MIMESIS
INTERNATIONAL

PSYCHOLOGY
n. 5

Fabio Benini

UNUSUAL PSYCHOANALYTICAL NOTES ON THE ORIGIN OF MONEY

הקקי או הערלה ?

Translated by Adam Elgar.

© 2020 – MIMESIS INTERNATIONAL
www.mimesisinternational.com
e-mail: info@mimesisinternational.com

Isbn: 9788869772801
Book series: *Psychology*, n. 5

© MIM Edizioni Srl
P.I. C.F. 02419370305

CONTENTS

Prologue	7
Introduction	11
Introduction to the English Edition	17
1. Once Upon a Time	23
2. Money and the Coin: Things Hidden Since the Foundation of the World?	35
3. Eric Fromm and the Property	57
4. Evil: in the World and in Psychoanalytic Theory	63
5. Sigmund Freud and Professor Alfred Jeremias	81
6. Naïve Barter and Chess	117
7. Merx,-cis	123
8. Numbers and the Knife	139
9. The symbol and Some Transformations	147
10. A Half-Serious Half-Facetious ante-Diluvian Digression	157
11. The Ur-bs not the Polis, Πολισ	165
12. Ur-Nammu	177
13. Bernhard Laum	181
14. Added Value or Original Capital	193
15. Wealth, Money, Coin and Gift Are They All the Same Thing?	199
16. Rome	209
17. Bulla: i.e. Money	215
18. The Epic of the Ring	221
19. A Very Quick and (Almost) Canonical History of the Coin	229

20. THE SACRIFICE AND THE NEW FAITH	235
21. ON WHY, IRONICALLY OR MAYBE NOT, IT IS DIFFICULT FOR THE COIN TO HAVE HAD A GREEK (OR ROMAN) ORIGIN	255
22. AVRAHAM, אברהם, FROM UR OF THE CHALDEANS, OR AVRAM, אברם, THE CHALDEAN OF UR	261
23. ARE WE REALLY SURE ABOUT THE CUNEUS?	275
24. THIS UNKNOWN FORESKIN	281
25. THE INGOT	289
26. FREE ASSOCIATIONS: THE BUTTER CURLER AND OTHER MATTERS	293
CONCLUSIONS	297
BIBLIOGRAPHY	301
NAME INDEX	319

PROLOGUE

The Hebrew letters in the title page are intended as a slightly mysterious announcement of this book's subject. Mysterious because the title has been partly expressed in a cryptic language and letters as mysterious and cryptic as the very subject of Money and Coins. Mysterious because of the mysterious links that are sometimes created between diverse elements, such as the two in question, but also the links between languages.

And mysterious because in using an "unfamiliar" alphabet, the Hebrew, I wanted to provoke a feeling of surprise in the reader about the material I will be presenting, generating a moment of puzzlement and suspense which could be a good travelling companion for the reader.

There is always the risk of helping to maintain an anti-Semitic prejudice when terms such as 'money' and 'Hebrew' are brought together: in the pages which follow, the concepts of Money and Coin will be associated with important episodes in the history and tradition of the Israelites who then became Jewish; a history and tradition to which the West owes an important part of its own roots, and which I see as incessantly transitional elements in a process begun by the Sumerian civilisation, continued by the Akkadian, and still present today. The Israelites drew heavily on the Sumerian and Akkadian civilisations – albeit indirectly via Babylon, but first of all through Ur of the Chaldees – and having in the meantime passed them on to the entire West, maintained many characteristics of these cultures, at least as far as my understanding and elementary study of Hebrew language and culture have thus far enabled me to see.

And it is in homage to this ongoing tradition, which has been and remains one of the important bridges between the ancient and modern worlds, that I thought it would be significant and amusing to begin this study by playing with the sense of mystery which "unfamiliar" signs like as the letters of the Hebrew alphabet can generate in those who try to read them.

These letters compose words: two nouns, a conjunction, and a question mark.

Hebrew is read from right to left and in transliteration the phrase becomes:

haqaqi or haorlah?   הקקי או הערלה ?

Which in translation becomes the very simple:

Shit or Foreskin?

Which are the elements around which the reflections being developed in this book will revolve.

A Giovanna, המורה שלי, hamorah shelì, la mia maestra.

INTRODUCTION

The pages which follow are the result of my deep personal and theoretical dissatisfaction with the subject of Money and Coins as it is commonly addressed, dealt with, discussed in general, and especially in psychoanalysis, and they are an attempt to find a less abstruse meaning for matters which have often become mere dogma.

Disregarding the siren voices of "modern" psychoanalytic conceptualisations and using "old" tools that are nowadays considered rusty, I will address the psychoanalytic dogma which wants Money to be associated with Faeces; arguing at length and skipping to and fro, I will try to find the relationships which link Money and Coins to genital features, and specifically to the foreskin, looking at this complex topic in a different light from the traditional view established by Freud and accepted, directly or indirectly, by whole generations of psychoanalysts – and not only by them – for over a century.

These pages are my effort, my personal attempt, to revisit – without a precise historical, economic or philosophical direction, but certainly with considerable puzzlement – concepts, commonplaces, and elements accepted as obvious, but which in my opinion have lost (if they ever had any) their vital significance.

Concepts such as Value, Price, Trade, Coin, Work, Barter, Slave, Master, Sacred, Profane, Scientific, Sacrifice, Right, Law, Need, Desire, Grace, Weight, Equity, Justice, Ethics and many others have dissolved into a thousand rivulets of meaning, often losing their meaning and their vitality along with it.

From a certain point of view, I think it is inevitable that any concept, after a bit of use, tends to lose that charge and energy which had characterised its birth, even without invoking the idea of entropy. Sometimes a concept is reified, becoming crystallised and giving rise to immutable dogmas and almost religious convictions; at other times it becomes fixed, though in a marginal position, within the sphere from which it emerged. Most often it melts like snow in the sun, revealing all the limitations of its origin. More rarely, it happens that, dismantling old, worn-out traditions, it founds a new paradigm, but immediately afterwards it inevitably returns to the path of what had preceded it, as Thomas Kuhn might say.

It seems to me that this applies both to grand, noble concepts and to cartoons, to an important scientific discovery and to a joke, to Galileo and to Walt Disney.

Habit, fashion, the unconscious, fatigue, misunderstanding, and excessive, superficial or simply repeated use are all reasons why so much dies and gets forgotten.

So, paraphrasing Goethe, we need to reacquire what we have received in order to possess it genuinely. And then start all over again.

Perhaps humanity, the human condition, is indeed "alone" in retracing – inwardly above all – the same path that millions of other individuals have already travelled, and millions more will; and also perhaps first establishing – an absolute novelty – that the atom is indivisible, as its name pompously declares, then finding the way to divide it and having to find another "idea" for another atom, a divisible one this time, and imagine it as a kind of string that someone will perhaps try to tie up or untie, and so on ad infinitum. Or, more "simply", declaring that what matters is the principle of "non-contradiction" and then understanding that, even if it might sound rhetorical, there is no life without contradiction.

This means going back over roads that others have travelled, everyone naturally in his own way with his own orientation, interpretation, and sensibility, roads which, sometimes seem to overlap. And this is how I will review some of Freud's observations on monotheism, for example, or Karl Abraham's on Amenophis IV, but in quite a particular light (different from Prof. Alfred Jeremias).

The same will apply to Dr Ilana Schimmel, Jacques Attali, and Keynes, but also to Herodotus, Xenophon, Aristotle, and others.

In any case, for my own part, when hearing a joke, reading a comic strip, or watching an animated cartoon, I have found myself thinking about, reflecting on, reading writings about Money and Coins.

I really don't know why I am so strongly focused on Money and Coins. To be controversial? Out of curiosity? Greed? Because of my thirty-year business relationship with my accountant? Or the mountain of tax I have to pay? Or because I never managed to ask out that girl at high school since I didn't have enough money for petrol to drive her to the lake? I really don't know. There can be any number of reasons why someone wants to be a gardener and work with plants and flowers, but I'm happy if he does it with passion and I'm grateful to him, even though every gardener has his

Introduction 13

"hinterland", his own tastes and sensibilities, and his own idiosyncrasies about the weather and which species and procedures to favour. Maybe the reason is just that: passion, and wanting to understand. Or else, a vague and diffuse feeling of irritation that comes over me when I detect the "smell" of moralism. I don't know.

I'd need to ask a psychoanalyst; or a gardener.

Just as we would perhaps need to ask a psychoanalyst why a psychoanalyst is interested in Money and Coins instead of sticking to Fantasies, Dreams, the Transference, and the Unconscious. Maybe we should also ask a psychoanalyst why some psychoanalysts aren't interested in these things at all.

And maybe I should ask myself why the idea of the gardener came into my mind in the first place, but maybe I have half an idea about that myself.

Imagine someone venturing into the reading of works written by psychoanalysts in the Twenties and Thirties. Imagine the taste for adventure, a challenge, a thirst for discovery, for desecration, and the reconstruction of elements taken as carved in stone by the pioneers of psychoanalysis, using some "crude" pickaxes. A creative, impassioned, fertile *Skandalon*.

Sometimes these seem like titanic undertakings: three concepts bring age-old pillars tumbling down and new perspectives are cast on ethnography, sacrifice, religion, sadness, love, and the tribes of Papua New Guinea and Irian Jaya!

Maybe my interest has arisen out of that same spirit of adventure, or more simply by that spirit of adventure I felt as a little boy when I read Verne, Stevenson, and Salgari, and which I found in those writings by the pioneers of psychoanalysis, often despite the fact that I often did not share their methodologies, conclusions, observations, or proposed theses; but I seemed to feel the fascination, the desire, the scent and splash of the sea on the shore of Robinson Crusoe's island. As a little boy, like so many little boys, on a distant shore I bought a conch to "hear" the sound of the sea and bring home in some way a little piece of Robinson's adventure. In the Pioneer's writings I felt the wish, the desire, the will to dismantle and reassemble ideas rather than make do with the ready-made product.

The wish, the desire, the will to look beyond the protecting wall of their own reassuring horizon.

The wish, the desire, the will to get entangled in infinite discussions and elaborations of sometimes frankly subjects, even though they we retracing steps already retraced a thousand times and in a thousand different contexts, but revitalised in those years by new blood, new challenges, new panoramas, new seas. Perhaps this is what impelled me: the topic of Money as a new South Sea; the wish, the desire, the will to start building my own cartography different from those I already know and which I find too reductive and stale.

The geographer in *The Little Prince* has explorers tell him about the territories they have explored and which he will then map, and he tries to assess the truthfulness of each account by weighing up the morality of the explorer who tells it, so that in this bizarre way he can ascertain the correctness of the information he receives.

Unfortunately, this bizarre way of proceeding – by Morality – is not exclusive to him. There is nothing like the topic of Money and Coins for stimulating terrifying nuclei of Morality, Moralising, Ethics, Idealisation, and full-blown Projection, of fury and bitterness which break out in any discussion that so much as touches on these themes, like bullets and sword thrusts which reveal how complex, conflict-ridden, and bizarre the inner world can be of people who move in and around Money and Coins.

I have never liked *The Little Prince*; personally, I think it's a terrible story, a perfect example of indulgence in Death masquerading as Life. Similarly, I've never loved geographers who don't do field work, who don't get their hands and shoes dirty, for whatever map-cartography-geography it might be. Clearly, though, there is a sense in which whenever we cite authors, texts, sources in general, we are to some extent acting like the geographer in *The Little Prince*. Someone else has gone somewhere and reports back, often in good faith, what he believes he has seen. In a way, this is also the problem of the dwarves standing on the shoulders of giants, given that is "almost always" the giants and not the dwarves who choose the times, methods, direction, and the goal to be reached, and even the routes and the vehicles for getting there. And so the dwarves often see only what the giants decided to see; they often look only for what the giants decided to look for, and find only what the giants wanted to find.

But it is a pity that the geographer in *The Little Prince* does not even know if there are oceans, mountains, rivers, cities, or people on his own planet, somewhere beyond his table.

This is why, given that I am concerned directly and concretely with the Money-Ocean and have no desire to get other people – explorers, cartographers, or giants – to do for me what I think I should do myself, I started to build a "boat" to enable me to set off towards the southern seas and distant oceans.

When the design of the boat had almost been completed, I "came upon" – thanks to a colleague and friend – the writings of Giovanni Semerano, a great but little known scholar, whose work immediately found a place

Introduction

alongside the sextant and compass, thus completing the set of instruments to help me find my way on the high seas. I feel great admiration and respect, albeit posthumous, for Giovanni Semerano, even though many eminent scholars find him as appealing as pepper spray in the eyes.

And so, joking a little about serious matters and taking comical things quite seriously, I have written these "travel" notes.

What I have seen and heard, the difficulties I have encountered and the flashes of light I have glimpsed, or think I have glimpsed, are described in the pages that follow.

And also what I have not been able to see.

INTRODUCTION
TO THE ENGLISH EDITION

Before continuing, I would like to make a few small clarifications about the English version of this text. Obviously, as the title page says, the text that follows is a bit unusual, and there are at least three reasons for this: the first is that, while trying to modify some passages so that they could be better understood in English (a problem faced by every translation and one that Adam has coped with well) I have tried to maintain a certain "provincialism", and even a certain "localism" here and there. Though it might seem bizarre, this is (second reason) a small homage to pre-Roman European history, and I hope that in its bizarreness it may be an invitation to some colleagues in Brno, or Bern or Brenner or Brennenburg/Brandeburg or Brenieux or Baroña or Beuron, or in Wales or Scotland or Ireland to search their own heritage of language and tradition for those shards of the "ancient" that lie, undetected, deep in their provinces, in their own isolated valleys, right inside their little fortified hilltop villages, and that they will play around a bit with the official languages which, over the centuries, have tried to enforce their supposed superiority, often by force. The third reason is that I like to jump back and forth, from the general to the particular and back again, from odds and ends to massive systems, from the apparently insignificant detail to the complete broad fresco, and vice versa; in the end, it's not very different from what goes on during sessions of analysis.

Beginning.

One day, a colleague who honours me with her friendship, Dr Gabriella Blesio, told me about a conversation she had had, long ago, with Musatti, one of the founding fathers of Italian psychoanalysis, already founded in 1925, which was later suspended by Fascism.

We were having lunch – Gabriella, her husband, my wife Giulia, and I – and Gabriella began her story. A long time before, she had asked for a consultation with Musatti in order to start an analysis with him. The Master, now at a great age, had welcomed my colleague and sat her down in his consulting room near a table piled high with books.

After listening to the then young aspirant analyst, with the sardonic smile that was so characteristic of him, he explained very graciously to her that he was now too old, touching gently on the topic of the analyst's possible death; and still smiling sweetly (my colleague's word) he thanked his young colleague in Veneto dialect, she also having roots in Veneto, for her faith in him, but suggested she try another, younger analyst, which my colleague did.

At the end of the consultation, before she left, Musatti said, still in Veneto dialect, "Wait a moment, Doctor, I want to give you an article to think about." He spent a long time looking through the chaotic heap on the big table but couldn't find the article. With the regretful air of someone who hasn't been able to say goodbye in the way he wished and alluding to the pointlessness of using stubborn insistence as a way to force the search of the Unconscious, and almost paraphrasing Bion's "without memory and desire", said, still in Veneto, "I'm sorry, Doctor, but the table will let me find it in its own good time!" The two of them said goodbye, and never saw each other again.

The then young colleague has still not received that article, but she preserves the great sweetness of the memory in this little anecdote as a precious, wise, and gentle farewell.

Clearly I'm joking, at least partly.

I don't know if it happens to every writer, but I imagine it does, even though I'm not much acquainted with Writers: after the work is done and the book has begun its own autonomous life, you realize you've left out, neglected, overlooked this or that detail, author, quotation, reference, explanation, note....

It sometimes happens, as with Musatti, that the table hasn't wanted to give back the one article which would have been just right for explaining that point, that concept, or that passage; and then in the sometimes spasmodic striving to bring the work to a close, in the face of assorted hassles, incomprehensions, and difficulties, the process gallops to a close and it's only afterwards that you realize how many things the spiteful table has kept hidden from you.

Here then, in this quasi-introduction to the introduction, I'd like to try and mention at least some of the sneaky tricks my table has played on me.

The first "thing" my mischievous table gave back to me, but only after the Italian version of this book had gone to press, is called Popper. In developing my argument, I've had a bit of fun with Popper and his *Logic of Scientific Discovery*. I wasn't kind, I admit it, but it's not my fault: it's the table's fault. Yes, the table's! And the reason is very simple: under a pile of books, right underneath Popper's *Logic*, there was *The Open Society and its Enemies*. And the table didn't want me to read that, so it only gave it back to me after my own book had to gone to press. This book of Popper's is a really beautiful piece of work in which he analyses Plato, Marx, and Hegel in a masterly manner, and much better than I have.

Of course, I didn't have the same aim as Popper (or his learning) but I would at least have been able to cite him and acknowledge that powerful breath of freedom of which he is the bearer.

Which doesn't mean I've changed my mind about the Popper of the *Logic*, but if any of my readers have the opportunity to meet Popper, please let him know that I don't only have an ironic attitude towards him, but also great respect and admiration.

And if any unlikely and bold-spirited readers of this book should happen to have an email address for Baruch Spinoza, I'd be very grateful for it because I owe him an apology too: not exactly an "Apology-Apology" but a lower-case apology. Oh all right, upper-case Apology!

In this case, the problem wasn't the table but my own presumption: at the age of 18 I picked up a copy of Spinoza's *Tractatus* with the result that I didn't understand a word of it. And I never wanted to pick it up again. The *Ethics*, however, I have read in its entirety with an adult's attention, but it bored me just as much. Then my "inner table" suggested that I re-read the *Tractatus* and, after a brief resistance, I started to: partly because I've studied a bit of Hebrew, partly because in writing this book I've addressed many of the things Spinoza writes about, partly because I happen to be older, but also because what I've just said about Popper is also true for Spinoza: the *Tractatus* is a beautiful book (there's the odd gross error, as in equating the Giants with the Damned, but overall it's really beautiful).

Another omission for which I feel I must make amends applies to Keynes: being all taken up with trying to grasp R. F. Khan's coefficient, 2% inflation as the earthly paradise, the theories of debt and deficit, "in the long run we are all dead," and paying men to dig holes and then fill them in again, I got lost in a theorizing which is still a bit undigested.

Until, reluctantly leafing through a book about the correspondence between Keynes and his good lady (*Lidia and Maynard*, edited by Polly Hill and Richard Keynes, Charles Scribner's Sons, New York, 1989) my eye happened to fall on a sentence of his which alluded to "my Babylonian madness". "Babylonian madness"? What does this mean, I wondered. What was this "Madness" that Keynes later abandoned? The only reference I initially managed to find was in Ingham, who is slavishly quoted by numerous authors (Graeber and many others) who make no effort to cite the sources, which Ingham himself actually does. And initially I found nothing to satisfy my sudden and peremptory curiosity: did Keynes know Babylonian? Could he read cuneiform? Did he go on a dig in Mesopotamia? Agatha Christie wielded a pick, but did Keynes? Did he perhaps have a personal correspondence with Hammurabi or Darius the Great that no one ever told me about? Did nobody write about this? And how did they communicate? Was there a Pony Express to deliver the missives? And what did Hammurabi and Sargon the Great think about full employment, infrastructures, and banks minting coins? And hundreds of other questions, a torrent of them....

None of my economist friends knew this part of Keynes's work; certainly, what he called "my Babylonian madness" was a marginal part of his vast output, and one he later abandoned, but a vitally important one for me since the "Babylonian," *lato sensu*, constitutes a large part of my own work's fabric!

And for weeks I tried to get access to this part of Keynes's work.

Certainly it's shortcoming of my own, and not the fault of the rascally table, that I spent too much time down the proverbial rabbit hole, and so it was only after my book was finished that I got hold of Keynes's volume XXVIII. And this is how I discover that some ideas which seemed to have emerged out of my imagination and reflections were sometimes thought, even if later abandoned, by the imagination and reflections of Keynes himself.

So it really is true that *"Nihil novum sub sole!"* And I think that the *Qoheleth* might be able to mitigate my Vanitas just a little, since it's been given a knock by the discovery of such noble ancestors.

However, if some of Keynes's observations preceded mine by almost a century, others (and here my Vanitas is less dented) are entirely mine and, unlike Keynes, I don't regard the work that follows as "Madness".

Or maybe it would be better to ask Erasmus what he thinks.

Anyway, for me it isn't and wasn't.

I hope I have been able to transfer at least part of these authors' enormous work into the fabric of my own, updating it a little. It shouldn't have been hard, given the deep consonance I have often noted between some of their positions and mine.

I also hope that all this doesn't become a further burden for the unwitting reader who has to gird his loins to read my text which, though not as expansive and complex as the *Kritik*, is definitely not light.

Probably, or rather certainly, the table is now giving me back what it wants to give.

Probably, or rather certainly, the table will give other things back to me, in its own good time.

Introduction to the English Edition

There are dozens of other writers whom I realize I have sometimes misquoted, for whom I've given no precise references about pages and/or texts I've cited, couple with the fact that translations from one language to another do not always respect original sequences; for dozens of others, I have obviously chosen extracts which support or challenge my theses (but everyone does this). Malinowski, Pareto, Marcuse, Schumpeter, Innes, but also Saraval, Fachinelli, Lacan, and many, many others could/should have found a place more befitting their stature. As I have also written in my conclusions, I've played the game as fairly as I could/should; as fairly as the "table" let me.

And since *"Hic Rhodus, hic saltus,"* this – oh! *Vanitas Vanitatum* – has been my jump.

The same argument – that it's not my fault, but that of my mischievous, malevolent table – applies to Prof. Garbini (neither he nor Giovanni Semerano is well-known outside Italy, but there's nothing my table can do about that): between the printer and the ashtray, right by the Hebrew dictionary that I've consulted thousands of times, his books have stayed hidden away, buried by the table's ostracism. I don't know exactly what my table had in mind with this particular game of hide and seek, (a kind of peekaboo!), but the fact is that it keeps offering me texts by a range of popular modern authors who I don't want to quote for obvious reasons of *tædium* (if Piketty or Graeber or Harari and/or a couple of Nobel Economics Laureates – who shall we say? – Kahneman or Thaler, or someone else, came to mind, you wouldn't be far wrong; about Thaler I'd just like to add that if anyone, in any way, saw fit to give me a 'nudge', however politely, it would provoke an immediate reaction of severe annoyance and I immediately say, in the Gospel sense and not the psychoanalytic: Noli me tangere!, or if you prefere, μὴ μου ἅπτου! or in Hebr., אל תיגע בי, or better in Aram. אל תיגעינא בי!).

Prof. Garbini wrote extremely interesting and profound things about ancient Israel, the Bible, and the Ancient Middle East, and I think it's appropriate for me to add that the great admiration I have for the work of Prof. Pettinato, Prof. Liverani, for Bottéro and Kramer, must also be extended to Prof. Garbini whom I thank profoundly, even though in his case, my thanks can only be posthumous.

Besides retouching a few small errors and adding a few more precise references, I won't be adding a chapter about Prof. Garbini's work to this edition (the bold letters in the root B-R-N at the start of this preface, are intended as a small reference to his work since the root b.r.n. indicates persons, such as Brennus who conquered Rome, and toponyms important to the Celts. His reflections on the Philistines-Peoples of the Sea, on Anakiti/Anaqiti and Anaq/Collare/Necklace/Ring, and on the Biblical "Tohu Vavohu", are also very important, and in my opinion are connected to the English *peekaboo*/ Italian *bau cettete*; but I won't be going into those here. Unfortunately, I have only found a couple of English translations of his splendid works).

Hoping to fill, at least partly, that void which my mean-spirited table is only now letting me address, I will add for this edition a brief incursion into another, still disregarded, aspect of the etymology of the term Moneta-Coin-Cuneus.

Let's hope for the best.

1.
ONCE UPON A TIME

His speech was substantial, and its contents extensive.
The messenger, whose mouth was heavy, was not able to repeat it.
Because the messenger, whose mouth was tired, was not able to repeat it, the lord of Kulaba patted some clay and wrote the message as if on a tablet.
Formerly, the writing of messages on clay was not established.
Now, under that sun and on that day, it was indeed so.
The lord of Kulaba inscribed the message like a tablet. It was just like that.

(Enmerkar and the lord of Aratta, 500-514)
From The Electronic Text Corpus of Sumerian Literature,
University of Oxford (etcsl.orinst.ox.ac.uk)

I found these lines (quoted in Pettinato, 2005) extremely beautiful in their essential simplicity. Breathtaking.

"בראשית באר אלוהים את השמים ואת הארץ", "*Bereshit barà elohim et hashamaim ve et haeretz*","In (*one*)[1] beginning G-d created the heavens and the earth," are written words which portray the Genesis of the Universe, while those of the Sumerian poem quoted above, describe a little more understatedly the genesis of (human) History.

The Word, in the very act of being inscribed, is transformed into a thing

---

1   "This world inhabited by man was not the first of the things created by God. He had already made other worlds [nine hundred and seventy-four generations prior to the creation of this world], but He had destroyed them one after the other because He was not satisfied with any of them until He created ours. Nor would this one have lasted if God had kept to His original plan of governing according to a rigorous principle of justice. Only when He saw that justice on its own would have brought the world to destruction did He ally it with mercy and make the two govern together." (Ginzberg, I, p. 24)

– concrete, tangible, third – it takes on a life of its own, generating a change which has still not ended today.

And as Pettinato very astutely point out, the word moves from the domain of the ear (oral transmission, and also proximity, fusionality, symbiosis) to the domain of the eye (distance), which is also thirdness, a break with symbiosis. He may be recalling the same concept expressed Isidore of Seville: *"Verba enim per oculos non per aures introducunt"* (Liber I, III,1).

In a general, superficial way, we could also hazard the hypothesis that, at that time, as the sun set on the world of Gaia, the Father decisively asserted himself. As a result, it is not so much the Word that is third – or rather, it is not so much the *spoken* word that is third, lending itself as we see every day to misunderstandings, projections, and the most diverse interpretations. Instead it is the written word which, perhaps rendering the flow of information poorer and perhaps a little colder, at the same time – without preventing it entirely – drastically reduces a piece of information's scope for, or margin of, ambiguity-ambivalence-interpretability.

I don't know what Lacan would think of this, but it will do for the moment.

These inscribed words come, more or less 3,000 years BCE, from a King, Enmerkar, of the first Sumerian royal Dynasty, "some" years after the Great Flood.

The absolute precision of the date is not of much importance, since at the moment nobody is in a position to go to the stake over it, Carbon 14 notwithstanding.

In any case, even if we wanted to date the Flood to 4,000 BCE, or even earlier, it would not make much difference to my thesis, given that it would simply mean lengthening the distance between and significance of the elements to be expounded.

Enmerkar inscribed words, not drawings or pictograms or simple logograms, but words which were already moving in the direction of the syllable and can be considered as the forerunner of alphabetic writing; signs which no longer have a direct figurative relationship to the object they "describe". Bottéro (1987, pp. 75-88) gives us a very clear and interesting description of these transitions between archaic and "late" Sumerian, which would fuse (or perhaps was already fused) with Akkadian.

But, setting aside questions of the date, why did Enmerkar 'pat some clay and write the message as if on a tablet', something nobody had ever done before?

Because, explains Pettinato, he *wants* the Lord of Aratta, a city 1,000 miles away in Iran, a city poor in grain but rich in minerals, to send precious materials to *Uruk* in recognition of the fact that Royalty, granted by the goddess Inanna (who would later be called Ishtar in Babylonia) had its seat in *Uruk*. Enmerkar *wants* gold, lapis lazuli, and other precious stones for the building and embellishment of a temple dedicated to Enki, the Sumerian god of the waters, and is offering wheat and barley in exchange.

It is not just an exchange, not only a king calling his own colony, his own vassal, to order. It is not simply a political, military or religious annexation, assuming that anything "simply" religious exists, and assuming that religion itself can be "simple", *pace* Voltaire and his "*Écrasez l'Infâme*".

Enmerkar *wants* primacy, considers himself supported by the gods (as did the Lord of Aratta, however, at least at first), solves riddles, proposes an exchange, establishes a relationship, is subjected to a swindle, reacts, threatens reprisals, brandishes his royal sceptre, decides to give battle and then accepts a "symbolic" battle in which two champions, one from each city – not the Horatii and Curatii but very like them – will challenge each other; a battle which will settle the destinies of the dispute that has lasted for decades. He offers "gifts".

Enmerkar *wants* precious stones, gold, and silver. And he wants them for the Temple of Enki, after requesting support and comfort for his sister, the goddess Inanna, since they are both "children" of UTU the Sun god.

I have emphasised the term "wants", that is Desires, perhaps Claims, or even Demands, because "wants" has very different implications from "needs". On this apparently small distinction whole systems of thought will be built in philosophy, sociology, economics, and even psychoanalysis, which – while looking at the same data – are profoundly different and often antithetical both in their presuppositions and in the conclusions derived from them. In psychoanalysis, this distinction relates directly to the concepts with which, clinically but also from an emotional and "ideological" viewpoint, the analyst approaches the patient in terms of "Need" vs "Desire", "Lack" vs "Conflict", of "Primary" vs "Secondary" aspects, and many, many others.

And in order to have precious stones and gold for the temple of Enki (Enki is not mentioned explicitly, but the temple he wants built is called "the great Chapel of the Abzu/Apsu," in other words the great Chapel of the Abyss, a term to which I shall return; and the Abyss is ruled by Enki, ergo Enki is the divinity to whom the temple will be dedicated) Enmerkar invents, creates the "Word on Clay" so that Words (sounds) take on a body and become visible (signs), as Pettinato acutely observes.

A "little Genesis", but this time created by men.

Enmerkar will also make use of a small aid, using the power of an Apkallu, a bizarre creature we will encounter shortly. In doing so, Enmerkar puts his seal on such technological and cultural superiority over the World that the Lord of Aratta, despite his proximity in rank and learning to Enmerkar, is only able with great effort to understanding the "miracle" he has before his eyes. But who is Enmerkar?

Following the chronology of the Sumerian King List expounded by Pettinato, Enmerkar would be the first, or one of the first, postdiluvian sovereigns of Sumer, and was preceded by antediluvian Kings who governed the world, some for 15,000 years, others for 28,000 years, and so on, like the great Patriarchs of the Bible; and just as in the Bible, the further we move from the Creation the more the lives of men are shortened.

The King Lists – whether Sumerian, Akkadian, Assyrian, Hittite, Gutians, Emorite, Chaldean or Babylonian – were revised and corrected every time the Royal House changed: that is, every time a new dynasty or population takes the place of the previous one. Hence, it is difficult to "historicise" a medley of information that is constantly being re-elaborated by conquerors over millennia in order to valorise their own dynastic right.

In any case, I will here place Enmerkar, a figure as epic as he is mythical, very close to the immediately postdiluvian era.

Enmerkar is the son and successor of Meskiaggasher, himself the son of Utu, the Sun god, who reigned for 324 years. Enmerkar, who founded *Uruk*, the Erech of the Bible, reigned for 420 years and according to the lists compiled by various scribes, is to be considered the father or at least the grandfather of Gilgamesh who in turn ruled *Uruk* for 126 years.

Enmerkar is also the last sovereign to avail himself of help from one of the *apkallu*, the Sages, who had appeared until that time, mythological beings who acted as counsellors to the king:

After the flood, during the reign of Enmerkar, Nungalpiriggal was apkallu.
He who made the goddess Ishtar come down from heaven into Eanna...
During the reign of Gilgamesh, Sinleqiunnini was ummannu.
(King List, quoted in Pettinato, 2005)

With Gilgamesh the *apkallu* disappear, and in their place as royal counsellors appear the *ummannu*, Teachers, human rather than divine counsellors. And Sinleqiunnini, the counsellor or Teacher, of Gilgamesh, will compile the Epic which will celebrate the deeds of his own King.

In its Hittite version (Pettinato 2004, p. XIX), this Epic readopts the Sumerian term "Song" (*šir*), while in Hebrew Song is שִׁיר, *Shir*.

And it is thanks to the *apkallu* Nungalpiriggal, and to his advice, that

Enmerkar invents writing. The *apkallu* (Sum. *abgal*: that is, *gal*, Great, and *ab*, Father, Sum. *abba*, sometimes *pap*; Akk. *abu*) were divine beings connected directly to Enki, the god of the waters and one of the principal gods in the Sumerian pantheon (and whom I shall re-encounter later and about whom I shall have much to say because, whether he intended it or not – but since he was a god, who can say? – he has assumed an important role in the history of psychoanalysis) along with An, the Sky, and Enlil, god of the air and of what lies between the earth and the sky, and Ninhursag, also called Ninmah, the sublime Lady also called Nintu, the Lady who gives life, the mother, in still earlier times also called Ki, the Earth.

Berossus (IV century BCE), the Chaldean author and priest of Bel in Babylonia, also quoted by Pettinato (2001, p. 430), described the first appearance of an *apkallu* called Oannes or Uan, during the reign of the first Sumerian king, Aloros, alias Ajjalu, alias Alulim, according to the various King Lists, who governed for 10 *saroi*. Saros is a Hellenization of the Akkadian *Sar*, a term which designates the number 3,600 – in this case, years; so Aloros reigned for 36,000 years) in the antediluvian era. Berossus gives this description of Oannes:

> There was a great multitude of people in Babylonia, and they lived without laws, exactly like wild animals.
> In the first year a beast called Oannes appeared out of the Eritrean sea [the name given to the Persian Gulf at that time] in a place adjacent to Babylonia. His whole body was that of a fish, but a human head had grown under the fish's head, and human feet had likewise grown from the fish's tail.
> He had a human voice and his image has been preserved until today. This beast spent his days in the company of men but did not eat any food. He gave men the knowledge of letters, sciences, and every kind of art.
> He also taught them how to found cities, erect temples, formulate laws, and measure the fields.
> He revealed seeds to them, and the harvesting of fruit, and in general gave them everything that is connected with civilised life. Nothing new has been discovered since the time of that beast.
> But when the sun set, that beast Oannes dived into the sea and spent the nights in the abyss, since he was amphibious. After him, other beasts appeared.

I add a small aside, also taken from Pettinato (2005) who notes a Greek writer Aelianus (165-235 CE) – in other words, very late in terms of a "chronicle" of the facts, but important because he re-adopts concepts now made "canonical" in order to give a better account of Enmerkar and then Gilgamesh, and the significance of the stories devoted to them:

There was once, long ago, a king,
his name was Enmerkar, lord of the city of Uruk.
To him the seers had prophesied:
"He to whom your daughter will give birth,
will take your royal crown from you."
The king was then struck with fear, and so that this should not occur,
shut away the virgin in a tower;
he had her watched day and night.
Nevertheless, she gave birth in secret a son of nobody
because the decision of the gods is immutable.
The guardians, in dread of their sovereign's anger,
threw the infant from the tower.
But an eagle, having espied him with its sharp sight,
seized the infant in its claws before it could be smashed on the ground,
and carried him into a palm grove where it laid him gently down.
The gardener discovered the child there
He loved him and brought him up: Gilgamos he named him.
Grown and become an adult, Gilgamos dethroned Enmerkar
the father of his mother.
And so the divine prophecy was fulfilled.
(Aelianus, *De natura animalium*, XII,21)

This seems to be the first version of the myth of Acrisius, who shuts away his daughter Danaë in a tower or a cave, or a watertight chest cast onto the sea, because in her womb would grow Perseus, who would kill his grandfather.

An anticipation, maybe only a variant – or rather, the antecedent of Oedipus. Many people are familiar with the myth of Oedipus, but I wanted to dwell on Enmerkar at the start of this book less to emphasise Freud's intuition that in Oedipus's conflict – conflict and not complex, as Davide Lopez has often reiterated (and actually Dr Ilana Schimmel, to whom I shall return, also writes about conflict) – we glimpse an important crossroads in human development, both for the individual and for the "human species", than to highlight the unacknowledged, or perhaps simply unknown, debt owed to the Sumerians by the West, both now and in antiquity.

The Sumerians, the *Sag-Giga*, the *Men*, the Black Heads (in Akk. ṣalmāt qaqqadim), the inhabitants of the Biblical *Shine'ar*, also known as *ki-en-gi* or *ki-en-gir*, the land of the Civilised Lords, have been subjected to almost two millennia of oblivion, an oblivion which is all the more mysterious since Cuneiform (Akkadian-Assyrian-Babylonian, though based on Sumerian in the form used for the training of scribes) was used

Once Upon a Time

as a lingua (and scriptura) franca right up to the first century A.D., and was only supplanted by Roman conquests and the Latin alphabet. It is such a complete oblivion that even at the start of the twentieth century there was discussion in the academies about whether something that might be called "Sumer" could actually have existed, or whether all the historical, archaeological, and epigraphic material was simply attributable to some variant of "Babylonian". This latter opinion was shared by Prof Alfred Jeremias, whom we shall meet later on.

For a more detailed study of Sumer and Akkad, certainly fuller and more precise than I am making here, I refer to the splendid, exciting, and passionate research of Giovanni Pettinato, and the careful, minute and profound work of Mario Liverani. Their writings have kept me company through these pages, composing the warp and weft on which I have woven much of my story.

I also refer to the original elaborations of Giovanni Semerano mentioned earlier, which together with the reading of Samuel Noah Kramer, a more traditional but no less brilliant scholar considered the founder of modern Sumerology (although the initiators of such studies between 1800 and 1850 were called Rawlinson, Hincks, Grotefend, Oppert, and Smith), have contributed to the growth of my curiosity for this splendid page of human history. This is obviously not to forget Bottéro and his 1954 work, *Le problème des ḫābiru*, which points out with great precision, and ruling out any connection to a specific ethnic group, the Babylonian term *ḫābiru*, derived from the Akk. *ḫabbātu* and *šaggāšu*, which are in turn connected to the Sum. LÚ.SA.GAZ – that is, "*malandrins, brigands, filibustiers, bandits, voleurs, vagabondes*" (p. 196), literally "Head-breakers" or "Cut-throats" – to which I shall refer shortly, using an argument from the observational viewpoint adopted by Fromm. (Henceforth, I shall incorrectly use the transliteration *ḫāwiru*, only because I like it more from the phonetic point of view.)

In the few lines I quoted a little while ago about the birth of Gilgamesh, we meet – indeed we almost crash into – the kernel of Sophocles' Oedipus, 2,500 – 3,500 years in advance. He is likewise present in the concept of "No man's son" who, thanks to Fate, will become the founding hero of an entire culture like Sargon (the Great, founder of Akkad), son of a sacred prostitute, entrusted to the waters like (*nota bene*) Moses, also the son of an unknown father; or like Cyrus, destined to dethrone his grandfather, or indeed Perseus, grandson of Acrisius, also entrusted to the waters with his mother, Danaë. Osiris too was cast upon the waters, as were Romulus and Remus, sons not of the Vestal Rea Silvia but a of a She-wolf (*lupa*), a "Bitch", as was Cino/Kyno/Spax for Cyrus (Herodotus, Book I; 110, 1), the "She-wolf" (Acca) Larentia, wife of Faustulus (or of the shepherd Tiberinus, or of a wealthy Etruscan – who knows?), also understood to be a prostitute (*lupanar*). This reminds us of ritual, sacred prostitution in Babylonia (Herodotus again, I; 199, 1) and the orgiastic rites present in so many cultures, not only the Greek. There is a further similarity to another sacred prostitute, Šamḫat who through a sexual initiation "civilises" and "humanises" the wild Enkidu, also no man's son, first the adversary and then the friend of Gilgamesh; and nearer our own time, the "virgin who bears no man's son" cannot fail to suggest Mary and Jesus.

Such is the power of the "myths" and culture developed by the Sumerians.

Freud, writing in his *Moses* about the birth of the Hero (see also O. Rank), cites very similar examples to these, but begins with Sargon and leaves Gilgamesh among the rank and file, an understandable approach for his time. In my opinion, however, he was following the path laid by Alfred Jeremias whom I mentioned earlier.

Not only is the West indebted to the Sumerians for mythopoeic and religious elements that are still fundamental to its own cultural roots, but also for a huge number of other things, including the wheel, the sexagesimal system, many features of mathematics, astronomy, geometry, elements of technology, agriculture, and botany, metallurgy (with all the alchemical, magical, and esoteric valences connected to its practice, so well indicated by Mircea Eliade), the firing of brick, the techniques of smelting, the creation of standard weights, algebra, architecture, the City, wine, and beer (Bottéro, 1986; Ascalone, 2005). Bottéro, I think rightly, considers the Sumerians, a non-Semitic population, as absolutely interconnected with the Akkadians, a (proto-)Semitic population. From an initial position of supremacy, the Sumerians were slowly absorbed by the Akkadians, who were able to benefit from a constant supply of "fresh blood" from adjacent populations, while the Sumerian population seems not to have been able to make use of "new sap". Bottéro even prefers to speak of Sumero-Akkadians. Although as this book progresses I will continue to use the term Sumerians, Bottéro's thesis must be considered implicit in what I have written.

And the West is also indebted for the great narratives which then became those of the Bible, from the Creation to the Flood; from the Books of Laws to the invention of the primal Eden and of the private Gardens that would explode centuries later, with Babylon becoming one of the wonders of the world, the concretisation of the original *Gan Eden* (Garden of Delights); from the division of space into cardinal points to the study of the sky; from cartography to topography. And much more besides. Not to mention writing, of course. (Equally obviously, I think, the Chinese and Egyptians would not agree.)

What I am concerned with here, trying to retrace its origins and developments, as a gardener with psychoanalytic interests and not as Historian, Economist, Numismatist, Anthropologist, Epigraphist, Archaeologist, Linguist, Ethnologist, Sociologist, Philosopher, Orientalist, etc. etc. etc., is Money and Coins, in order to try and understand if and in what way I am also indebted to that great nation and its great culture for these last two elements, before continuing my journey towards nearer shores.

Being a gardener, I am not so sure that Enmerkar did not use Money. And Coins too. But the commonly held view is that at the time of the Flood, before the Flood, and for a while after the Flood (assuming that there was only one Flood), mankind did not use Money and, still less, Coins. Nevertheless, in the horticultural circles I frequent there are those who claim that Enmerkar might sometimes have solemnly said, "*ana* KUG. BABBAR *nadānum*," while in private yelling, "*ana kaspim idin!*" ("Pay for it in silver", in Sumerian and Akkadian respectively.)

"And Queen Anne's dead!" someone might retort. "They used barter, didn't they? I need a goat, you need cabbages, so let's swap. What's new about that?"

And in fact, there isn't anything really new about it: but barter didn't work exactly like that, or at least not so simply and naively, given that it is not all that simple to work out how many baskets of cabbages are worth a single goat. Or on another occasion, how many goats are worth a single basket of cabbages. Just try it, even with cigarette cards.

Is the goat old or young? Healthy, sick, small, big? How much milk does she make? Or is it a kid? Or a lamb? Or an old billy goat? And what colour is it? Does it possess the very strict qualities required for being sacrificed, or not? And how much does it cost to bribe the officials to accept my offering of a sacrificial animal and pay me for having provided the temple with the perfect sacrificial animal, even after I've given it "a little aesthetic adjustment" to make it truly "perfect"?

And are the baskets of cabbages big? How many cabbages do they hold? And of what quality? At what time of year are they being exchanged? Are they an early crop? Have they come from a long way off, or are they grown just outside the city walls? What do they fetch at the market? Has there been a shortage somewhere which affects the price, or demand, or the "intrinsic value"? Or is the diet of some tribe changing for mysterious reasons? Has there been a drought? Or over-production? Has *Pieris rapae* (the Cabbage White) done a lot of damage to the crops this year? And *Cœnurosi*s, given how many lambs *Taenia multiceps* (tapeworm) has killed this year, along with wolves, lions, and raptors. And the Gutians, "*accursed monkeys*", how many incursions have they made?

And if someone who needs a "goat" finds himself hundreds or thousands of kilometres from any "cabbages", as happened to Enmerkar, what does he do?

Someone who would like to look more deeply into this aspect and find out for themselves how reductive it is to think of barter in the naïve terms of "Goats vs Cabbages" could glance at Peter Temin's excellent *A Market Economy in the Early Roman Empire* (2001). It is about Rome and not Sumer, but it is highly interesting nonetheless because it allows us to understand in detail, what Joseph, for example, should have taken into account when making provision against famine, the "lean kine", and sheds some light on the much-abused concept of "redistribution" to which so many scholars have made abundant recourse. In any case, I will deal with some of these aspects later on.

For the moment I shall only reiterate that neither Enmerkar, nor Lugalbanda, nor Gilgamesh used Money – perhaps. Nor did Prince ZI.U.SUD.RA (which means "Life of prolonged days" in Sumerian), also known in other poems as Ut(a)napishtim (which means "He who has found eternal life" in Ninevite, a dialect of Assyro-Babylonian): that is, the Sumerian Noah who built an arc 120 cubits in length [1 cubit is about 44-50 cm] and not 300 like Noah. Perhaps Noah himself did not use Money either; certainly not Coins – or did he? Atra(m)ḫasis – another name for another, palaeo-Babylonian Noah – definitely (maybe) did not use Coins, though the scribe who compiled his story certainly did.

The exact attribution of these characters to precise cultures and periods is only amenable to a general overview, given that the various accounts of the Flood are copied by different scribes in different times and different cultures, as a result of which we go from the *Poem of the Great Sage* to the *Songs of Gilgamesh*, from fragmentary tablets to descriptions of precise measurements which overlay and cut across each other (for more precise details see Pettinato, 2001, 2005; Bottéro, 1992; Ascalone E., and Peyronel, 2006).

In any case, it would be very easy to read the dispute between Enmerkar and the Lord of Aratta, or the letter which Gilgamesh wrote to an unknown king, also "demanding gifts" to celebrate the funeral rites of Enkidu (Pettinato, 2004, p. LXXXIV), in moralising, ethical, religious terms, or in terms of disillusionment about human nature, or in terms of conflicts caused by the modes of production of goods, which are the ways in which I have heard Money being spoken of all my life. Simplistically believing that *c'est l'argent qui fait la guerre*, uttered in disparaging terms, explains everything once and for all, does an injustice to the fact that if a complex, global, and tragic phenomenon like war, but also like culture, and in the end History itself, were really governed by Money, it would in any case mean that Money is something much more serious, complex and multifaceted than mere faecal matter, which can be discharged with a simple hand gesture, pushing a button and sending down a bit of water.

Before moving on to a brief description of my point of view about Money and Coins, I think it will be important to set aside for the moment the "Money = Faeces" equation which I will address later, and to remember that Freud was being neither moralistic nor "politically correct" when he first made this interesting observation about money (1913, *On Beginning the Treatment*, pp. 131-2.)

> The next point that must be decided at the beginning of the treatment is the one of money, of the doctor's fee... powerful sexual factors are involved in the value set upon it. He can point out that money matters are treated by civilized people in the same way as sexual matters – with the same inconsistency, prudishness and hypocrisy.

Then, in a somewhat ingenuous, rational, Enlightenment manner, as if it were an easy matter to talk about sex, he goes on to say

> The analyst is therefore determined from the first not to fall in with this attitude, but, in his dealings with his patients, to treat of money matters with the same matter-of-course frankness to which he wishes to educate them in things relating to sexual life.

and lastly adds a very clear passage that is often ignored today:

> It seems to me more respectable and ethically less objectionable to acknowledge one's actual claims and needs rather than, as is still the practice among physicians, to act the part of the disinterested philanthropist – a position which one is not, in fact, able to fill, with the result that one is secretly aggrieved, or complains aloud, at the lack of consideration and the desire for exploitation evinced by one's patients.

He was not being "politically correct" in this, any more than he was when he looked at complex, important, and disturbing phenomena such as the unconscious, sexuality, incest, pleasure, hate, love, transference, Oedipus. Nevertheless, it is hard to postpone the "Money = Faeces" question any further. Let's read Lowenkopf (2003, p. 4):

> Unfortunately, over the years, psychological theories, by and large, have not departed too far from equating it [money] with feces, and this has proved crippling for current- day understanding, leaving the literature on the subject and the role it plays in the lives of children (as well as adults) sparse and not particularly enlightening.

"*Unfortunately*"? A really curious choice of adverb.

Since I don't believe in "misfortune" I think it is necessary to try to understand what is around me, including things written by the giants, without using them as an alibi, and using my own eyes instead.

And apropos the giants, I would like to add that if it is certainly true, as I suggested in the introduction, that a dwarf sees further on the shoulders of giants – in other words, if it is certainly true that Gulliver sees further if he sits on the shoulders of Glumdalclitch – it is equally true that a dwarf like Gulliver, on the shoulders of a giant or giantess making the stressful crossing of an all too human, fragile, wobbly rope bridge in the pouring rain, is surely destined for a ruinous and fatal fall.

So it is sometimes better for Gulliver to manage things for himself, even if he is a dwarf, a *Grildring, Nanunculus, Omino, Mannikin*, as Jonathan Swift says of Dr Lemuel Gulliver; even if he is only one small man.

Although I am a small man, I weigh less. Maybe it will cost me more time and effort, and maybe I will make many more mistakes, but I will be the one who decides where I go and how I get there.

2.
MONEY AND THE COIN: THINGS HIDDEN SINCE THE FOUNDATION OF THE WORLD?

I begin my partial and highly personal panoramic general history of Money and the Coin with a first slightly peppery and certainly overgeneric statement which should nevertheless contribute to the *polemos* that is needed to help us reflect on the thesis I am expounding: in the history of philosophy I have never come across any of the great writers investigating Money. I have encountered a great many who say nothing at all about it and a very few who take note of it, but almost always do so in disparaging terms.

And here we find something strange. Remarkably, it seems to me, not even the economists have much to say about Cash, Money, the Coin. And when they do, then like the philosophers, they often exude Moralism, Ethics, Scorn or Condescension (all of these, naturally, are solemnly Upper Case). Here is a brief quotation from John Stuart Mill (*Principles of Political Economy*, 1848, Cosimo: New York, p. 465):

> There cannot, in short, be intrinsically a more insignificant thing, in the economy of society, than money; except in the character of a contrivance for sparing time and labour.

And another from David Hume (1752, "Of Money", p. 1) who, though he addresses many important reflections to various economic topics, holds that:

> Money is not, properly speaking one of the subjects of commerce; but only the instrument which men have agreed upon to facilitate the exchange of one commodity for another.

Let's make an exception for Georg Simmel, who I sometimes find astute but at other times rather tedious given that – despite the likely disagreement of numerous scholars, many passages are awash with moralism and, in any case, he is not really an economist; and for Adam Smith who remains a giant, but also a grandee writing about and for grandees, with highly refined observations about many aspects of the human mind. However, neither of them dwells on the origin of Money in anything other than the usual way: that is, with a vision that is always highly "reasonable". Instead, I want to record my admiration and respect for Serge Viderman who, as both a psychoanalyst and a "gardener-floriculturist" and retaining a "classical psychoanalytic" approach, wrote a fine book on money, a fine, very dense book, so dense sometimes that it leaves a messy, even chaotic impression with all the references and reflections it stimulates about a *Homo* who cannot be considered solely *oeconomicus*.

I pass over the various Moralists who, wherever they come from, have exerted themselves in the most fantastical exercises of rhetoric which nevertheless come to the same conclusion: that Money is a Bad Thing.

I could almost understand (I say "almost", given that Religion is the *primum movens* of Money, going back to the gods Inanna and Enki) that Religion, all Religions (at least in theory), and theologians, all theologians (at least in theory) want nothing to do with Money, at least officially. Two questionable exceptions are the Reformation – in particular, John Calvin – which, from mini-schism to mini-schism, has led to the TV evangelism of Prosperity Theology, and a small Japanese "Buddhist" sect which can only with difficulty be called "religious".

After Annikeris remitted the debt owed to him for having redeemed Plato from slavery as a result of the "joke" played on him by an annoyed Dionysius the Elder (one of the reconstructions of the first voyage to Syracuse), I can almost accept that Plato would use that very money on his return to Athens, despite his hardly modest social background, to build a house and studio – twenty *minae*, apparently, 8-10 kilos of silver, or perhaps 30 *minae*: that is, ten to fifteen times the value of an adult male slave, which was only 180-200 *drachmae* in the 5th-4th centuries BCE (for the moment, I will use the terms, Cash, Coin, and Money as synonyms) – and yet with the snobbery of the intellectual, Plato would despise that same money. To take two examples: "nor does holiness suffer them to mingle and contaminate that heavenly possession with the acquisition of mortal gold, since many impious deeds have been done about the coin of the multitude," Republic, Book III, 416e-417a; or, condemning money-makers' habitual "neglect of everything except the making of money, and as complete an indifference to virtue as the paupers exhibit", Republic, Book VIII, pp. 550ff.; but any number of further examples could be cited. In this connection, a 1983 work by Massimo Venturi Ferriolo is of great interest, with much to say about the meaning of Money in Plato, going on to examine the concept of Chrematistics in Aristotle, which I will address shortly. In places, Ferriolo's book inevitably reflects the climate in which it was written, but it remains original, fresh and intelligent, and a stimulating read. The upshot is that, without money, Plato would not have become Plato but a vagabond like Socrates or Diogenes.

The same applies to Lycurgus, later venerated as a divinity in Sparta, who actually abolished money but did not, it should be noted, abolish the slavery which allowed the Spartans to devote their own time to war, keeping the helots, *perioikoi* and other slaves who, under a rule of iron, provided for their masters' daily needs. And since money is needed to pay for goods and services, if I already have someone to provide me with goods and services, then obviously money is no more use to me, at least in everyday life, than it is for children.

Though Athens was open to the outside world on the commercial front, under Dracon and Solon its internal affairs were managed much as they were in Sparta, at least until the 6th century BCE.

I am somewhat puzzled that Spinoza should not go beyond the identification of a generic greed, "to be satisfied sufficiently," stating that "they who know the true value of money... live content with little" (Ethics, IV, 29). It is by no means clear what he means by "true", "little" or "sufficiently". And I am thoroughly surprised that Kant, writing about the "realm of ends" (*Groundwork for the Metaphysics of Morals*, ed. Allen W. Wood, Cambridge, Mass. Yale University. 2002. chap. 10, p. 93), declares:

> In the realm of ends everything has either a price or a dignity. What has a price is such that something else can also be put in its place as its equivalent; by contrast, that which is elevated above all price, and admits of no equivalent, has a dignity. That which refers to universal human inclinations and needs has a market price; that which, even without presupposing any need, is in accord with a certain taste, i.e. a satisfaction in the mere purposeless play of the powers of our mind, an affective price; but that which constitutes the condition under which alone something can be an end in itself does not have merely a relative worth, i.e. a price, but rather an inner worth, i.e., dignity.

Or maybe, Kant being Kant, it isn't so surprising after all.

But I find it frankly difficult to comprehend that Marx (and he is not alone), should claim (I am quoting approximately) that "at a certain point barter was replaced by the coin" and "the coin is a form of money", "money becomes a commodity" and "capitalists use this commodity in the same way that they use all other commodities", albeit seasoned by Commodities-Money-Commodities, or Money-Commodities-Money, without taking the trouble to explain why and how it would make this "very marginal" passage between Goods, Barter, Coinage, and Money (Money coming after Coinage, but given his presuppositions, that is the only way it could be). To be clear, I have always liked the fairy tale of *The Sleeping Beauty*, and I adore expressions like "... and then Prince Charming arrived, they married and lived happily ever after," but that does not mean that the brothers Grimm or Perrault or Andersen or Collodi, for all their heavy moralising, were advocates or founders of a "materialist fabulism" with claims to being considered "scientific". Expressions similar to some of those I have quoted and attributed to Marx, and many others which he presents as his own, or which others present as his, actually existed before him, for example in Aristotle (the concepts of value of use and value of exchange), in Ferdinando Galiani (the concept of labour: though, as a Neapolitan, he calls it "toil" in *Della Moneta*, p. 42) for whom the *Quality of the Work* and not its mere Quantity) is a determinant of the Value of a Commodity. Some maintain that Work as a base element of Value had already been introduced by Locke, but I believe this is only partly true because it seems to me that Locke, who certainly introduces this notion before Galiani, tends to refer to it as an ethical, not an economic element. For him, it is an ethical element aimed at emphasizing the historical priority of and entitlement to private property – in this case, property of any man's individual labour.

Locke writes (I am quoting from memory), "The chestnuts I have gathered on the common are mine." ("Common" is closer to the Latin concept of *res nullius* than to the concepts behind community or State ownership.) But I think this element mainly concerns physical, manual labour of a material and agricultural kind (*Two Treatises of Government*, 1690; I think Karen I. Vaughn, 1978, gives a good summary of this work) which precedes the element of "community" that was invoked in the seventeenth century, before Morgan and Engels, and perhaps as early as Plato, as a primordial state, the condition of humanity in a state of nature. In my opinion, this invocation was aimed at maintaining, even then (though it had been present "from time immemorial", albeit in a variety of forms) the legitimacy of the "Leviathan's" intrusion into the lives of individuals. This *querelle*, expressed in different ways, is still with us today, and I will meet it again; as in Galiani's concept of Supply and Demand for determining Price, although Verri has been suggested as the originator of the concept (while others, not without good reason, suggest Locke, who also applies it to Money, considering it a commodity like any other). Verri is indicated as the first to call Money a "Universal Commodity"; others, Anglo-Saxons obviously, go back to Locke as "inventor" of the concept of Money as "Universal Commodity", but his view that Money (in the 1691 text *Some Considerations of the Consequences of the Lowering of Interest and the Raising the Value of Money* [*The Works of John Locke in nine volumes*, London, 1824] is "as necessary for trade as food is for life" (p. 7), later adding that "the price of money would be raised, as it is of any other commodity in a market, where the merchandize will not serve half the customers" (ibid., p. 10), makes him a forerunner of Krugman, but certainly not the originator of the concept in question. And many things are present in Ricardo, from whom Marx "borrows" some ideas about land and property, turning them into Concentration of Capital, as if the land and money were the same, concrete thing. But many other ideas whose paternity was later attributed to Marx are also present not only in Locke but in Galiani, Verri, Ricardo, and even in Adam Smith, the "template" of many of Marx's "intuitions", as well as in Hobbes, Malthus, Sismondi, and many others. But not the oddity of the falling rate of profit: perhaps that is his own; "perhaps" given that, in this case too, it does not seem very different from the law of diminishing returns, still in the sphere of agriculture and landed estates, earlier hypothesized by Ricardo and still connected to the constant diminution of the rate of profit hypothesized earlier and certainly questionable even then. In his substantial, posthumously published *History of Economic Theories*, Marx devotes only one line to Galiani [ed. Kautsky K. Papamoa Press, 2018]; and gives Pietro Verri just one page out of 1500, purely for the purpose of underlining a detail related to a note of Verri's about the poverty of country people; but nothing on Money, Supply and Demand, or the Concept of Universal Commodity. Another notable Neapolitan, Genovesi, with his splendid and thoroughly politically incorrect intuitions about the concept of "full employment" (pt. 1, chap. 12, p. 174), expressed in an eighteenth-century language and style which anticipate Keynes by two hundred years, is entirely unknown. Obviously Scipione Maffei does not exist. Malthus is Marx's *bête noire*, but one who nevertheless does exactly what he does himself: colludes with one of the "Hegelian" dyads. Sismondi is apparently considered to be "on the ball" but not "collusive" like Marx, and he says slightly different things, being in fact somewhat tainted with cowardice by Marx himself.

Obviously we find in Marx no trace of the concept of Subjective Value, not so much because this concept had not yet been elaborated by the Viennese School, but because – even though it originates, as we will soon see, in Ferdinando Galiani – it is hard to reconcile with the immense work of collusion/confusion which Marx aimed to construct. On the other hand, the point of origin for Marx's work is Quesnay's *Tableau économique*. But that's enough of this because I have no interest in saying any more about Marx and the economists.

In any case, the fact remains that all the giants to some extent concur that from Predation, (we don't know how or why) we passed onto Barter, and from Barter (we don't know how or why) to Money and/or to Coinage. And these "transitions", which actually seem rather to be hiatuses in the way they are conceived and proposed, are still present, though in a variety of ways, in practically all the economists, historians, and philosophers. I should just have said "all" and left it at that. Servet's reflections are interesting in this context, as are those of Felix Martin (2013), who observes how a thoroughgoing myth has been built up around the concept of Barter with the aim of highlighting the frenetic mercantile activity which begins in the seventeenth and eighteenth centuries. This myth "helps", not consciously, to assist the introduction of all those cultural elements which will flow into the great bloodbath (this observation is mine, not Servet's) called the French Revolution (Liberté! Égalité! Fraternité!). Servet notes that until the seventeenth century there is no historical reconstruction which introduces the concept of Barter:

> Ainsi, grâce à l'économie classique et à son mythe du troc il est possible de concevoir une société sans monnaie et donc sans Prince (p. 30). (Thus, thanks to classical economics and its myth of barter, it is possible to conceive of a society without coins and therefore without a Prince)

The meaning of the construction of the myth of Barter would therefore lie in the need to free oneself from the sacred and hierarchical values which underpin monetisation. And Alary also concludes his book by claiming that:

> *La filiation entre troc e monnaie ne semble pas évidente* (p. 29).
> (The filiation between barter and coins does not seem evident.)

However, like all myths, Barter has a basis, as we shall see later, for example in Herodotus. Personally, I think that many "ancient" authors dealt with Barter, long before the seventeenth century, but Servet's idea certainly has an intriguing flavour.

The element which explains this universal phenomenon is a feature that von Hayek calls constructivist and I would call functional-rational: in his opinion, it starts with Descartes, but I seem to find traces of it in Herodotus. To clarify: for almost all the giants, "at a certain point", "historically", "in an earlier time", Man first gathers, then steals (or the reverse, or both together) and then finds himself using Barter; later on, Money appears, and after that the Coin. According to the constructivist-functional-rational approach, this means that, more or less simply put, men "decided" that Barter was better than Predation, and then that Money/Coins were even more effective, rational, simpler than Barter. And this effectiveness, economy, and flexibility would have justified its origin and its adoption. Like me, Paul Enzig (1948, p. 26) also says he is greatly puzzled by this reading. And even before Barter, how we passed from continual, mutual Predation to Barter, likewise remains a mystery on which I will do no more than pause in passing.

In any case, to adopt the "rational" argument to explain the transitions which occur is like saying, 'One thing is better than another because it is more practical or simpler or quicker, therefore let's adopt the better, more effective one.' The same would then be true of the Coin; and again for the Banknote/Coin; the same again for the Credit Card, for Bitcoin and so on. I think that from this constructive-functional-rational approach we may deduce at least one thing with certainty: almost all intellectuals, economists, historians, scholars, philosophers, and anthropologists live, as a fixed rule, in their own homes and/or in their ivory towers, Descartes and Herodotus included. My conclusion is certain and absolute. And it cannot be otherwise since it is a falsifiable claim (if not actually false), therefore absolutely scientific: and certainly approved by Popper. And Popper says so himself!

If these scholars had lived in a block of flats with its own "democratic" decision-making bodies, they would have known very well how few decisions of a residents' association are constructive-functional-rational, bearing in mind the fact that apartment blocks generally speak the same idiom, agree about the maintenance of shared property, depend on the decisions that are taken, and must fully respect the decisions taken as well as those not taken.

If we "project/transpose" what happens in a block of flats and in a residents' association onto the scale of a nation state, we will – scientifically – obtain the proof that, given the claims of the scholars, none of them can ever have taken part in a residents' association meeting and haven't the slightest idea what goes on in them. As a result, they can only have owned and lived in private houses, undoubtedly with big enough gardens to avoid the noise of possible other houses and the racket from blocks that would certainly be irritating but far away.

In other words, they are completely detached from what they are pronouncing judgements and opinions on.

Naturally, Plato, Aristotle, St Augustine, St Thomas, Kant, Hegel, and many others have not shied away from this position with obviously

different nuances, arguments, perspectives.
They too were always the owners of isolated, luxurious detached villas.
Here is a brief quotation from St Augustine (*Commentary on John*, Homily 40, 9: ed. Fitzgerald AD., *The Works of Saint Augustine: a Translation for the 21st century*. New City Press, New York. 2009) to clarify his interest in money:

> We are God's coinage, coins that have gone astray from the treasury. What had been stamped on us has worn off because of our straying; someone has come who will do it over, because he had done it in the first place. He too is looking for his money, just as Caesar looks for his. That is why he says, Pay back to Caesar the things that are Caesar's, and to God the things that are God's. To Caesar money, to God yourselves. So that then is when truth will be stamped on us.

See also Discourse 335/C, 7-9 where Augustine describes money as a frivolous and wicked lover (at http://www.augustinus.it/italiano/index.htm) the terms Money/Coin/Solidi/Sesterces/Gold appear 485 times, always in this vein): beautiful terms, but hardly pertinent to the language and dynamics of a resident's association meeting in an apartment block; and just as predictably as every Psalm ends with "Gloria", Augustine, in his Exposition on Psalm 131 writes equally predictably that:

> It is because of private property that among men there are quarrels, hatreds, scandals, sins, crimes, murders. What is the reason for all this? Because of property possessed in private.

These words, which echo Plato, though not word for word, recall the phrase attributed to Augustine by Campanella in *The City of the Sun* (quoted in Croce, 1899, p. 183n.): "*Amputatio proprietatis est augmentum caritatis.*"

Sometimes even the Greats of antiquity look at the finger pointing to the moon instead of at the moon itself; Lycurgus, Plato, Augustine *in primis*.

The old story that tells how taking something from someone inevitably brings an advantage to someone else must be deeply rooted if, after centuries, it continues to be trotted out. On the other hand, even *Genesis* shows how deep-rooted some nasty habits are, but despite being so deep-rooted, they still do not become good, useful, and intelligent.

That scoundrel Scipione Maffei, who chooses not to know his illustrious descendants, writes:

We say the knife cuts, and yet it does not cut unless it is moved, and to good purpose: this is how Money bears fruit. Without human labour, even the field is sterile. Coin does not give birth to coin, just as the earth does not give birth to earth; but worked earth gives grain, and invested cash gives profit (1744, III, I, p. 194).

And the *knife* must first be invented, then we need to know how to construct it, and after that have the skill and desire to *move it*. In this passage Maffei puts forward exactly the same argument that Locke (1690) used a little earlier when, speaking about Labour, he claimed that it is human (the individual's) Industry-Intelligence-Attention which makes the soil fertile, making it richer, allowing it to better satisfy the demands of those who work it, thereby enriching the whole collective, including that collective which does not want to involve itself by investing its own labour, given that if my land produces more for me when it is worked, at the very least, there are many more chestnuts which other people can gather on the common; *pace* Aristotle and/or his epigones who often have their own axes to grind.

Ferdinando Galiani also makes some amusing observations about "philosophers", those whom I could, in today's terms, call *maîtres à penser*, opinion leaders, or experts who pontificate on every subject from every pulpit; the people Hayek (1949) calls intellectuals:

> Perhaps it will be said to me by many philosophers, even if it is true that the value of jewels and rare things is founded on human nature, as I have demonstrated, that they never cease to regard my opinions as ridiculous, miserable delusions. To such persons, I reply that I do not know if they will find anything human that does not seem so; and nothing in the world can make them change this opinion. But I would wish the good philosopher, after he has divested himself of earthly deceptions and, almost dehumanizing himself, has risen so far above other people that he can laugh at us wretched mortals and find us amusing, when he later detaches himself from these thoughts, descends once more and mingles in society (as the necessities of life compel him to); I would, I say, wish to see him return as a common man and not a philosopher. That laughter which, when he is a philosopher, has given health to his spirit, could, if he engaged in it now, perturb the spirits of those dear to him and of others too (Galiani, 1750, p. 40)

These observations recall Nietzsche's famous and provocative "invitation": "Aboard ship, ye philosophers!" (*The Gay Science*, § 290, in *Complete Works of Friedrich Nietzsche*. Hastings, Delphi Classics, 2015.) By "association", it also occurs to me that I do not recall a single Saint who does, or rather did, the regular duty of a Husband or Wife. Like the intellectuals cited above, they are sometimes married, but the odour of sanctity is only detected in them when they go back to being single. Strange. And yet Marriage is a Sacrament. Can there really be no Married Saints, elevated to the honour of the altars precisely *when* they lived in couples, precisely *because* they lived in couples, during the sacred bond of marriage, precisely because of living in the married state, in a family?

Married people are sometimes canonised, but for what they have done outside marriage: wars in the service of the faith, charitable works, benefices, cessions of territory or concessions in politics; never for what they have done in marriage.

Saints are virgins, martyrs, ascetics, celibates, young women, priests, popes, children, old people, sometimes widows and widowers, sometimes intellectuals and sometimes illiterate. Always single. I can't think of any married person canonised *for* being married.

Given the Church's great historical and emotional intelligence, I think there must somewhere be a Saint, or a pair of them, created *ad hoc* to refute claims of this kind, but it is quite obvious that such a Saint is less important than Ambrose, Augustine, Aquinas, Frances, Anthony, Benedict, Clare, Catherine, etc.

After this digression, from my love of truth, I must note that Aristotle (384-322 BC), drawing heavily in my opinion on a text by Xenophon (430-355 BC), the *Œconomicus*, is alone in having writing a brief work on economics, or rather on the management of Money, or better still, the administration and management of the Home – and to be absolutely precise, the direction of the *Oikos*, οικος (hence the term *Œconomia*): in other words, the "single house" of which he speaks. But he says nothing about the nature of Money except as *Chrematistica*, often associated with Usury – although I have never liked this commonly accepted translation of the Greek, which seems a bit forced to me since the original term is much broader, deriving from the root χρη, 'necessity', 'need', 'scarcity', and I would prefer to call it Financial Activity, even if this distinction seems irrelevant to many modern ears. *Χρη* is from the Akkadian (w) *aqru*, 'scarce', 'rare', cognate with the Latin *carere*, 'to be scarce", 'to be needed', hence 'costly', and also with the root χρ of χρυςός, gold (see Semerano). Χρ is also very close to Christ, the Annointed (Greek, *chrio*, χριω, to anoint), the Messiah, He who has the rare gift of the *Chr*isma (not charisma, which is linked instead to *charis*).

I also add that in Hebrew "an occasional, rare, chance event" is *aqrai*.

And where there is any reference to Money/Coin in the *Nicomachean Ethics* (V, 5, 1133a 30) and in the *Politics*, it is again always in terms of Ethics (and always obviously Upper Case) and some rhetorical-philosophical exercise about what the Coin "would be": i.e. "nonsensical" and existing "altogether by convention", then slipping via the moralistic metaphor of King Midas who is an example of the uselessness of coins-money-cash-wealth itself (*Politics*, I, 1257b 8-17) along the lines of "This is how you'll turn out if you don't eat your greens."

Calling the Coin, as Marx does, an "agreement altogether by convention" does not seem very far from Aristotle. Aristotle is also the first to make a distinction between Value through Use and Value through Exchange, a distinction which will have its day many centuries later and, as I have already suggested, will in its commonest use be attributed to Marx, a distinction which I will attend to later. In any case, the *Peri-Oikoi*, the "free" citizens of Sparta, are 'οι περιοικοι, and the metics of Athens, the "extracommunitarians" are *Meta-Oikoi*, 'οι μετάοικοι, which tells us a great deal about democracy, work, the concept of a Free Man, the *Oikos*, houses, and money; as do the terms *household* and *familiar* in more recent times.

Still on the subject of *Chrematistics*, I find Scipione Maffei's (1675-1755) observations on the question interesting and pertinent. He maintains that it is the translation from Greek to Latin (profoundly influenced by a certain *Chr*istian vision – Augustine's in fact) that brought about a misunderstanding of Aristotle, who would not have given *Chrematistics* its modern pejorative sense of Usury (a term which occupies more than half the 350 pages of Maffei's text). For Aristotle, *Chrematistics* is an Art, that of creating money out of money, an art he definitely dislikes, but it is not Usury; and Maffei believes that, by Usury, Aristotle means that particular, hateful type (from the Latin *usare*, already in Cicero and still present, not necessarily in a negative sense, in Aquinas) called *obolostatica*, ὀβολοστατική, which consists in a "loan" at 1% interest a day, charged daily (Maffeit, 1744, III, chap. 1, p. 197). He also makes interesting reflections on the subject of the *Weltanschauung* of translators (and here I add the old saying that translators are traitors) regarding their fundamental preparation-training-disposition with regard to the term Usury. Maffei reveals how very easy it may be for a translator who – consciously/unconsciously, I might add – is "partial", or "parti pris", or somehow inclined towards a certain view of life and the economy, to confuse, for example, *usura* with *foenus*, an institution of Greek origin referring principally to maritime law and connected to the risky financing of overseas expeditions for which the investor took on the risk of losing the entire investment, but in the case of a positive outcome required a very high rate of interest on the original sum.

In the same way, translators may superimpose, confuse, blend εκτοκίζω, I charge interest, with pure usury; or ανατοκισμός, compound interest; or δάνειον, money lent; and χράω, to need (I, I, 11). *Chrematistic* comes from χρήμα, something that is needed, a helpful thing, a commodity for use, substance, goods, money, wealth.

Maffei's 1744 text is in fact a "defence of usury" of usury, or rather it is an attack on any moralism lurking in assessments of it, an attack which brings him into conflict with the Catholic Church, to the point that it suffered the most violent ostracism, was placed on the Index, and both its publication and dissemination were prohibited. I do not know if Bentham knew this text, but it is a fact that he published his highly celebrated *Defence of Usury* (1787) forty-three years later and that he uses the concept of *foenus* ("maritime trade and maritime loan") just as Maffei does. Sarcastic in tone but, nevertheless acute and stimulating, although made in quite a different context – that of Writing, which is just as relevant to Translating (apart from the Newspeak of *1984*) – George Orwell's observations in his essay 'Politics and the English Language' of 1946, anticipates many of those on the use of so-called "politically correct" language which Robert Hughes, for example, would make in *The Culture of Complaint* (London, Harvill. 1999).

Even in the age of the so-called scientific, aseptic, detached, and icy "Illuminati", from *Candide* and *Émile* onwards, it seems that the whole world feels newly obliged to pour scorn on Money; after a millennium *of* and a millennium *from* the "Dark Ages", in the era which could at last enjoy the Light – theirs obviously – Money starts being attacked again with renewed emphasis but now in a "lay" manner, "democratically".

Naturally, as Seneca's were before them, their pockets are filled with the money they despise. However, the claim "the whole world.... after a millennium" should be slightly revised, since it is a bit narcissistic and Eurocentric. What does "the whole world" mean? The whole of "our" world!

In the millennial Byzantium which is no longer (only) Rome, though it had lost its African coasts, with the Macedonian dynasty at the peak of its solidity and power, functioning also as an "aircraft carrier" of that formidable drive for evangelization, literacy, and acculturation of the Slav nations and the immense Russia. Islam extends from Cordova to the Indus with a fantastic flourishing of inventions, culture, trade.

In China, the Song dynasty reaches sublime peaks in the arts and achieves the unification, leadership, and administration of an Empire of enormous dimensions, setting up a definitive, complex, and elaborate "Exam

System" for selecting the legions of imperial officials needed to make the vast administrative, bureaucratic, and military system function, a system of selection that would last for another whole millennium (see Miyazaki, 1988). It also invents the Encyclopaedia, by which I mean the concept of "summarising" all that is known in a certain area of human activity, and applying it systematically to all areas of knowledge: that is, ignoring works like the *Naturalis Historia* of Pliny the Elder). It introduces paper Money (while China's geopolitical situation remains enormously complex over the centuries, it is the Song dynasty which first prints banknotes, and not, as Murdin (2012, p. 92) suggests, the Han) which made the Venetians smile some time later when Marco Polo returned from the Celestial Empire. With regard to this, Sismondi's observations about paper money and gold made in the early nineteenth century are amusing, interesting, and in their way prophetic.

In Cambodia, the Khmer culture, which intersected first with Hinduism and then Buddhism, both of Indian origin, conceives and creates splendid cities; likewise in India which at the same time is fiercely resisting the Islamic invasion; and in Africa, where the wealth and opulence of Ghana and then of Mali unleash Islamic "cupidity" well before the discovery of American, well before the European obsession with Eldorado. Given the geographical distance, I will not dwell on what is yet to become Central and South America (*auri sacra fames*!, Vergilius, Aen. III, 57)

All this is going on while "our world", still to a large extent pagan, is licking the wounds caused by the dissolution of the Roman Empire, but "only" the Western half, at least for a time. And with regard to the time needed for licking those wounds and to the so-called "Crusades" judged "imperialistic" by some historical interpretations which would like to be considered politically correct, we should keep alive the memory of the fact that, while the origin of Arabic writing, based on Aramaic-Nabatean, is indicated in 512 and was introduced, it is said, by two Christians, Zaid Ibn Bammad and his son (see also Caetani, 1914, p. 159, for whom the origin of Arabic script is Syriac; while Gaur, 1984, p. 97, seems to incline towards Aramaic-Nabatean), we have Mohammed's Hegira only a hundred years later in 622 (the Lesser Hegira happened eight years before). And yet the Prophet's body is scarcely cold before Islam is already looking towards India and treating the waters of the Indus as a mere footbath, while only thirty years later Byzantium has lost almost all the Middle East and Zoroaster and the Sassanids have been brushed aside. Another hundred years, and peaceful, tolerant Islam arrives "on foot" with the edge of the sword and exempting new converts from taxes, reaching as far west as

Poitiers as early as 732; and "our world" is only able to get itself organized with the first "imperialist" Crusade in 1095: that is, 365 years after Charles Martel, 365 years after terrible devastation. In any case, Poitiers represents only a first brief halt in an expansion, military before it was cultural, which has been expanding for at least a century and does not stop the Saracens from making incursions deep into France and making themselves at home in Spain. *Hic manebunt optime* for a long, long time, given that it will take another 781 years to reach the end of the *Reconquista*; and even if it is more legend than history, it will be 317 years of terrifying blows since Roland at Roncesvalle, which happens in 778. After swallowing up Byzantium, reducing it to a glowing ember in 1453, Islam continues to exercise a fearful pressure by land and sea on the very heart of Europe, controlling almost the whole of the Balkans, even besieging Vienna in 1529 and again, in spite of their defeat at Lepanto in 1571, in 1683: in other words, 588 years after the first "imperialist" Crusade. But at least the failed siege of Vienna seems to have given us croissants (croissant de lune, crescent moon, *yareah/sahar*, as I will show below).

"What the caterpillar calls the end of the world, the world calls a butterfly," says an old Chinese proverb (why on earth are Chinese proverbs always "old"?!); and here the allusion is to Jacques Le Goffe, another giant, who seems to consider the history of Money mainly from a European viewpoint, or rather from a *côté* that is just a little bit French, when in fact the butterflies have already flown around the entire globe millions of times and billions of new, highly vigorous caterpillars have been transformed into the same number of multicoloured butterflies.

As I was saying, the eighteenth century which saw the birth of Adam Smith and records the beginning of interest in the subject of economics is also the century that witnesses a far-reaching surge of moralising which aims to deny and condemn, almost more forcefully than in previous centuries, the importance, the potency, the meaning of Money. Will this be an inevitable consequence of the same old English industrial revolution and/or of what Servet refers to? Well, who knows? In any case, in "this" world of "ours", after the "chrysalids of the Dark Ages" (?) new butterflies keep being born: naturally from the Code of Justinian to Charlemagne's monetary reform, from Maimonides to Averroes, from the debates in Bologna and Paris about a "fair price" to the Aquinian concept of *Communis Aestimatio*, or the reflections of Albertanus Brixiensis (Albertano da Brescia) to whom the great Geoffrey Chaucer is somewhat indebted for his *Liber consolationis e consilii*:

ex quo hausta est fabula gallica de Melibeo et Prudentia, quam, anglice redditam et The Tale of Melibe inscriptam, Galfridus Chaucer inter Canterbury Tales receipt

(from which was extracted the Gaulish tale of Melibeus and Prudentia which, translated into English and called "The Tale of Melibe", Geoffrey Chaucer adopted in the "Canterbury Tales"), [J.Thor Sundby, 1823]

from the foundations of individual right posited by Accursius, from Irnerius and the Christian commentators, long before even the *Magna Charta Libertatum*, to the uprising of the Commons against the Empire, laying down the basis for the Reformation and the Renaissance, from the School of Medicine in Salerno to Dante, these are all obscurantist side-shows.

And among the countless species of almost entirely equivalent Lepidoptera which come to life and take flight today, the only "recent" exception, it seems to me, apart from Adam Smith and Ferdinando Galiani – and it should be treated cautiously – is the butterfly represented by the Austrian School: Carl Menger, Ludwig von Mises, Friedrich von Hayek, and others. Since it was von Hayek that I mentioned earlier, I think he must be the most congenial to me. This school has also been called the Psychological School, which will be one reason why I like it and probably also the reason why some "old school" writers do not. And it is to this School that we owe the concept, or rather the modern rediscovery, of "Subjective Value" mentioned a little while ago, which is so interesting and suggestive. Subjective Value was also introduced by the great Ferdinando Galiani, cited earlier, who (1750, p. 39) for example, reflecting on Count Ugolino who would have paid "a thousand grains of gold" for an egg, who highlights an assertion by another writer, Davanzati:

> "A rat is a most disgusting thing; but during the siege of Casilino one was sold for two hundred florins, an enormous sum; and yet it was not expensive because the seller died of hunger while the buyer lived." Here, thank God, we see one occasion where [Davanzati] acknowledged that expensive and *good value* are relative terms. (My italics)

This School, like all economic disciplines, nevertheless tends to adhere to "conscious" data, but even so it comes up with interesting reflections. And my reason for calling it the Only One is because it seems to me that no other School stays so anchored to the roots of Man without an over-reliance on mathematical formulae and without forgetting, in its examination the Economy, that there are unconscious factors which motivate people in their relationship with money; that the consequences of their economic decisions and actions are by no means sequential and rectilinear; that the reasons and aims of economic actions are often not knowable or predictable; and that not everything, indeed very little, is attributable to rational choices.

In other words, it is the only school which, at least sometimes, frequents condominiums and not just Frank Lloyd Wright's *Falling Water*. And it is recent because it is barely more than a century old and its most prominent exponents spanned the twentieth century. But it is to be handled cautiously because in any case all, truly all, these butterflies – the Austrian School, the London School of Economics, the Chicago School (even though they do make some interesting reflections on the Condominium), to say nothing, for heaven's sake, of the Frankfurt School, and ignoring the craziness of Five Year Plans, or Agnes Heller and her Bisognism and of the various variables considered "independent" (such as Salary and Work), or Great Leaps Forward and Sugar Cane Economics – in the end all talk, write, think, discuss, propound theses in terms of Macroeconomics (even when they say or write Micro-), and hence of Economic Policy, Fiscal Policy, Monetary Policy, and hence *de facto* of Ideology, Ethics, Morality (obviously always, rigorously, Upper Case). And even in their most fully worked out and clearly expressed form they reiterate exactly the questions posed by the "classical" economists, which they essentially and provocatively summarise within a range which might be as follows: Is it better to sell our grain to strangers who will pay us in gold and thus have much gold and little grain in our country, or better to buy grain abroad, which costs less, and enrich foreigners by paying for the grain in gold and making them rich?

These words also recall those of Pliny the Elder (*Naturalis historia* XII, xli, 84):

> Minimaque computatione miliens centena milia sestertium annis omnibus India et Seres et peninsula illa imperio nostro adimunt: tanti nobis deliciae et feminae constant.

> (At a minimum estimate, India, the Chinese, and the Arabian peninsula relieve our Empire of a hundred million sesterces a year: that is how much our luxuries and women cost us.)

If, as I believe, Pliny is referring to the "Great" Sesterce, a unit of accountancy rather than current coin – given that for important transactions the neuter *sestertium* was used, as in *sestertia milia*, rather than the masculine *sestertius*, and Pliny does indeed use the neuter – this would give us the tidy sum of 150 billion euros (1 Great Sesterce = 1000 Sesterce; 1 Sesterce = about €1.50. Today, only the LVMH Group 50 billion euros; and it is also possible that Pliny simply means "70 times 7": an infinite number of sesterces. Then he adds a moralistic touch:

> *Quota enim portio ex illis ad deos, quaeso, iam vel ad inferos pertinet?*
>
> (How much of this sum, I would like to know, is dedicated to the gods above and how much to the underworld?)

All possible intermediate and imaginable positions lie within this range: that is, they are always the same; yes to tariffs, no to tariffs, maybe tariffs; print a lot of money, a little money, some money (with all the bestialities connected to the famous concept of the circulation of coins, a concept introduced, in this case correctly, by Hume in his *Political Discourses* of 1752); a lot of gold, a little gold, some gold, and so on; yes to colonies, no to colonies, a few colonies; yes to war, no to war, a bit of war; treaties yes, treaties no; the State yes, the State no, the State maybe a bit; transformed products versus semi-transformed products, prime materials versus other prime materials to be bought in turn with our gold.... *et coetera*.

Though they are hardly contemporaries, the observations of John Locke and John Stuart Mill on coinage and foreign trade are interesting in this regard, and almost always in the same vein; the great Galiani is aligned with them in this respect too.

In truth, a great many studies of economics, not only those of the Austrian School, acknowledge at some point in their work that there must be hidden somewhere something "irrational" or "non-rational" in economic choices: whether Animal Spirits for Keynes (a term picked up by Akerlof) or the Invisible Hand for Smith, or the Hand of God for Scholasticism. But straight afterwards they dive back into some reassuring formula or some bold general idea which always seems to have been taken from Descartes, where on the one hand there is an external, mechanically predictable, concrete, "objective" reality and on the other a rational and sentient Man, *Homo Œconomicus*, beautiful, strong, and reasonable, rationally planning his own destiny and investments; stable, even if he sometimes changes his hairstyle and the colour of his shirt; male on the whole, though in fact it was women who made the Phoenicians' fortune. And never – I say again, never – does any emotion prompt him to do something stupid, because these commentators believe that Irrationality is entirely marginal and ineffectual, even as they extol its importance.

In reality they entirely gloss over, or rather repress, the problem which had raised its head; for all these scholars, the Human Being continues – annoyingly – to be not entirely rational, not totally a matter of tables and formulae. Really very irritating. We must do something to eliminate this anomaly!

Which brings me back to the non-distinction between Macro- and Microeconomics, since these terms are used in markedly different contexts and with equally varied, sometimes even contradictory, meanings by different authors. In any case, it seems to me that I am not alone in having made these observations since Behavioural Economics also reflects on the matter. For example, writing about economists and psychologists, Furnham and Argyle in *The Psychology of Money* (1998, pp. 3-4) cite Lea and Webley (1981) who, after calling Money "a neglected topic", write:

> We do not need to look far to understand this negligence. Psychologists do not think about money because it is the property of another social science, namely economics. Economists can tell us all there is to know about money; they tell us so themselves. It is possible, they admit, that there are certain small irregularities of behaviour, certain deficiencies in rationality perhaps. This, psychologists can try to understand, if this amuses them. But they are of no importance. As economic psychologists, we disapprove of both the confidence of economist and the pusillanimity of psychologist.

As far back as 1986, Krueger, though he is not part of this trend, edited a book about Money, centred more in fact on the "pathology" than the "physiology" of Money, with a highly explicit title, *The Last Taboo*. For convenience, I will not dwell on Behavioural Economics and its exponents (Arrow, Simon, Tversky, Kahnemann, and others such as David Tuckett; and more recently in Italy, Egidi, Motterlini, Guala, and others) who since the nineteen-fifties have been making interesting studies of Micro-(Mini-)economics, since such studies, often carried out *in vitro*, concern the subjective perception of Money according to cognitive-Gestalt criteria, game theory, problem solving, statistics, probability, etc., but nothing on the genesis-nature of Money. Kahneman's *Thinking Fast and Slow* glances at these researches from a general psychological or sociological viewpoint. While this approach draws on elements that are (perhaps) quite remote from psychoanalysis, such as concepts of bias or WYSIATI (What you see is all there is) or Priming or Loss Aversion, it has the merit of not considering Money as the sole preserve of economists. Behavioural/Cognitive Economics poses, although not directly, a series of problems for psychoanalysis too, and I will not address those here.

As for Neuroeconomics, sometimes called the daughter/stepdaughter of Behavioural Economics, I would rather hold off discussing it, despite being largely in agreement with Damasio's (1995) criticisms of Descartes (apart from the fact that, in my opinion, he falls into the same error I have just highlighted). This is because, when they hear about a certain type of "scientific" observation, a certain type of theory, a certain type of approach – the kind that John R. Searle calls "neuronal chauvinism" (1992, p. 38) [*The Rediscovery of the Mind*. Cambridge, Mass; MIT Press.] – my "mirror neurons" immediately go on strike. And even when some researcher mentions that Freud himself had great faith in the developments of biology, attributing para-gnostic and divinatory qualities to the Master, I cannot help wondering if Freud's faith in biology's marvellous progressive destiny had more to do with the Master's frustration over the gonads of Triestine eels in 1876 than with his awareness of a radiant biological future or with the Aphasias so dear to Solms (2000). So, I will confine myself to citing – only to emphasize how slow psychoanalysis has been to address these studies – certainly not the observations of Sacha Gironde, but the words which Matteo Motterlini, though he is culturally close to Damasio (words which in this case, thank God, do not bring into play those close relatives of the Cartesian pineal gland, the amygdala, hippocampus and cerebellum) uses in his introduction to Dash's book about the Dutch financial crisis of 1636 which originated in the speculative moves around tulip bulbs. That crisis was a forerunner of the present one, and its main features are well conveyed by Schama (and as ever, the schema is the one given by John Law1 and largely identical to the South Sea Bubble in which Isaac Newton, no less, lost a packet, and used in more recent times by Charles Ponzi, a

---

1   John Law (1671-1729), a gambler, founded the Mississippi Company after persuading the Regent of France to entrust him with running what would become the National Bank of France, and promised enormously high rates of interest to those who purchased "guaranteed" shares in plots of land in America. The South Sea Bubble (1711-1721): in a gamble largely identical to Law's, in this case the English public debt was taken on by the South Sea Company, to which the Government paid annual interest and granted certain monopolies. In order to finance the project, the South Sea Company began to issue its own shares which immediately increased sharply in value with the promise/ illusion of "certain" future profits. Charles Ponzi (1882-1929), an Italian immigrant to the USA convicted of postal fraud, promised his shareholders that he would turn a five-dollar investment into a capital of five million dollars in two years. He accumulated a personal fortune by means of a financial "ploy" involving international postal services, which gambled on/ exploited the different values in different countries when buying and selling "stamps". Despite their apparent differences, in all these cases there is a so-called St Anthony's Chain or, if you prefer, a Pyramid Structure:

model in which the Futures and Derivatives of the time, as always and as they do today, played an important role). In his introduction to Dash (1999) in the Italian edition, Motterlini writes:

> Contrary to what we are used to believing, and as this book testifies, financial decisions do not necessarily have anything to do with money. They involve intangible motives such as avoiding losses or not feeling regret. They are to do with anger, frustration, envy, pride, honour, fear, and panic, or with lack of optimism and trust.

Overall, apart from numismatists (an interesting term, from *Nomisma*, νόμισμα, *Nomos*, νόμος, *Nummus*; No-mos, directly connected to *Mos*, which I will address soon), the only people who seemed to have taken some kind of interest in the subject of "Money" seem to have been sociologists, anthropologists, and ethnologists, each of these being interested in paying attention – rightly, I think, from their point of view – more in Money's function than in its nature and genesis: its function as an intermediary between men, on the same level as rituals, beliefs, institutions, totemic animals, systems of kinships, etcetera. We could add to this approach that bizarre text by Karl Marx. Perhaps also Silvio Gesell, with whom I won't be dealing in this book; just as I won't be considering Karl Polanyi and many others. It seemed to me that the sole exception in the attempt to understand something of the nature of Money is Bernhard Laum (1924), a scholar whom I can only categorize as an antiquarian, and whom the literature makes an object of heavy irony, when not subjecting him to repression and oblivion, and who "together" with a few others has tried to say something different from the mantra we constantly hear in various permutations. I

---

an expectation of high earnings in a short time is created (and it would be interesting to understand how and in what circumstances). It starts by paying high rates of interest to the first subscribers with the money paid in by the second wave of subscribers; then these are paid with the money collected from the third wave, and so on. There is no genuine investment and no real production of wealth or goods, but a financial vortex of money paid for with other money, maintained solely by the subscribers' wish to obtain very high financial gains without apparent risk. In Italy, not long ago, someone using mass communications media built up a small financial empire in exactly this way. He was later convicted on the same basis. It is a pity that the entire, supposedly redistributive, Italian pensions and welfare system has for decades been based (for financing its current account and not pensions) on the same principles as the business founded by that person: the principles of the St Anthony's Chain, or if you prefer, the same criteria as Law and Ponzi, throwing "some fear" into the system's resilience: the welfare system, and much of the system of so-called Futures, Derivative Swaps, etc.

will not refer to Marx now, but perhaps later on. And perhaps also some of Max Weber's intuitions might be numbered among the "heterodox" approaches to Classical Economics, at the very least for his attempt, which is interesting in its aims though highly debatable in its content, to make a connection between Faith/Religion/Ethics and economic matters.

Obviously, many other authors could be cited along with those I have recalled, but I think these are enough for the task in hand. The "exceptions" must also include many of Freud's observations which form the central fulcrum of my entire study. In these pages I will try to understand something about.... I don't know what to call it, given the Tibetan bridge and the giants to whom I referred earlier and given that this is not a work of Economics, History, Religion, Anthropology, Philosophy, or Sociology (again, all of these in Upper Case) but aims to consider elements on a minute scale. In homage to Jonathan Swift, I could coin the ironic neologism *Quinbus Flestrinian Non-Œconomics* – the Non-Œconomics of the Man Mountain – referring to the name given by the Lilliputians to Gulliver, who will shortly afterwards become Lilliputian himself. And in our house, the house of psychoanalysis, suspending reflections I will make later about Freud, since this book should have the ambition of mainly, if not exclusively, addressing a "psy-something", it is better to draw a merciful veil over the Fromm of *To Have and To Be*. But look, I've made a "mistake". In fact, I chose to make a nice lapsus: I wish it was *To Have and To Be*! Instead it's much worse: *To Have or To Be*! Given that Fromm represents the point of an iceberg much larger than the one that sank the Titanic, I will indeed draw a merciful veil. Or rather, I will try to lift this veil, just a little.

3.
ERIC FROMM AND THE PROPERTY

Erich Fromm is a giant and held in high regard by huge numbers of people; but sometimes even giants say things that are, to say the least, debatable (I mean my use of the adjective "debatable" to be politically correct). Here's just one example: in *To Have or To Be ?* (London, Bloomsbury. 2013, p. 59), speaking about property, he writes:

> This kind of property may be called *private* property (from Latin, *privare*, "to deprive of"), because the person or persons who own it are its sole masters, with full power to deprive others of its use or enjoyment ["sic"].

Fromm's *Kultur* is certainly vast and his nuanced reading of some aspects of history and the human spirit is undoubtedly interesting, but *privatim*, *privatus*, and *privus* mean no more than One's Own, Personal, Individual, Singular, Personally, On One's Own Account – ie. Autonomous – as well as Particular, Special, Distinct, Different. *"Privos privasque antique dicebant pro singulis,"* wrote Sextus Pompeius Festus in the 2nd century AD., quoted by Valerius Flaccus in *De verborum significatione* and handed on to us by Paul the Deacon (8th century) in his *Historia Langobardorum*, as Semerano emphasizes, adding that *"Privus* was glossed as *prei-u-o-s*: that is, prior, isolated, detached, alone," and goes on to make the striking point that the adjective "proper" derives from *pro privus*. The verb *Privo, -as, -avi, -atum, -are* means to detach, take something away from someone, but in the sense of "split off", "make isolated", "sever": for example, take away love, or take away pain. It does not mean "make off with something belonging to someone else", rob, appropriate another's goods, except occasionally in a metaphorical sense, as in steal someone's soul, or life, or cut off their light. Of course, "steal something" also means isolate that something from its legitimate owner, given that "stealing someone's life" makes that particular good unavailable for that person, as when I steal a soul, or my soul is stolen. But this would mean acknowledging that if I deprive someone of life (literally, not metaphorically), that life was previously theirs! In other words, it was a state, a "property", or an exquisitely *private* quality before being taken away, before the subject was *deprived* of it; and it was not previously taken from anyone else, but was a characteristic *proper* to the subject, and *private*. One's own *property*, indeed. (This aspect will be revisited in relation to a possible reading of original sin.)

Furthermore, to deprive someone of life has its own names: *killing, murder, etc.* In everyday language, in the Sumerian codes, in those of Noah and of Moses, and in all the world's penal and civil codes, and has come to be considered a different kind of behaviour from theft: indeed, it is considered as "shedding blood". All the more because, in this case, the thief – i.e. the murderer – definitely does not gain an increased lifespan by "stealing" life from someone else: the exception being the removal of organs from the victim for a transplant, either on behalf of the thief or as a commission, but Fromm is not referring to this specific context. Therefore, his book should not 'be' (!!!) entitled *To Have or To Be*, but *To Be Autonomous or To Have Autonomy*, or rather *To Be Autonomous and Have Autonomy*, but in this case it would have turned out to be a very different book written by a very different Fromm, although, even at the dawn of 1982, some Marxist writers graciously deigned to concede that "*l'avoir est une métonymie de l'être*" (Aglietta and Orléan, 1982, p.34). So, Having and Being are not mutually exclusive alternatives. It took 142 years to move on from the work of Proudhon to which Fromm seems to refer, but we've managed it. Perhaps. In any case, committing Theft (*fur, -is*), Stealing (*furtum, rapina, expilatio, direptio*) and Robbing (*furor, -aris, -atus sum; furtum facio; eripio; abigo*; and also *surripere; subducere; aufere*) are different from the original *privare* and it is not "*to deprive of*". And Fromm should have known it.

I realize that when you quote a sentence that someone has spoken or written there is always the risk of decontextualizing the meaning of which the sentence had been part. However, this is not the case here, since Fromm's sentence clarifies and exactly summarises – as we gather from a reading of the whole book – the meaning he wants to attribute to the term 'deprive', and the whole meaning which leads him to force, *pro domo sua*, the meaning which the Latin term does not have, thereby clarifying and explicating the forced ideology of the concept. As does the entire Frankfurt School. Fromm should have known that it is only under the impact of Christianity, beginning with St John Chrysostom (344-407) who defined Wealth as injustice, with Augustine (354-430) and later, I think, with Boethius (480-524), that the term begins to acquire – without losing any of its original meaning – the nuance of negativity proclaimed by Fromm and to which he clutches so fiercely. Christianity makes its full impact with the complete collapse of the Empire, for which it certainly in part responsible, though "only" in the West. Christianity was then experienced as *communitas*, in relation to which the single individual cannot propose to be autonomous but must form a part *in toto*. St Cyprian (210-258) had already written, *"Extra ecclesiam nulla salus."* The Catholic – that is, universal – *communitas* was necessary even for mere biological survival: in other words, people had to stay together so that they could stand up to persecutions, barbarians, famines, and invasions in those days when *mala tempora currunt*: the Catholic and Ecumenical *communitas* as *Oikos*, Home, Universal Community. Someone has said that *ecumene* indicates dry land but, given that mankind can live in, inhabit, name Home – that is, *Oikos* from *oikein* (οίκος from οίκειν) to inhabit, to have something on which to reside/pitch one's tents, *wašābum* in Akkadian, *lashevet* in Hebrew. So, *ecumene*, *communitas*, *oikos* and home are, semantically, the same thing. Erich (Augustinus) Fromm is evidently much more Mün(t)zerian than Mennonite in his origins and to a far greater extent than his liberal declarations would have us understand. There would be nothing negative about this if only it were at least acknowledged; in practice, however, this approach says more about his sympathy for Proudhon and/or the Anabaptists than his etymological rigour. I wish, and hope, that Marx's troubling text on the Jewish question (1844) will not be numbered among the roots of Fromm's reflections, even though many of Fromm's concepts, including the admiration for Mün(t)zer which I have speculatively attributed to him, appear in that terrible work.

We cannot rule out the possibility that, as it was for Weber, the model of Augustine (*"amputatio proprietatis est augmentum caritatis"* quoted earlier) so profoundly present, though elaborated somewhat differently, in the education of Luther and Calvin, is still there, and "active" in Fromm's theorizing. It is still there and still "active" like something else that I will talk about later. Nor can we rule out nominative determinism, given that, in German, Fromm means pious, calm, devout, peaceable, subdued, tranquil, religious, docile, tame (with a name like mine, Benini, Little Goods, this is a joke I could have done without).

On the other hand, the mental schema which Fromm uses is fundamentally the one used by the Babylonian Kings or the Canaanite authorities when – in the El Amarna Letters, for example – they apply the term *ḫāwiru* (man/ free man) to the pastoral nomads, both Hebrew and otherwise (according to some scholars, the term Hebrews may derive from this, i.e. *ivri*) who did not answer directly to the Imperial authority and did not endure the pressure either of taxation/levy or the obligation to render *corvée* or other services to their Lords. *Ḫāwiru*, simple pastoral nomads and semi-nomads who, not being part of the Imperial organization since they lived, often voluntarily, on the margins, were considered "free" as a matter for scorn, "free man" being understood as "man alone, solitary, isolated, homeless, a vagrant" not belonging to the community/ organization / Empire. In other words, *ḫāwiru* has the sense of "wild", "barbaric", "vagabond", "thief". More simply, Freud also cites the term in his Moses (pp. 128, 151), likewise making *ḫāwiru* the etymon of *ivrì*, Hebrew. Jacques Attali too notes in passing the possible origin of the term "Hebrew", but relates it to Ever, the great-great-grandson of Noah.

In Hebrew, "wild" is also rendered as *bar*. In many languages there is no absolute superlative for adjectives: thus, for example, "very many" is rendered as "many-many"; and the plural is often expressed by duplicating the adjective. For example, the Sumerian term *kur* (mountain) becomes *kur-kur* in the plural; while, in Hebrew, *gal* (wave) becomes *galgal* (wheel). In the same way, *bar-bar-(ian)* could derive from savage-savage, wild-wild, i.e. very wild, most wild (in its Semitic origin); and "barbarian" in Hebrew is in fact *barbàri*. There are those who hold that this Hebrew term *barbàri* is an import from Latin, while others derive it from the Sanskrit *bàrbaras* or *vàrvaras*, meaning *woolly-headed, shaggy*, when not *babbling* because they do not really know the new language. However, given that a foreigner, a stranger, a barbarian is called URBAR/UR.BAR.RA in the Sumerian language, and *ubār* in Akkadian, and that intensifying duplications are very common in these languages, I find it very hard to see the root as being Latin. More crudely, Liverani renders the term *ḫāwiru*, or *hābiru* in his transliteration, as "exile, enemy." Ishmael, Abraham's eldest son, born to Hagar, and himself a *ḫāwiru* by definition, is called by God – in the English Standard Version – "a wild donkey of a man, his hand against everyone and everyone's hand against him." Rashi di Troyes calls him "savage", and in the Torah he is "*pere/arod*", an "*onager*", a wild ass (*Genesis* 16:12); and I do not think he can have been, even metaphorically, half man half beast as Bernal (1987) suggests. Instead, the descriptions of Enkidu "seem" to indicate a "primeval man", *wild*, uncivilised, sent by the gods to punish the arrogance of Gilgamesh; and it is curious the Gilgamesh, who mourns his death in several passages and versions of the epic, calls him an onager, a wild ass. Maybe *bar* was initially associated with *hābiru*, with those brutish wild thugs who don't abide by the Rules. In other words, I believe that there is a further element in the term *ḫāwiru*, an element of mockery, if we think that among the Babylonians the term *awīlu*, applied to a Babylonian, referred to a free and noble man fully incorporated into civil society (see also Saporetti, 1984, p. 16), while the phonetically similar *ḫāwiru* scornfully, even sarcastically, indicates a "free" individual, one who does not pay "taxes", but because he is solitary, a poor wretch, a wild ass, a "bandit", and not because he is "noble". This is rather like the phonetic-linguistic game played by Gramsci when he tries to derive *Christian* from *cretin* in French. We could also consider the fact that in old Akkadian *barbaru(m)*, an almost completely Sumerian borrowing, is the word for *wolf*, which would strengthen still further the concept of barbarian-*ḫāwiru*-savage.

In modern terms we could call this, as much current and widespread ideology does call it, Private versus Public; and everything that is not Public – that is, everything that is Private – becomes something the Community is Deprived of: that is, Robbed from the Community; Robbed in fact from the totalising and omnivorous bulimia of the "Community". Therefore, everything that is Private deprives the Community and, as a result, the Private Subject, the *ḫāwiru*, necessarily becomes a thief by definition. Even today, as in the past, it seems to me that those who do not pay homage to the "King" or the societal Superego are considered *ḫāwiru* and not *awīlu* – i.e., thieves, as Fromm seems to say – and must be "re-educated" into life and society since they are considered delinquents by the Sheriff of Nottingham, John Lackland, and naturally, Sir Hiss.

I add that, in other circumstances, *ḫāwiru*, *hābiru*, *apiru*, *hapiru*, *ha'iru* also acquire the sense of Husband, Man, Strong Man, Hero [Chicago Assirian Dictionary]; it is also interesting to emphasize that this term *ḫāwiru*, which originates from the Sumerian syllable IR, male (in this case, male slave), and scornfully indicates the state of non-submission to the "imperial community" and stigmatizes the fact of independence, gives rise to the Latin term *vir*, as Semerano suggests; and the term *eviratio* (emasculation, castration) is pretty clear.

Vir, *viri* indicates a concept quite different from the Babylonian one. It is a term suggesting Strength, Virility, and Courage, exquisitely *manly* and *individual* characteristics which will combine to form Rome and the Romans, a Rome – as we are reminded by John Kenneth Galbraith, whom no one can accuse of jingoistic, fascist rhetoric – which is the often discounted cradle of modern Law. *Vir*, *viri* closely connected to *Virtus*, *Vis*, *Vigor*, *Robur*, and also *Virga* and *Virgo*. And *Vira* which "via India takes on an identical meaning" (De Gubernatis, 1899, p. 123). Irony of History.

After this digression which implicitly concerns a large chunk of psychoanalytic thought, I close this panorama. In the end, the idea which I have unearthed is that Money and the Coin seem to belong to the "things hidden since the foundation of the world, to quote Girard: hidden things that not even he takes into consideration.[1]

---

[1] René Girard seems uninterested in Money and the Coin, but some of his epigones – i.e. Aglietta and Orléan – seem on the contrary greatly interested: so much so that they have written a book called *La violence de la monnaie* with an Introduction by Jacques Attali. This work is picked up several times by Dr. Schimmel. The authors, to whom I will later on refer as "advocates of "the Penny Black" attempt to bend Girard's reasoning to a Marxist logic, going so far as to claim openly (p. 28), "À partir de cette hypothèse que Girard appelle mimesis, il est possible de fonder une conception générale de la valeur d'usage dont une conséquence importante est de soustraire la théorie marxienne des formes de la valeur aux critiques des partisans de la théorie subjective de la valeur" ("On the basis of this hypothesis which Girard calls mimesis, it is possible to found a general conception of use value, one important consequence of which is to remove the Marxist theory of the forms of value from criticism by the partisans of the subjective theory of value"), using, as Fromm used Proudhon, Girard's words to support somewhat outdated theses; a further proof of this rather partisan approach can be found in a note on p. 220 where the authors report a contribution by Young who explains how Weimar overcame hyper-inflation by licensing an enormous mass of public dependents who, being inserted into the state accounts both as the outcome of the war effort and as a "social shock-absorber", along with war reparations, weighed enormously, along with war reparations, on the national balance of payments. But straight afterwards, only a line later, on p. 221, they go back to their first true love, maintaining that the hyper-inflation was caused by the bourgeoisie exporting capital and by the dirty tricks of the commercial banks. In a spirit of patriotism, no doubt, the authors make not the least reference to a substantial additional cause of the destabilisation inflicted on the already highly fragile German economic situation: the brutal and treacherous invasion of the Ruhr, the only viable economic pole in post-war Germany, by the French army which, unilaterally and ferociously, decided to occupy the region from 1923 to 1925 as "compensation" for losses incurred as a result of the war, a move opposed by all other European governments. This is beside the fact that they present themselves, with questionable taste, as among the few who have really understood the Theory of Value in Marx (perhaps only they and Althusser), yet they do not even take the trouble to cite Aristotle as the founder of the distinction between Use Value and Exchange Value. Instead, they cite him (p. 158, but only, I think, in reference to the *Politics*) to support the usual thesis: that Chrematistics, to which they add the introduction of the Coin, kills off the old, sweet, harmonious, and redistributive arrangement (obviously considered καλός καί αγαθός), thereby killing off the primordial, paradisal social solidarity. With no right of reply, they then make the distasteful claim (p. 160), "quant au coeur des empires perse et babylonien, il a résisté au monnayage jusqu'à la conquête d'Alexandre » (« as for the centre of the Persian and Babylonian empires, it resisted the introduction of coinage until Alexander's conquest"). Thus, I gather from these learned savants that Alexander fought the Babylonian empire, something I was entirely unaware of: but I especially thank them for having shed light with such great precision, clarity, acumen, and historical authority on the origin of coinage in Mesopotamia. Faced with such wisdom, it would be better if everything yet to come in this book, given that we are only on p. 62, were thrown straight into the garbage bin or that the book be returned to the vendor: with a full refund, of course.

4.
EVIL: IN THE WORLD AND IN PSYCHOANALYTIC THEORY

Money, its genesis, and its nature simply do not exist in any fully worked out system of thought, and when they do, Money is "simply" Evil – when viewed favourably. When viewed unfavourably, it is EVIL ITSELF. The same conclusion holds for the Coin which is sometimes "the shape of Money".

Is this why money becomes excrement? In Greek κάκκη, from the verb κακκάω, evacuate, defecate, undoubtedly connected to κακός, ugly, κακως, bad, and κάκη, malignity, vice, wickedness, baseness; in Latin, *cacca*; in German, *Kacke*; in French, *cacà*; in Spanish, *càca*; in Portuguese, *cocò*; in Brazilian, *cacà*; in Romanian, *càca*; in (dialectal) English, cack; in Irish, *cac*; and in colloquial Hebrew *Qaqa* or *Qaqi*, as appears in the title of this book. As for the various forms of *"merde"/"merda"*, they come from Latin *merda, -ae*, which in turn echoes the Sumerian *Mudra* and the Akkadian *Mardītu*. "The Vedic *mr'd*, soft earth.... The Latin *merda* can have no other original meaning than soft" (De Gubernatis, 1899, p. 181). Droppings, sheddings: Sumerian UR:A; Urinate: Sumerian A.SUR; Defecate: Sumerian SUH, Akkadian *zû* (see Semerano and ePSD).

Money is evil/EVIL and, therefore, in the hope of keeping at a distance from it, it is called excrement: or is it the other way around, that being excrement it becomes evil/EVIL from which it is necessary to keep our distance?

Is the case that *Pecunia non olet* (Vespasian): or instead, *olet* (Ferenczi)?

Sometimes, more for the sake of argument than from conviction, someone claims that Money is the absolute GOOD (a Hebrew proverb copied by Marx), but here too we can smell a rat.

The point I want to emphasize, getting back onto the rational paradigm, is that if we take as our starting point the poem with which I began, and play with the entire philosophical, economic, and scientific output of at least the past 2,500 years – starting in other words, from Plato – it would be as if a Sumerian scribe had left us this:

> One day the Lord of Kullab got up from his bed
> the lord of Kullab got up
> he got up and wrote a letter to his friend in Aratta.
> But the letter was never received because we had not yet invented stamps.
> O Inanna, O Inanna, invent stamps for us.
> O Inanna, O Inanna, invent stamps for us.
> The great Inanna harkened to the prayer
> the great Inanna harkened
> harkened to her brother Enmerkar and blew in the ear of Gilgamesh
> who hastened from the land of the Cedars with a new Penny Black produced by the printer in Kur.
> Praise Inanna, praise Inanna, who has given us stamps.
> Praise Inanna, praise Inanna, who has given us stamps.

(Taken from the – thank God, non-existent – poem *Gilgamesh the Postman and the Lebanese Printer*)

This would certainly make a horrible poem, but as a poem it would nonetheless be interesting (fortunately none have been found that are as terrible as this) since it is a poem about the distinction between the public-spirited civil servant (*awīlu*) and the outlaw (*ḫāwiru*) which I would like to explore further.

Obviously I am not referring to THIS non-existent "poem".

However, what would be neither interesting nor acceptable would be for me to believe the story of the Penny Black, the Lebanese Printer, and a postman called Gilgamesh, whether Sumerian or Babylonian. Which is, however, exactly what we would read next according to the constructivist-functional-rational paradigm:

… and at a certain point, voilà, because we're tired of bartering we'll move on to Money, to Coinage, because it's better!

which sounds exactly like:

O Inanna; O Inanna, invent stamps for us because we're fed up with not being able to post letters and having to carry them to the recipient personally, on foot!

And it's just as well that an invisible hand, or the hand of God, or some Spiritus Animalis regulates things: but even so, Adam Smith, and St Thomas too, and likewise John Maynard Keynes would be banned, to some extent, by this invocation, this paean.

Incidentally, we should observe that in Keynes, Spiritus Animalis could without much difficulty also be called Intuition/ Preconscious, and not Animal Spirit as someone influenced by a hasty reading of Hobbes and to much reading of Fromm would have us believe. In any case, Keynes's

Spiritus Animalis seems very different from the Spiriti Animali[1] of Descartes, even though, ironically, he gave them the same name. Instead, but it is only my "impression", it is the conception of the pineal gland in Descartes which most resembles the cerebral localising of Envy, Jealousy, or Anger "identified" today with CAT scanning, magnetic resonance, or the "God Helmet", rather than the electrodes of a few years ago. Maybe,

---

1   Descartes, *The Passions of the Soul and Other Late Philosophical Writings*. Oxford, OUP; 2015. Article X, 'How the animal spirits are produced in the brain': "But the crucial point here is that all the swiftest and most subtle parts of the blood that have been rarefied in the heart by its heat are constantly flooding into the cavities of the brain. And the reason they flow there rather than anywhere else is that all the blood which has left the heart via the main artery heads straight for the brain, but since it cannot all enter there, because the openings are very narrow, only the most agitated and subtle parts of the blood make their way into the brain, while the rest is dispersed throughout the body. Now, these very subtle parts of the blood form the animal spirits. And for this to happen they do not need to undergo any further change in the brain, except being separated from the other denser parts of the blood. For what I call 'spirits' here are only bodies, and their only properties are that they are very small and fast-moving, like the parts of the flame that comes from a torch. As a result they are never stationary, but while some are flowing into the cavities of the brain, others are simultaneously flowing out through the pores of the substance of that organ. Through the pores they are conveyed into the nerves, and thence into the muscles, by means of which process they move the body in all the various ways it can be moved." Descartes really lets himself go in this work, an amusing text in some ways, and in some ways frankly pitiful. It is amusing because in Descartes' supporters, some philosophers of science and, as always, some of Descartes' relations we can see an attempt, outdated but still interesting, to confirm the existence of mysterious phenomena. Perhaps in those "subtle bodies" that are "very small", half way between Speedy Gonzales having convulsions and a "flame that comes from a torch", we can see, or hope to see, a poetic anticipation of the neurotransmitters. It is frankly pitiful, and not in the Latin sense of *pietas*, because the picture which emerges is one of presumption, wishing to apply one's own conceptions to spheres entirely outside one's own, and, turning to more recent times, philosophy (Philosophical Consultancy, for example) is not unfamiliar with this kind of "overflow" (even psychoanalysis cannot always resists this temptation), resulting – for Descartes – in a series of features which it is almost a compliment to call "bizarre" since constructing a system of this kind and in such a manner is a lot closer to the delusions of President Schreber than to a comic metaphor. I will just add that I think Animal Spirits here means Spirits of the Soul and not Bestial-Animal-Beastly-Primordial Spirits, given that they come and go through the pineal gland which, as we know, is the seat of the soul and in its "oscillations" regulates all the perceptions, sensations, and movements of human beings. On the other hand, Descartes' passion for "very small particles" which explain Light and Magnetism is familiar from other works of his. However, it wouldn't be a bad idea sometimes to say, "I don't know", as Newton seems to have done (on just one occasion, apparently).

to the great satisfaction of Franz Joseph Gall and Luigi Ferrarese (1836) we are on the threshold of a new era in phrenology? From the paradise of Barter, so New Age and ecological, voilà, we move on to the foul age of "The Devil's Excrement", an expression attributed to Luther,[2] while it may possibly be attributable to St Francis, or as someone suggests, St Filippo Neri, or perhaps Francis Bacon: and for others, me included, simply to folk

---

2   In his Table Talk (quoted in Haight, 1977, Martin Luther expresses himself thus: "Gelt est verbum Diaboli, per quod omnia in mundo creat, sicut Deus per verum verbum creat" ("Money is the word of the Devil, through which he creates all things, the way God creates through the true word.") Ergo, Gelt verbum Diaboli, non stercus! In Discourse DLXXIV of the Tischreden, [*Table-Talk of Martin Luther* (trans. Hazlitt, W.). Philadelphia: Lutheran Publication Society.] Luther writes that "*The devil vexes and harasses the workmen in the mines. He makes them think they have found new veins of silver which, when they have laboured and laboured, turn out to be more illusions. Even in open day, on the surface of the earth, he causes people to think they see a treasure before them, which vanishes when they would pick it up. At times, treasure is really found, but this is by the special grace of God. I never had any success in the mines, but such was God's will, and I am content.*" This passage clashes with the general, perhaps catholic, Vulgate. On the other hand, Luther's indignation about the sale of indulgences (and the economic vexations caused by many Bishops, including the Pope, wielding temporal power and possessing great personal fortunes) and the lucrative activity linked to the possession of "authentic" reliquaries prompted his rebellion, at least initially. Hence, it should not surprise us that, having been educated as an Augustinian, he has it in for money (see my earlier comments on Augustine's position. Tomasso Campanella's position – "*amputatio proprietatis est augmentum caritatis,*" is almost a Wealth Tax *avant la lettre*; a position which runs smoothly from Plato and Lycurgus to Marx, and then in our case, to Fromm) though we should not forget that his protector and saviour, Frederick the Wise, owned a substantial and highly profitable collection of "authentic" relics, including an authentic drop of Mary's milk. These were "authentic relics" which Luther never found reprehensible and against which he never hurled his thunderbolts of indignation. We should likewise not be surprised that in the *Turmerlebnis*, the Experience in the Tower, the "visions" of the German Hercules often played out among toilets, demons, and excrement, clear elements of scorn presented incarnate in the figure of the Pope seen as the Anti-Christ who accumulates into himself everything that Luther despises. Since the Pope is involved with Money via the sale of indulgences (and via "authentic relics") the Pope is the Devil (that is, the Anti-Christ; as are the Turks and, often, the Jews) and Money is in league with the Devil: that is, the Pope. So, at least as far as I can discover, Luther certainly has the idea of a relationship between the Devil and Money (mediated by the attack on the Pope, and its metaphor; Erikson's work is still very interesting, speculating that, among other things, for Luther the Pope, or *Papa/ Papà/ Dad/ Daddy*, was an object onto which to deflect the anger he nursed towards a Papà, i.e. Hans Luder); someone wrote that he changed his name, Luder in Luther, because Luder, in German means Carrion/Rotten. Even here, though, we do not come across Faeces, except "on the rebound", in the sense that when we wish to make an unflattering judgement about

"wisdom". In other words, we move from Barter to the root of all evil, as the English have it, borrowing from St Paul (first letter to Timothy, 6: 10). Actually, St Paul says that "the love of money is the root of all evil", the subject rather than the object, a consideration also taken up, though in a "classical" context, by Haight (1977).

For St Augustine, however, as I noted above, the beginning of all iniquity is private property but deep down, although Sirach does not use these words (*Ecclesiasticus*, 10: 7-13), its true starting point you will find in the Scripture which says, The origin of every sin is pride (Augustinus, *Commentary on St John's Gospel*, Homily 25:16).

John Donne (1572-1631) indicates a different "root of all evil":

> In this rebellious part, is the root of all sin, and therefore did that part need this stigmatical mark of circumcision, to be imprinted upon it (Sermon CXXX, quoted in Calcagno, 2009, p. 147).

John Donne is obviously not speaking of Money but, tying in remarkably with what is to come, the foreskin. In the end, everyone finds the root they're looking for.

Nevertheless, it is interesting to observe the Calvinist – or, for convenience, Protestant in general – development of the idea about the "Devil's Excrement", a development which would lead Calvin to "sanctify" this same "Cack/Shit", albeit in the guise of successful work, go so far as to sanction it as an indicator of divine Grace and Benevolence, with all the polemical significance which Predestination had acquired, which produced another singular effect that, having become a further dogma, it would cause the Reformation to give rise to (modern?) Capitalism (see Max Weber, *The Protestant Ethic and the Spirit of Capitalism*).

Unfortunately, whatever the case, I seem to have read so many people, both Giants and little people, making the Penny Black manoeuvre: out of the black box of Barter, with a backflip and death-defying double

---

someone, in many languages and with various, often creative nuances, we apply excremental attributes to them. I would add that, depending on the circumstances, such insults may also refer to anatomical parts which have little to do with defecation, but are instead associated with reproductive and/or sexual functions whether male or female. Nor should we forget the fact that they very often refer to genealogical and professional characteristics, specifically the mother's profession, often the sister's, sometimes the father's, and (surprisingly) hardly ever the brother's. (In general, the sons and daughters don't turn up either.) All of which has little to do with Money, but a great deal to do with the "stuff" of the person being abused, and always with the anger of the abuser. But this too has little to do with Gold.

somersault – hey presto! – comes the rabbit, Money! And on one side we have Gilgamesh the public functionary, as I mentioned, a splendid fellow, being a postman and therefore honest by definition, since his costs are met up front; and on the other side, the *ḫāwiru*, obviously criminals, nasty, filthy, wicked, who stay outside the Community and steal from it.

Whether this comes about so that we can adopt a "scientific" explanation founded on the constructivist-functional-rational paradigm, or by alchemy or magic or sleight of hand, or else is a mental state created by aliens, the recipe is always the same, and the claim is reiterated absolutely and categorically. There was no monetary and financial system worthy of the name in Antiquity: only a series of transactions involving Goats and Cabbages. Or else, if we had found one – a civilisation which, making allowances for wars, was capable of functioning internally by means of barter: Cabbages rigorously seasoned with abundant Redistribution, two drops of Primeval Communitarianism, and a dash of Oriental Despotism. Naturally the tremendous inflationary phenomena of ancient Rome (going back to Servius Tullius around 560 BCE, or – if you prefer – dating from the First Punic War and running through every other war, both the Republic's and the Empire's, Punic or otherwise, through the dawn of the Empire itself and going on until its fall – in the West, not to mention the East, and to say nothing about Ancient Babylon) are mere trifles.[3]

And, following the same recipe, it is claimed that the monetary and financial system is a modern, indeed contemporary, invention introduced by capitalism. End of story.

And woe betide anyone who says the contrary; that once the state of innocence that was intrinsic to Barter (?) had been lost, Man had to reckon with Excrement and create a new ethic for himself – oops, sorry: Ethic (upper case).

---

3   Paul Einzig (1948, p. 213) notes the changes which occurred between the last dynasty of Ur and the start of Hammurabi's reign. The "official" rate of exchange had been 1 silver Sheqel = 300 Sila of grain (= 1 gur), already becoming in Ur III a rate in real terms of 1:240, and straight afterwards, with Hammurabi, a rate of 1 silver Sheqel = 150/180 Sila of grain, and later, during a period of famine, 1 Sheqel = 20 Sila. There is a striking relationship (devaluation? inflation? Sooner or later we'll have to understand something about this!) between Sheqel and Sila until the Assyrian conquest of Babylon: i.e., 1 Sheqel = 3 Sila of grain, and all the more so on the black market; whereas, in 539 BCE, the year of the Persian conquest of Babylon, we pass in a single year from a ratio which had returned to 1:234 to one of 1:15-40. Einzig adds that the same situation applied to dates. Bottéro (1987, p. 175) also makes interesting observations about real salaries compared to the official ones in the Code of Hammurabi.

A "state of innocence" that lasted until the Renaissance? To me that just seems weird.

In fact Max Weber said no such thing: the received wisdom is that his ideas changed everything (naturally if my child does something wrong it is always the fault of the bad company he keeps) but the truth is that he dipped his toe into this as well; and many other writers have pushed much further than Weber, blending the Penny Black, the Luddites, and Science Fiction, and forcing Hegel – who had earlier got his feet thoroughly wet – and Girard beyond any reasonable limit, dribbling their way up the wing and systematically, intentionally removing all alternative viewpoints and carefully weighed utterances, starting with the grievously abused Benedetto Croce (who had read a bit of Hegel, after all). Anyway, I think Ernesto Sestan's Introduction, from the now far-off year 1963, to an Italian edition of Max Weber's work mentioned earlier, makes the general terms of the question very clear, giving unto Caesar what is Caesar's and unto Weber what is Weber's. Neither Sestan nor Weber, however gives unto the Cathars what was the Cathars' "some" years before Hus, Luther, Zwingli, and Calvin. The Cathars are not only a French cultural and religious movement, as some historiography, especially transalpine, tries to assert, but a movement which reached from the Black Sea to Catalonia, generating Occitania, the original Occitania which was not merely a piece of southern France but for nearly three centuries the incubator of much modern thought.

However, if we assume that Calvin is the founder of the modern capitalist *ethos*, it would be fun to argue that the inhabitants of Brescia and Bergamo, as is attested by certain extremely earthy and *crude* linguistic elements – that is, differently cultured, differently graceful elements – present in the language of those regions, are the direct descendants and continuation of the Sumerians, or at least the Chaldeans: as we might deduce, if we wished to, from certain conclusions in the present work.[4]

In any case, weirdness like this, the idea of a lost "state of innocence", the lost virginity intrinsic to Barter, generates in turn, or perhaps simply endorses, another dogma: the equation that Money = the Demon's Excrement, with all that this entails. (I will not dwell on the etymology of the terms Demon and Devil which, in this case, represent a negative reinforcement of Excrement since "shit" is most certainly not a compliment

---

4   On the other hand, as V. Kutra notes, the Celts, κελτοι, really did reach Babylon in 324 B.C. where they encountered Alexander the Great, and in 280 B.C they undertook the Great Expedition which brought them to Delphi and then to the creation of the region of Galatia: the κελτοι who became γαλάται in central Turkey, and to whom St Paul wrote an Epistle: perhaps *baita*, house (Akk. *bitum*,

when it comes to be used to address something or someone. Moreover, the situation isn't greatly improved if the Excrement is produced by a foul or ambiguous individual, the spirit of Evil, bringer of conflicts.)

This dogma has been laid down so deeply and extensively that even psychoanalysis, its Metapsychology, and the lines along which libidinal development is constructed are profoundly infused with it. And I am not referring solely to Fromm. There are countless writers, psychoanalysts and otherwise, who have accepted this equation Money = the Demon's Excrement: that is, Money = Faeces.

But this begs the question, is the Excrement Demonic or Human? If they are the same, it would equate Human with Demon: a strange and hardly secular conclusion, though maybe St Augustine with his belief in Original Sin would not consider it a bestiality. Freud himself seems to accept an equivalence of this kind (Money = Excrement and Devil = Human), although with some ambivalence, I think: more because of the need to systematize the "Metapsychology Witch" than from personal conviction. Or maybe it is just my idealization of Freud that makes me see the matter in this way. Who knows?

Nevertheless, the letter which Freud writes to Lou Andreas Salomé, as we will see more clearly later on, seem "truer" than the attempt to create a systematic taxonomy of psychoanalytic Metapsychology or the spirit of folkloric research which seems to motivate it.

But there is no point denying that, not only Fromm, Ferenczi, Reich, Róheim, Jung[5] and many others, but almost all psychoanalytic intellectual, theoretical and clinical production has been aligned to this approach, to the equivalence whereby Money = Faeces: as the previously cited Löwenkopf (2003) also says.

Only *privatim et rarissime* (*privately/ in camera caritatis*, and not "by theft") do we hear things put a little differently: for example by Eissler, Glover, Saraval, Bleger, and Etchegoyen who, while officially staying

---

Heb. *bait*-) and *cuz*, a traditional mutton dish in some Brescian valleys (Turkish *kuzu*, lamb; Sumerian UZ.UN, goat) come from there? It is also interesting, in the context of this book, that St Paul's Epistle is about circumcision and that one object in particular distinguishes the Celts: the Torque. This term connects back to the famous *Anaqiti*/Giants of the Bible (Numbers 13:28), as Garbini acutely observes on p. 88 of his *I Filistei*.

5   C. G. Jung, Collected Works, vol. 5, *Symbols of Transformation*, § 276: "We might also mention the intimate connection between excrement and gold: the lowest value allies itself to the highest. The alchemists sought their *prima materia* in excrement, one of the arcane substances from which it was hoped that the mystic figure of the *filius philosophorum* would emerge (*"in stercore invenitur"*)."

orthodox, treat the topic of Money very pragmatically, allowing us to glimpse positions "inside" the analyst himself, and a long way from any anal references. Likewise, John Gedo, whom I had the pleasure to meet a long time ago, used to speak about money in a decidedly less "orthodox" way, as did Johannes Cremerius, whom I met during my youthful training and who expressed, at least on the other side of the Alps, positions very little aligned with the orthodoxy about Money.

And in a note to this passage, Jung adds: "De Gubernatis says that dung and gold are always associated in folklore, and Freud tells us the same thing on the basis of his psychological experience. Grimm reports the following magical practice: 'If you want money in the house all the year round, you must eat lentils on New Year's Day.' This singular association is very simply explained by the indigestibility of lentils, which reappear in the form of coins. In this manner one defecates money." Some observations: first of all, the dates. This is a text from 1912 when Freud and Jung were still (more or less) in agreement: the break between Jung and Sabina happened around 1911, and the break with Freud (officially) came to a head in 1914. In spite of it, the definitive edition of this book dates from 1952, as we know from the chronology of Jung's works: and so, break or not, the equation Faeces = Money remains a fixed point for Jung too. As we will see in the next chapter, the claim "Freud tells us the same thing on the basis of his psychological experience" more reflects the idealization of the Master than a clinical approach or a real knowledge of the sources. The *filius philosophorum* is often called the Philosopher's Stone, and since the alchemists spent centuries looking for it all over the place, it is no surprise if they also looked in excrement. As for Angelo De Gubernatis, he was a highly cultivated Italian scholar who travelled the world in the second half of the nineteenth century and studied for a couple of years in Berlin around 1863, the same Berlin which would soon witness the glories of Panbabylonism. He wrote many works beside *Zoological Mythology* (1872) cited by Jung. A quick scan of this book shows that in volume 1 he refers to the fable of *Fearless Giovannino* in which Giovannino goes down to Hell and comes back with an ass which "shits" gold from under its tail (p. 388). However, it seems to me that Hell, the Devil, and the ass only appear in the (as he says, unpublished) "Piedmontese" version which he cites, and that there is nothing of the kind in the variants from the Po valley, nor the one adopted by the brothers Grimm, nor in Italo Calvino, nor in the original sixteenth-century version. At best, we would have to turn to Bertoldo to find a farce in which we are presented with a "gold-shitting" ass, or to the Grimm brothers' fairy tale, "The Wishing-Table, the Gold-Ass and the Cudgel in the Sack," in which the ass expels gold coins, but "from front and back". The other element that we need to take with a grain of salt comes, I think, in the way De Gubernatis reads the myth of Midas turning everything he touches into gold. For De Gubernatis, this transformation would happen after Midas had eaten "things", digesting them and turning them to gold (p. 383). But this interpretation would not account for how Midas runs the risk of dying from hunger and thirst. As far as lentils are concerned, I think any comment on their digestive properties is superfluous. I would rather point out – though I will leave

There are whole passages in which Karl Abraham (Abraham – interesting name! I have a feeling we'll meet him again), but he's not alone, cheerfully expatiates on the inherently anal character of concepts which, to him, are clearly "mediaeval" in origin – they may or may not be – and which he is using to shore up the edifice he is constructing: that of "The Demon's Excrement". His implicitly accepted reasoning merely expounds something that it should explain.

I do not question the anal phase. I believe it to be one of Freud's most brilliant intuitions.

And I certainly think Money may become highly significant in the articulation and development of the anal phase, but it is the reduction of the entire subject of "Money" to the anal phase that I find absolutely inadequate.

It would be just as inadequate and meaningless to claim that Sexuality is a characteristic element of one phase of libidinal development.

Libidinal development is the story of the entire evolution and transformation of Sexuality itself.

In the same way, I believe Money is a dimension that is articulated and enriched step by step, and takes on characteristics which may be an integrative part of a certain phase of development but not exclusively so.

As a "synonym" for Wealth-Wellbeing-Ease-Security, Money is an integrative part of all developmental phases and, like Sexuality, it takes on features, valences, and meanings which change with the changes brought about by maturation, experience, and by the ways in which a person works through it: obviously reflecting all the possible permutations of which human nature – and pathology – are capable; as in so many other things.

---

this mysterious for the moment – that "lentils" (*lenticchie*) are like "freckles" (*lentiggini*) in that both are *round*. Moreover, in volume 2 we find foxes coming out of their lairs with gold, ants which separate grains of wheat from gold, some cockerels and magpies involved with gold, girls rewarded with gold, dragons guarding gold rings and various other things, but nothing connecting Gold with Excrement. On the subject of the ass, in another of his works, from 1899 (p. 151) De Gubernatis goes into more detail: "While May is the month of roses... the fact that it is also called the month of asses indicates the relationship of the ass, a phallic animal, with Priapus and his cult, a relationship most evident in the myth of Silenus, and also in Lucian's story of the *Golden Ass*, reworked by Apuleius.... In my *Zoological Mythology* I tried to illustrate the Greco-Latin proverb of the *asinus in unguento* and the Hellenic myth of the onocentaur." In the whole of his chapter on the ass De Gubernatis constantly pays attention to the ass's ears (and not to its tail) as the animal's characteristic feature in its various mythological transformations.

So it is frankly difficult for me to accept, for example, that Money is simply a desexualized partial object that is destined uniquely to serve as a means of exchangeas a colleague seems to believe, summarising Freud's position in a work from some years ago in a sentence very like one I previously quoted from John Stuart Mill. I first encountered the sentence in an article by Gilles Arnaud who attributes it to Dr Schimmel. Having read Dr Schimmel's complete article, as well as her very interesting book on money, it seems to me that Arnaud rather misrepresents her, partly misunderstanding what she has to say in her summary of Freud's thinking.

However, it is also true that whether I sum up Tolstoy by saying he is old, ugly, tedious, prolix, and writes badly, or that he is tall, handsome, blond and blue-eyed, I am bringing something of myself to the description.

On the other hand, much of the psychoanalyst's work consists in just such "misunderstanding" of what is said or happens during sessions, and so I too will "misunderstand" Dr Schimmel's work, using it to draw attention to a concept dear to my heart: that is, my belief that her summary of Freud's position in 1915 is part of something which a great many people/ colleagues think, or at the very least take for granted, and going straight on to apologize to my colleague for having pulled her hair, and thanking her for her collaboration, however involuntary.

The observations which follow are not intended as a critique of that fine clinical work of Dr Schimmel, but a spur to reflection about a very widespread way of thinking which characterizes much of contemporary psychoanalysis.

I do not agree with the sentence I quoted above, that money is "a desexualized partial object that is destined uniquely to serve as a means of exchange," first of all because Money is by no means desexualized. Quite the reverse. Next because, while Money does have a function in exchange, this is not its only function: it can take on other configurations, as I was recalling earlier, guaranteeing security, ease, wellbeing: matters of very important internal valence both for individuals and for nations. (Even Kant was convinced of this: see the *Metaphysics of Morals*, I, II, III, 31.) And lastly, because I really do not think it is possible to consider money as a partial object, unless we think of symbols as partial objects, which I do not. As it happens, I do not think that even Kleinians hold such an opinion, speaking instead of symbolic representation, symbolic equation, or "real symbol".

In her fine work, Ilana Schimmel gives an excellent summary of the current state of theorizing about Money in psychoanalysis, also noting the historic positions which the Master passed through, and she does not conceal their limitations as she expounds them. She also dwells on the concept of Symbol: whether it is viewed as the result of a libidinal shift from a motor discharge to a sublimated cathexis or whether libido is considered an element present as a motor discharge and also in its withdrawal and deferral in sublimation, in either case the Symbol is generated by a shift, and this shift relates back to secondary elements of abstraction which, if successful, should have the upper hand over the instincts. At least, so Freud hoped: in this case, I would say, from an *un*Enlightenment perspective. In any case, the Symbol seems to be something other than a partial object.

In her work with a young patient who sought an analysis and immediately raised the problem of the analyst's fee and how to pay it, Dr Schimmel continued to operate in the sphere of the equivalence between Money and Faeces, although this fundamental theoretical given did not stop her doing admirable clinical work. At the end of the article she quotes a witty fragment from the correspondence between Fliess and Freud in which Freud writes:

> Happiness is the belated fulfilment of a prehistoric wish. For this reason wealth brings so little happiness. Money was not a childhood wish. (16 January 1898)

(so how could Money be a partial object?). Freud's is an appealing observation which shows that the work has reached a good conclusion because his patient has accepted the meaning of "true reality" entailed in the exchange of money for therapy, along with the acceptance of reciprocity, the limits of his omnipotence, etc.

Dr Schimmel's acceptance not only of a much reduced fee but of the whole "classical" stance in relation to the equation Money = Faeces does not, however, stop her adding that in any case (p. 537, my translation), "the Symbol stands in close correlation with the status of the Ego within the psychic apparatus," and also, on p. 549, that Money:

> *est un symbole qui peut appartenir à différentes organisations*
> (is a symbol which can belong to different organizations),

which means, if I have understood her correctly although she puts it in rather a complicated way, that the symbol comes to be used differently/ assumes different valences in the various maturational phases which the *Moi* addresses and constructs. So I find myself perfectly in agreement with her as far as this parallel reading of the customary equation Money = Faeces is concerned, the more so when she adds, taking the idea from Hanna Segal, that Money is a:

symbole vrai représentant du monde réel
(true symbol representing the real world)
("Revue Française de Psychanalyse", p. 539, 1990).

In 1999, Krueger pleasingly concludes an article which does not greatly diverge from his 1986 work by writing: "Money may be a myth, but it is never a fiction. And sometimes, the myth is greater than the reality."
I won't go into the concept of Money's relationship with the *réel* in Lacanian theory (Lacan, Miller, Arnaud, P. Martin) since I frankly do not understand much of it, though I hope that sooner or later we will grasp some of its aspects.

All in all, besides Dr Schimmel's indisputable clinical excellence and her capacity for summary – and I say again, I have only used her an example of a much more widespread orientation – it seems to me that what emerges from reading her article is that the work with her patient went well *in spite of* the theoretical equation Money = Faeces, and not because of it. In fact, the work hinged mainly on the need for and importance of Payment and the awareness of this necessity, rather than the Money itself. And given that the Payment is made in Money a certain confusion is often created, and not only in this case history. It would be interesting to make a further investigation of the problem of payment: session by session, end of month, in cash, by cheque or bank transfer, by third parties, by institutions.)

And it is a real and present confusion since, as I will try to show below, a great many writers continue to superimpose terms such as Payment, Money, Coin, and Gift onto each other and relate them to the most varied existential situations and to the pathologies of highly diverse patients, always drawing on the equation Money = Faeces as a fundamental theoretical standpoint: one might say, a basic assumption. The further point should be made that not just Money, but Payment (that is, the act of Paying) generates problems not only in subjects to whom we can attribute conflicts related to the anal phase, or borderline subjects like Dr Schimmel's patient, but also in subjects presenting autistic or symbiotic features, or – viewed from another perspective – narcissistic subjects, or from yet another perspective, those who are depressed, manic, bipolar, or even those with parasitic characteristics such as drug dependency, False Selves, the downright exploitative, and of course the "eternal sucklings" as Freud calls them ('A Seventeenth-Century Demonological Neurosis', SE XIX, p. 104). Problems of and with Payment very often arise when Thirds enter the consulting room, such as parents (as in the case of Dr Schimmel's young patient), tutors, institutions, tribunals, insurance companies, etc. And problems can certainly occur with invoicing and its multiple implications in terms of presence/ invasion/ irruption into the session by the Stone Guest of the State. And in any case, problems *always* arise with Payment when we are faced with a certain type of negative transference, whatever its origin and irrespective of the patient's neurotic character or personality structure. Requiring a capacity for acknowledging Otherness, Payment is a highly delicate and complex tool, the peak of a person's maturational capacity, and for this reason alone cannot be considered a pregenital element which only involves the anal phase.

I think I have taken Ilana Schimmel's article as an illustration of what I am saying because it is one of the few papers I have found which dwells on the emancipatory aspects of Money, for which I must sincerely thank her.

And even while she quotes some radical proponents of the Penny Black (Aglietta and Orléan), Dr Schimmel makes an interesting observation in the introduction to her book on the subject (*La psychanalyse et l'argent*), claiming that the Introduction of the Coin necessitated a major evolutionary leap forwards in the ability of the human species to symbolize. This too is an observation with which I find myself fully in agreement, without even having to bring Jean Piaget into play (I am thinking of his four transformation operations) or bothering another giant, Kant in *The False Subtlety of the Four Syllogistic Figures*, since the reversibility of human thought seems to have been established in Antiquity. We need only read the *Epic of Gilgamesh*, so richly expressive and dense with such profound humanity that it is hard to consider it "inferior" to anything "modern". If modernity does have a new property, this can only be its striving towards the achievement of Truth and Justice *erga omnes*: that is, *Kittum* and *Mīšarum*, as described by Bottéro (1987, p. 106) whom we will meet again later. And in this context, I think it is interesting to emphasize that Freud shared this opinion, at least to some extent, in his view that the distinctive features of monotheism as propounded by Akhenaton and subsequently by Moses, whom he saw as the founder of "complete" monotheism, are represented by the fact that they lived their lives in Ma'at (*Moses*, SE XXIII, p. 21): that is, within the concepts of Truth and Justice, concepts which I regard as entirely equivalent to those of Equity and Justice which we will meet later. And Ma'at is also the goddess of the Feather which weighs the heart of the deceased using the *bilanx*, and these too are elements we will meet further on.

Anyway, this assertion by Dr Schimmel that the appearance of the Coin necessitated an "evolutionary leap" will, to some extent, be the conclusion of this book, though with a small difference: I do not believe that modern man, understood as belonging to a certain historical period, can claim primacy in this regard, and as this book develops I will try to explain the reasons which have led me to make this claim.

If "modern" man, in the sense of man beginning to be literate, can boast of anything, it is simply that, in spite of the deviations and brutality of his History, he has continued the work of those who, "as if raving mad", had first laid down the bases of a demand for Equity and Justice, the bases for escape from Predation and the Talion. That being the case, Freud does not help us to clarify this by proposing his own rather contorted and bizarre system for understanding the concept of Gift, which includes Excrement, the evacuative function, adult expectations, and much else besides. While he had proposed the equation Baby = Faeces, soon afterwards he mixed everything up, giving rise to the unfortunate situation we are still faced with today. Dr S. Zucconi (personal communication) has pointed out to me that this fact, the equation Money = Faeces, is an inevitable consequence of the drive theory adopted by Freud, an approach within which this equation would make perfect sense.

However, dissenting from Dr. Zucconi's view perhaps for the first time, I hold that this is not the case and that whether we base our reasoning on drive theory, object relations, the field, intersubjectivity, Bion, Kohut, or Lacan, this equation is a mere ornamental fringe which has little to do with the Master's thought and with the brilliance of his other intuitions, but instead with other aspects which I will try to clarify shortly. While I detach myself *in toto* from the "faecal column" and "excrement", I will develop the equation Penis = Child during the course of this study. In any case, if Money "brings so little happiness" because it "was not a childhood wish", I still find it hard to understand the reason for all the attention, condemnation, hope, endless moralising, accusations, and desires which human beings have devoted to it, even when – and especially when – denying its importance. Perhaps some hypotheses will emerge later on.

And this appealing observation which arises out of the Freud-Fliess letters seems to be evidence of the difficult in calling Money a part object, specifically given that it has no roots, at least no "direct" ones, in childhood. And if it has no roots in childhood, I keep wondering how it can be connected to the anal phase? Simply through Condensation and Displacement? The Gift could certainly be the outcome of this concatenation, but why Money? In other words, why should Money be derived from faeces and not from the pacifier, or the penis, or the breast, or urine, or teething, or failure to suck, or bottle-feeding, or from premature or delayed weaning, or traumatic abandonments? Or from the mobiles that circle above the cot? Anything, so long as it can be somehow connected with childhood. Why not derive it from measles or scarlet fever? Or why not from the fantasy of a state of complete wellbeing that we wish never to lose? In this case, it would be more like the Age of Gold which existed for every individual, when milk and honey flowed and we didn't have to work because Mummy and Daddy took care of everything. A primeval Edenic fantasy to be reconstituted on earth, the Age of Gold flowing with milk and honey (and not *Qaqi*), which frees us from necessity and the "sweat of the brow". Or could it just be a piece of wisdom which enables us to understand that life is more complicated than it seems and that we need to keep something set aside in case we lose a tooth, or the washing machine or car breaks down, or we require more specialist treatment for an illness?

Or, if we focus on the difficulties in our relationship with money, why should this difficulty not be seen as resulting from such a multitude of secondary advantages that it puts understanding it "out of reach" of patients who are not at all obsessional but suffer from an entirely different type of "neurosis"? Or to put it yet another way: if we take it for granted that faeces can also assume the value of a Gift and be part objects, what does such a Gift have to do with Money, assuming that it has to do with something? Is it just because of the stubbornness associated with the anal character? And if the transformation of a sign into a symbol also happens thanks to the subject's ability to withhold something/ sublimate/ displace libido and to place this something at the service of the Ego, why should this "withholding" necessarily occur in relation to the anal phase? Why is something only withheld during that phase? Is breastfeeding not a Gift? Is a kiss not a Gift? And can't these be "withheld"? and does that mean they represent anal aspects?

And a lullaby, sung with love, is that not a gift? Giving a friend a pat on the back, even with no words: isn't that a gift? Having a cup of coffee ready when she wakes up?

When an adult calms a baby in a state of distress, and the daily attentions on which the relationship with the baby is founded, aren't these a gift? And seeking out a little something before one comes home: a flower perhaps? A loving goodbye before going out? A passionate embrace with one's beloved? A kiss full of desire, an intense, exciting, overwhelming shared orgasm: aren't these also gifts? And isn't all this "kept inside" one's own body, thereby cementing a profound and loving relationship between adults? Between adults and babies? Between friends? Between lovers? And what is anal about these gifts? And is all alchemical research, with its many consequences for chemical, philosophical, and mathematical research – the search for the Philosopher's Stone itself, for example – even when it often concerns Gold, only attributable to "cupidity" and anality? I don't think we can easily dismiss the interesting and well-reasoned observations of Mircea Eliade (1935, 1937, 1956, 1978) about this. And poetry? And music? Aren't these also gifts?

And paying one's debts, honouring a contract, constantly and laboriously making the burdensome repayments on one's mortgage? Paying an adequate and regular wage to someone who works in your house, knowing that she lives on this money and needs to be able to continue doing so on a stable footing? What is anal about any of this? Comparing prices in the supermarket and carefully choosing the "right" balance between quality and price, with all the possible variants this can involve: what is anal about that? Trying not to waste resources, without ideology and leaving aside global heating: wat makes that anal? (Caetani's 1911 observations about ancient and post-glacial Arabia and those of 1904 about Semitic origins and the Alps are of interest here.) What is it that makes Freud see such an obvious equation Gift = Faeces and then Gift = Faeces = Money? The equation Faeces = Gift often holds good, but why Faeces = Money? For Otto Fenichel (*The Psychoanalytic Theory of Neurosis*):

> What money and feces have in common is the fact that they are deindividualized possessions; and deindividualized means necessarily losable.

But he immediately adds:

> Actually anal-erotic persons who love money, love money that is *not* deindividualized; they love gold and shining coins or new bills, money that has still a "blue", individualized character.

Frankly, apart from an association with the gleam which attracts the thieving magpie, I for one don't find this very clear; and if I were to take into consideration Fenichel's 1938 work, republished later as *The Drive to Amass Wealth*, its ideological approach, based on the verb 'amass', which later becomes 'accumulate', would require me to write a longer chapter than the one I have written on Fromm. There would, however, have to be a brief reference to a pathological "prophylactic self-castration".... But not today.

A good anthology of the position, not only Fenichel's but that of most analysts, can be found in *The Psychoanalysis of Money* (1973) edited by Borneman, which though some non-psychoanalytic articles are added at the end, represents a perfect summary of the "orthodox" viewpoint.

And in an interesting article from 2012 (published in the book *Money*

Talks), Theodore Jacobs, who is no minor player, concludes with these interesting and problematic words:

> A collusion to avoid dealing more actively with money and its complex psychological meanings for analysts, as well as our patients [gives rise to...] one of the most important, and neglected, problems facing psychoanalysis today (p. 11).

And yet, a few pages earlier we read:

> For many analysts, as with much of the population, their personal sense of worth is closely linked to financial success. This connection, which often is only loosely tied to reality considerations, begins early in life with the unconscious association of money with bodily products, particularly faeces. In the child's mind these bodily products are linked with objects of value and often offered as gifts (p. 6).

Jacobs then goes on to pursue the orthodox line. Still, in 2012! Consulting other authors (Erle, Desmonde, Tulipan, etc.) I have found no greater clarity or diversity in their positions.

And the psychoanalyst who lives on this money, or should do so[6] (there are slightly ideological, but even so, stimulating observations on this by Klebanow, 1989; Dimen, 1994; Myers, 2008; and a little generic but also interesting, Lasky, 1984), where is he in this process? What does he do with those aspects? How does he use them? How does he live them? How does he consider them himself and for himself? Is it only a question of countertransference? Or of projective identification? Or of the analyst's own anality? So many questions. To one of these, I hope to find a 'partial answer'. Meanwhile I shall take a look at Freud; and at Prof. Alfred Jeremias.

---

6   In relation to 'should do so', it is probably my ignorance, combined with the fact that I live in Italy, and in the provinces, and am therefore excluded from important cultural movements given the weight of provincialism of the province where I live and work, but I really cannot think of a single publication in which the formative process of a psychoanalyst has been analysed with any concreteness. I have no memory of anything written about what, in economic terms, we could call an accounting system for analytic activity. Plenty of idealized chatter, even in the provinces, but nothing about Return on Investment (ROI). Words galore about dedication/ altruism/ love for one's neighbour and devoted commitment to one's profession: about Return on Equity (ROE), zero. To put it another way, how much should a psychoanalytic session cost in order, over time, to balance the costs (the real, not abstract/ abstruse costs) sustained, so as to be something different from the hairshirted philanthropy it sometimes resembles?

5.
SIGMUND FREUD AND PROFESSOR ALFRED JEREMIAS

In fairness to Dr Ilana Schimmel, since it was not she who invented the story of the equation Money = Faeces, and for reasons of my own, the moment has come to call upon the Father, given that within the body of doctrine he elaborated there is a series of assertions which are, it seems to me, in need of some reflection. I need to take a deep breath before immersing myself in a laborious topic, a game of Chinese boxes. To summarise very briefly: I shall start with the letters to Fliess, in one of which Freud, aged forty-one, declares that he has been thinking about the subject of Money, but in quite a general and confused way. In this letter, though without going into much detail, Freud explicitly shows that he has started out on the path which he will later try to organise using apparently "deductive" reasoning. The letter (24 January 1897) moves between Witches, the Devil, Shit, Hysteria, Cagliostro, and Alchemy:

> I read one day [*where? when? written by whom?*] that the gold the devil gives his victims regularly turns into excrement... Cagliostro – alchemist – *Dukatenscheisser* [one who defecates ducats].... So in the witch stories it [*the money, or the ducats, or the gold?*] is merely transformed back into the substance from which it arose.

Or take another, of a quite different kind, from 21 September 1899:

> A patient with whom I have been negotiating, a "goldfish", has just announced herself – I do not know whether [she has come] to decline or to accept. My mood also depends very strongly on my earnings. Money is laughing gas for me.

And passing through the correspondence with the Protestant Jung (specifically the letter from Jung dated 23 October 1906 about Sabina Spielrein and the way she defecates, in my opinion more interesting about the turn which events will take between Sabina and the young, good-looking Gustav, rather than on the level of the study of the anal phase, as

I have hinted), we will come to the letters Freud writes to Lou Andreas Salomé in 1927, which seem "truer" to me in terms of the topic being addressed here than the whole theoretical edifice built around the equation Money = Faeces. Some of the letters exchanged between Freud and Lou Andreas Salomé respond to an extremely concrete demand: that is, the survival and professional dignity of the analyst, exacerbated by the Weimar crisis, which was on the way to being solved in 1927 and resulted in appalling inflation. Moreover, these letters touch Freud very directly (see Tögel, 2009) and concern neuralgic aspects of the profession that are still alive today, and not so much Freud's supposed anal fixation, as has been suggested (Warner, 1989). The pressure to survive which emerges in 1927, is in fact only a further element in the chronological sequence, in which the difficulties of psychoanalysis as a clinical practice/profession, are given voice. For at least nine years – in other words, from the end of the First World War, everything that happened to Freud was bound up with the fall of the Austro-Hungarian Empire, and with the dissolution of the old political, economic, and social order of the whole of Europe. During the early post-war years, the level of desperation striking Austria was such that, after a kind of "self-annexation" to the Weimar Republic (even endorsed by art. 61 of the Weimar Constitution of 1919), the Bishop of Vienna, Ignaz Seipel, Austrian Prime Minister from 1922 to 1924, and Foreign Minister from 1924 to 1929, campaigned politically – while maintaining a markedly Germanophile position – for the closest possible alliance with Italy, his country's worst enemy at that time, given that Austria was literally being considered a kind of Italian protectorate. This "protection" lasted until the assassination of Chancellor Dollfuss in 1934 and came to a definite end a couple of years later with the withdrawal of the three (or perhaps four) armoured Italian divisions from the Brenner Pass, thereby opening the way for the Anschluss. Echoes of this position can be found, years after Seipel, in Freud's first Preparatory Note of 1938 to the third essay, *Moses, his People and Monotheist Religion*, written in Vienna shortly before his unavoidable and painful departure for London: in other words, shortly before the Anschluss:

> … strange to say, it is precisely the institution of the Catholic Church which puts up a powerful defence against the spread of this danger to civilization— the Church which has hitherto been the relentless foe to freedom of thought and to advances towards the discovery of the truth! (1939, 55)

A little later, disconsolate and embittered, Freud would write a second Prefatory Note in which resignation and grief at the loss of all hope in the face of unchecked Nazi power and arrogance made him express some very gloomy judgements.

From a professional viewpoint, the fall of the old order of Mitteleuropa after the Great War was, for Freud, reflected in a general drop in patient numbers, and particularly in a brutal re-scaling of his income (that smaller number of patients were paying less than they would have done in the pre-war years). It is during this period that Kardiner and the Americans arrive with the dollar which becomes not just "laughing gas" for the Master, but "oxygen". Indeed, in the letter which Freud writes in reply to Kardiner in 1921,[1] the Master asks for a fee of ten dollars per session in cash, not in cheques which he would only be able to change into Austrian crowns, and asks, it seems, that this fee remain "undeclared" given that he makes no allusion to invoices and so forth. (It would be interesting to make a deeper exploration of the topic of the fiscal regime Freud must have used under the Austro-Hungarian Empire.) According to Tögel, Freud goes from earning 17,700 dollars in 1911 to 6,250 in 1917 and 7,196 in 1920: that is, 60% less after ten years and a World War.

In any case, by 1927, in a pair of letters to Lou Andreas Salomé, Freud "threatens" his pupil with breaking off their friendship if Lou does not ask a fee of "at least 20 Gold Reichsmarks" per session.

I've made some effort to try an understand the value of 20 Gold Reichsmarks today. Let's see: 20 Gold Reichsmarks... As far as I can tell – and Freud makes no reference to it – those 20 Gold Reichsmarks would have to be tax free.

After a desperate, exhausting and frantic online search (lasting a few seconds) I managed to find on the website of Oregon State University a piece of research by Dr. Robert Sahr, who has developed, quite precisely I think, some conversion graphs for the value of the dollar from 1665 (although the history of the dollar officially begins in 1794-96).

---

1   On p. 13 of his book *My Analysis with Freud*, Kardiner records a letter from Freud himself: "In late April 1921 I received a letter from Freud, who said as follows, 'Dear Dr. Kardiner, I am glad to accept you for analysis especially since Dr. Frink has given so good an account of you. He is strongly confident of your chances as an analyst, and spoke highly of your character. Six months are a good term to achieve something both theoretically and personally. You are requested to be in Vienna on the first of October, as my hours will be given away shortly after my return from the vacation, and give me definite assurance of coming some time before—let us say, in the beginning of September. My fees are $10.00 an hour or about $250 monthly to be paid in effective notes, not in checks which I could only change for crowns. If you understand German, it would be a great help to our analysis, and you can work here in the Redaktion of the Intern. Jour. of Psa. With kind regards, Yours truly, Freud.'"

In around 1927 a dollar was being exchanged for about 4 new Gold Reichsmarks. The new Gold Reichsmark had just been issued, in 1924, by the Weimar Republic to replace the Papier Mark, in an attempt to stem the looming and atrocious inflation with the aim of stabilising the currency and thus reconnecting it both to the Gold Standard and to the old pre-war Wilhelmine Gold Mark. The new Reichsmark had the same value in gold as the old currency.

So 20 Gold Reichsmarks were worth about 5 *dollars* at that time.

The relationship between dollar and gold mark, not the Papier Mark, remained stable until the Second World War at a level of about 1:4 (1:4.2 to be precise; see also tab. II, p. 188 in Aglietta and Orléan, who cite Bresciani-Turroni, *The Economics of Inflation*, 1937; von Mises, 1979, p. 67, records the same ratio in the pre-war period). And yet, in the period between 1921 and 1927, the purchasing power of the dollar itself remained stable according to Dr. Sahr, always around the coefficient 0.083/0.081, which is the coefficient by which a dollar of those years is divided to obtain a value adjusted to its "current" value. In other words, a dollar from the period 1921 to 1927 was worth 12 dollars from 2009. This coefficient is obviously different for different years.

From Dr. Sahr's study, given that a dollar from that time had the purchasing power of 12 dollars "today" (though Tögel writes of a ratio of 1:26 in 1920), we can reasonably deduce that, if his conversion table are accurate, the 20 Gold Reichsmarks mentioned by Freud as a minimum fee for Lou Andreas Salomé in 1927 in Göttingen, and not in Berlin or Vienna, are equivalent to about 43-60-73 euros today, depending on whether we are taking respectively the lowest point in the euro/dollar rate – that is, weak euro and strong dollar in October 2002 – or the 1:1 rate, that is parity with the euro, or the highest point: strong euro and weak dollar (at the time of writing, July 2011). Always after tax. If we prefer to work by Tögel's calculation, the income roughly doubles, bring Lou Andreas Salomé 106-130-182 euros per session. Still tax free.

Hence, back in Vienna, we can deduce from Freud's letter to Kardiner that the fee requested by Freud is located – depending on whether we take the value indicated by Dr. Sahr (with a euro/dollar ratio of 1:1) or by Tögel – between 120 and 260 dollars today. Still tax free, and in any case double what Freud suggested to Lou Andreas Salomé.

Considering the enormous crisis and economic difficulty facing Weimar, all this seems to me to offer a very interesting point for reflection on the fees which psychoanalysts ask, or sometimes do not ask, today.

Having ended this digression into the possible value of the 20 Gold

Reichsmarks and 10 dollars requested by Freud, I shall resume my discussion.

I have developed my observations largely from six works by Freud, which are general rather than clinical in nature, although his references to Money/Coin/Gold are scattered through many other works.

The viewpoint from which I will try to look at these six works will be a little different from other observations which have over time been made by others.

Not wishing to add confusion to an already quite confused picture about Money, I will not risk even opening the case of the Wolf Man (*From the History of an Infantile Neurosis*), considering it as the development of a particular, singular, individual, and *private* situation, in which the patient's specific idiosyncrasies, and Freud's own I believe, predominate over more general clinical considerations. In any case, much has already been written about this.

I will touch on the 1917 work *On Transformations of Instinct as Exemplified in Anal Erotism*, which I think contains interesting elaborations, although they take for granted the equation Money = Faeces.

I shall only give marginal attention to the Rat Man.

So in the end, the texts I shall take into consideration will be:
Three Essays on the Theory of Sexuality (1905)
Character and Anal Erotism (1908)
Notes upon a Case of Obsessional Neurosis (The Case History of the Rat Man) (1909)
Dreams in Folklore (1911)
On Transformations of Instinct as Exemplified in Anal Erotism (1917)
A Seventeenth-Century Demonological Neurosis (1922)

I will not consider *The Interpretation of Dreams* since, *strangely*, it contains no reference to Gold = Faeces. Or rather, there is one reference, on p. 403 ("Dreams with an intestinal stimulus... confirm the connection between gold and faeces which is also supported by copious evidence from social anthropology"), but this is an addition made in 1929. It is a clearly attributable reference, as we shall see, to Professor Jeremias. I shall ignore *The Psychopathology of Everyday Life* (1901, p., n1) where Money is associated only with a general ambivalence about payment or with hysterical and narcissistic features, with play, and so on. And I shall calmly pass over Little Hans (1909) with his *"lumf"*, along with the girl and her eggs, her colours, and her lies (1913).

I shall use these texts to try and understand the source of Freud's conviction that Money is a derivative of Faeces, and no longer a golden fish, as he had put it in the letter to Fliess. As its scientific name *Carassius auratus* suggests, the Goldfish is precious in itself, very much so in China during the Song dynasty around 1,000CE, but it is also precious because it lays a great many eggs. So I do not believe I am incorrect in thinking that Freud's prospective patient was seen in the master's fantasy/ dream as a goldfish who would lay many golden eggs/ 20 Reichsmarks in her therapist's hand. As Freud himself states in *A Seventeenth-Century Demonological Neurosis*, "Five dollars in a dream can stand for fifty or five hundred or five thousand dollars in reality." This quotation is connected to another, from Dreams in Folklore. In the end, the patient/ goldfish is not very different from the Hen/ Goose which lays golden eggs and which Jack steals from the giant.

We could quibble that the Egg passes through canals not altogether unlike those of Faeces, but its symbology seems somewhat different. Ferenczi, however, would not entirely agree: see his Two Typical Symbols of Faeces and Children, 1915, and The Ontogenesis of the Interest in Money, 1914 (in Ferenczi, S. (1952). *First Contributions to Psycho-Analysis*. The International Psycho-Analytical Library, 45:1-331. London: The Hogarth Press and the Institute of Psycho-Analysis). It is also interesting that both these works come significantly after 1902, an important date as we shall see shortly. Nevertheless, pace Brahma, Eurynome, Pangu or Thoth, the Egg is born from somewhat different sources than those through which excretion takes place, and only uses that canal in certain species (the hen, for example), and only its final stretch. So, although inter faeces et urinam nascimur (although the expression seems to be St Augustine's, Augustinus actually says sub peccato, carnaliter, in carne peccati), birth is generated ab ovo (from the ovocyte), even if the new creature is "ex-creted". The Egg has origins, functions, and outcomes "just a bit" different from those of Excretions.

Moreover, the term "Goldfish" which Freud uses in his letter to Fliess may also contain reflections on Avarice, or Cupidity if you prefer, but the Cupio in Freud's letter is that of being able to work – ie. to live/ survive – at a difficult time, and very unlike the description by Fenichel recorded earlier, a definition more appropriate to the glint in the eyes of Harpagon. We often forget that for a long time Freud lived entirely on the actual fee from his actual patients, and could not count on other, guaranteed, extra-analytical earnings, as is sometimes the case for many analysts today!

I do not think that the chronology of the works listed above can be considered to have come about by chance, since it has the appearance of a crescendo by Rossini. It begins in 1897 with the thought/whisper that there may be an equivalence Money = Faeces (letter to Fliess in January). The *Essays* of 1905 speak/hint ever so gently that Excrement is a possible equivalent in infant fantasies of the birth of a child. There is a hint about the gift and the concept of gift expands from child = gift to faeces = gift. In 1908, Freud declares fully and officially that excrement and faeces are linked to *chthonic* symbolisations of wealth, and he looks at the equation wealth = money = faeces as a whole, without losing the connection to the Gift. In the 1911 paper he definitively reasserts that wealth = money = gold = faeces = gift, and that this is the whole story. In the study from 1917 there is a phantasmagoria of leaps, pirouettes, and jumps all performed by balancing on these equivalences, fully in tune with the understanding of the anal phase and its implications, but somewhat less with the understanding of money.

When Freud was 66, in 1922, this urgent musical rhythm took on a more "tender" motion, almost faded away, like the closing scene of a film where the characters ride off into the sunset; and, without modifying what he claimed to have discovered, Freud treats the poor painter of the story with an almost empathic, and in some respects identificatory, understanding, writing about the economic difficulties which might have led that possessed "poor devil" to seek refuge in the monastery. In other words, I think that this approach was substantially provoked by the fall of the Austro-Hungarian Empire and all the rest. I believe that if Freud had dwelt a little longer on what he had begun to suggest in 1905 – that is, if the equations baby = gift and faeces = gift had been a little less bound up with faeces = money, perhaps many things would have turned out differently.

But history is not made of "If" and "Perhaps".

Freud liked details, and as I was saying earlier, I too have been struck by a detail.

In 1902, before the *Essays* of 1905 and well before *Character and Anal Erotism* (1908), a certain Professor Alfred Jeremias published *The Old Testament in the Light of the Ancient East (vols. I and II)* in Leipzig, having a couple of years earlier published a book on the Babylonians: *The Babylonian Conception of Heaven and Hell.*

And earlier still, H. Winckler, who had already published *The Cuneiform Texts of Sargon* in 1889), and H. Zimmern published works on the same subjects.

This Professor Jeremias was convinced that everything derives from Babylon, and his 1902 work, relating to the Old Testament and motivated by archaeological interests but even more by religious concerns, attempts with great learning to reconcile, re-read, and harmonise the information which comes from the Bible with the reality of the archaeological excavations of his own time. He was part of an intellectual current of the period which was known as Pan-Babylonism, and Otto von Bismarck, had he not died shortly before this, would have been pleased with Jeremias, given Bismarck's obsession with the three Bs railway (Berlin-Belgrade-Baghdad) in constant competition with the colonial powers of the period (Britain, France, and Russia) for cultural, political and military pre-eminence in the Middle East and Africa, and "allied" with the moribund Ottoman Empire. Besides Pan-Babylonism, Jeremias's studies engaged with a characteristic cultural debate of the time called the *"Bibel-Babel Streit"*, which Pettinato give the fine literary name of "the war between Bible and Babel", describing the white heat in which the debates and discussions of the time were conducted (Pettinato, *La Saga di Gilgameš*, Introduction and Epilogue; an interesting overview is also offered by G.P. Basello, 2004, and to some extent also by S. Parpola, 2004; in my opinion, a more developed and complete view is given by the Epilogue to H. Winckler 1907 text, written by Pietro Mander, who also clearly brings out the vein of antisemitism which cut through the otherwise admirable attempt at exploration and discovery, faithfully reflecting the times. This antisemitism will also appear in the positions taken by Alfred Jeremias, as we shall see shortly).

Alfred Jeremias was a fervent "anti-Sumerian", going so far as to call everything which preceded Babylon simply and dismissively *"Euphratesian"*, a term I introduced earlier. This is an odd position, given that in the second volume of his *Old Testament in the Light of the Ancient East*, he very skilfully discusses the Hammurabi Codex (newly discovered in the winter of 1901-1902 by J. J. M. de Morgan) which he clearly knows in immense detail. Both in the Prologue and in the main text of the Codex we repeatedly read the phrase *"šarrum... ana mat šumerim u akkadim"*, "King... of the land of the Sumerians and Akkadians."

But there is something about this which Professor Jeremias really dislikes, and on page 2 of his *Old Testament (vol. I)*, clarifying his point of view, he writes:

"and since – in very late Assyrian records, it is true – there is mention of a "language of Sumer and Akkad", we speak of a "Sumerian" civilisation, inherited by Babylonian–Semitic people. Nothing can be said with certainty as to the character of this first civilisation, which we call in future "Euphratesian", to distinguish it from the later Semitic-Babylonian epochs. The hope of solving the problem by new discoveries of yet more ancient literary remains has been invariably disappointed. The oldest records known betray a Semitic character."

To sum up, he is so unwilling to speak of Sumerians that in note 2 on the same page 2, he adds, "The present author has recorded his "antisumerian" views in the Theologische Literaturzeitung, 1898, No. 19."

In the same years, H. Winckler and H. Zimmern were writing very differently.

As late as 1970, another scholar, Krauss, quoted by Pettinato (2001) claimed that "the Sumerians and Akkadians are evidence of a single culture, that of Babylon."

Rather like me, Jeremias believes that much of what we know today was generated by the Babylonians, while I think in terms of the Sumerians/ Akkadians. (Out of respect I ought to say, "Rather like Professor Jeremias, I too believe that…" but we need a little narcissism every now and that, dammit!) Perhaps we two are also quasi-Doubles.

He was a Pastor, a Lutheran, an Academic, and a Theologian. Definitely quasi.

Joking apart, Jeremias draws on the culture of his period – that of the Greater Germany which was funding research in Mesopotamia, competing with the British and the French not only in the fields of railways and war, but also that of archaeology: the same Germany that was filling Berlin Museum with the altar of Pergamon and the Gate of Ishtar, and which was undertaking cultural developments with the impetus of an express train. Schliemann had died only a short time before in 1870, and the echo of his discoveries was still being heard. In any case, Jeremias was a highly cultured, well-trained, and attentive scholar.

And Jeremias provided Freud with a formidable structure while he was getting to grips with the elaborations of symbols and meanings, having recently published *The Interpretation of Dreams* in 1899, *The Psychopathology of Everyday Life* in 1901, and preparing to publish *Three Essays on the Sexual Theory* in 1905.

I think it is worth pointing out that in Austria Felix and in Europe, the phantom of the "Devil's Excrement" had been doing the rounds for at least three hundred years, probably seven hundred, and well before other phantoms which were around at that time throughout Europe and Russia. Taking the 1897 letter to Fliess in which Freud was already thinking of money = faeces (of the Devil), the *Three Essays on the Sexual Theory*, as I said earlier, contain the first "official" sketch of the idea that excrement = baby = gift; but the interesting thing is that in all five of these works, when Freud refers to the meaning of faeces, the name of Alfred Jeremias is always quoted, either directly or in a note.

This is not to say that the reference to Cagliostro and the Witches in the letter to Fliess of 27 January 1987 mentioned above ("I read one day [*where?*] that the gold the devil gives his victims regularly turns into excrement") is not also connected to Jeremias, or Winckler, or Zimmern, or other authors/scholars of the period cited by Jeremias himself who tried out equivalences of the kind Jeremias would attempt.

I wrote earlier that I would be considering six works by Freud.

I lied, knowing full well that I was lying; on the other hand, I was born in a small town of "liars and imposters", and I attended high school in another little town which is a byword for fibbing and lying. I will in fact be considering only one element, a single detail which unites all these six works: and that is, Professor Alfred Jeremias, or perhaps I should say the "ghost" of Professor Jeremias. This is the detail which has intrigued me: the continuity of the presence of Professor Jeremias's ghost as it cuts across and flutters around this whole area of Freud's work.

The not so old proverb *"The Devil is in the detail"* could hardly be more apt than here in the case of the Devil/Hell and Details. (Some people say "God is in the detail", and this version might have appealed to Jeremias.) It's just that this detail and its examination is a job in its own right, and I don't find it easy to summarise in a few lines the elements which compose it, but I will try.

Leaving aside Jeremias's Lutheran background (a topic which it would, nevertheless, be interesting to investigate, along with Jung's cultural hinterland and the brief observations I made a little earlier about Luther and Calvin, and then Weber), the "revolutionary" thesis he proposes to demonstrate in *Old Testament in the Light of the Ancient East*, obviously based on the data he had at his disposal and, especially, on his type of interpretation, is – extremely, brutally summarised – *Gold is the Devil's excrement*.

And as I noted a little earlier, at least three hundred years after the Reformation and Counter-Reformation, and seven hundred years after good St Francis (and my goodness, don't we need him now!) this is hardly what we could call a scoop!

Jeremias begins with a Babylonian poem, *Nergal and Ereshkigal*, Babylonian but with Sumerian roots, in which a god, Nergal, goes down to the "Land of No Return" to expiate an offense given to an ambassador of Ereshkigal, queen of the "Land of No Return" during a ceremony of the heavenly gods: but then, in order to defend himself against an attack which would have resulted in his "death" and imprisonment in "Hell" like any mortal, he makes a pre-emptive strike and tries to kill Queen Ereshkigal, who is enraged by the wrong done to her and lashes out at "poor" Nergal, who had brought this upon himself; however, after the fight is over, she falls in love with him. So he marries her and settles down to become King of the Underworld.

Jeremias actually starts his book with the poem *Enūma Elîš* (*When up there...*) and an erroneous interpretation of the poem itself, which concerns a *Mummu*, understood as the first son of the couple Apzu-Tiamat (an error which goes back to Damascio, see Bottéro, 1989, p. 642, note 6. Winckler, 1907, makes the same mistake as Jeremias, and on pp. 96-102, fired up with literary enthusiasm, he goes on at great length, even worse than me, about how Mammu=Logos=Word=Tao....), but as far as Money-Coin-Gold is concerned, the important part is the way Jeremias deals with *Nergal*.

Simultaneously, there arrived in Babylonia from somewhere, though Jeremias does not say where or how, a divinity or *ilu* called Man-man, from whom he derives the name Mammon, the god of money, who he claims is in the Old Testament. *Man-man*, alias Mammon the god of money, is an epithet – again, so he claims – of the chthonic Babylonian divinity Nergal, god of the Underworld/*Unterwelt*, god of the dead and hence, in his opinion, of Hell. But according to Jeremias, Nergal is also the Sun, *Shamash*: "*Shamash and Nergal are one*" (p. 30); furthermore, still according to him, Nergal is also Saturn, and Saturn is the *star* of the Sun (*ibid.*); and Nergal is also Mars (p. 140), besides naturally being god of pestilence, terror, and storms, as he in fact is in Sumerian tradition. With respect to storms, there is some overlap with the god Erra, who I may encounter later on. Jeremias therefore "deduces" from all this that Money comes from Hell.

This is the source of the quotation in *Character and Anal Erotism* (SE 9, p. 174, n4), which we will see more clearly in a moment.

I think my summary is much too concise and risks being mistaken, but this is the problem: trying to summarise a truly gross misunderstanding in a few words. In SE 9, p. 174, n4, we see that Freud is referring directly to Jeremias and his *The Old Testament in the Light of the Ancient East*. Both in the body of the text and in the note Freud states that *Mammon = ilu Man-man*, a Babylonian deity, the very one cited by Jeremias. Shortly afterwards, in *Babylonisches im Neuen Testament*, Jeremias heightens the importance of this deity, endowing him with the additional epithet *kakkab*, luminous/shining/glowing.

In Jeremias's book, at least in the English translation I have, a reprinting of the translation of the second German edition of 1911, Jeremias writes of the Serpent as the representative of the Evil One and quotes Micah 7: 17, "They shall lick the dust like a serpent, they shall move out of their holes like worms of the earth: they shall be afraid of the Lord our God, and shall fear because of thee," and Isaiah 65: 25, "The wolf and the lamb shall feed together, and the lion shall eat straw like the bullock: and dust shall be the serpent's meat. They shall not hurt nor destroy in all my holy mountain, saith the Lord." In these verses the Serpent eats dust because it creeps (Jeremias, p. 234).

In the Hebrew text of Micah, however, the word is ילחכו, *ielachekhu*, "will lick", from *lelachekh*, ללחך, (though the Hebrew verb is actually closer to "graze") and they will lick *afar*, earth/soil, as in *Gen.* 3: 19 and not, as Jeremias would have it, *avaq*, dust.

In the effort to demonstrate the equivalence between Dust and Shit, Jeremias "forgets" that in *Gen.* 3: 19, "*Pulvis es et in pulverem reverteris*", reiterated in *Ecclesiastes* 12:7; *pulvis* is intended to translate *afar*, earth/ soil, עפר, but literally means dust, *avaq*, אבק, which we find in the term "abacus"; and also in *Gen.* 3:19, and again in *Gen.* 18:27, where we find *afar*, earth/soil and not *avaq*, dust. The Greek text of the Septuagint is more correct in using γῆ (root of γέα and γαῖα), the Earth.

At this point in his book, Alfred Jeremias, still on p. 234, n 2 of his *Old Testament,* cites other authors, including Winckler, who (perhaps) equated dust, the food of the Serpent, with shit: that is, he plays on Dust and Dung, because of which the Serpent-Evil One would eat dust: dust which would, without going into a detailed explanation, be understood as shit. He kindly prefaces the note with, "The literal eating of dust cannot be meant," but straight afterwards, still in the note, he asks "Or does the serpent eat dust?" But just a moment later he brings other authors into play on the question of Dust and Dirt, so that the Serpent-Evil One would again eat dust, dust which Jeremias intends, without explanation, to be read as dirt and then as shit.

In my opinion, the works cited by Jeremias would only serve to show that other writers, well before the 1897 letter to Fliess, had certainly enjoyed "playing in the sand" (or perhaps more plausibly, with "star dust"), and Freud could and should, have become aware of them, even those beyond ultra-Catholic Austria, especially if we consider that he would go on to draw so copiously on Jeremias. Among the authors of the *Bibel-Babel Streit*, Zimmern is never quoted by Freud, while Winckler is quoted only once, in the *Traumdeutung*, in a note added in 1909 on p. 103 of the German text, *en passant* in a reference to wordplay in ancient Eastern civilisations, and certainly not in connection with what follows below. There is another reference by Freud to a Pan-Babylonist writer, E. Stucken, in 'The Theme of the Three Caskets' (SE XII, p. 291) on the subject of Astral Myths; then another general reference concerning "the first beginnings of astronomy among the Babylonians" in *New Lectures on Psycho-Analysis* (1932, SE XXII, p. 173). Yerushalmi (1991) further reports that "according to an eye witness", in 1904 Freud held a conference at the B'nai-B'rith Association on Hammurabi, entirely in the spirit of Delitzsch, adding that the text of the conference has been lost. In any case, it seems to me that we have enough evidence to say that Freud knew the *Bibel-Babel Streit* to some extent. And although both Delitzsch and Jeremias held the same view about the Euphratesians, Freud chose not Delitzsch, the teacher who is never cited in his work and who with great intellectual integrity would change his opinion years later, composing the first Sumerian grammar, but Professor Jeremias!

Immediately afterwards Jeremias quotes himself, and rather hastily writes (p. 234 n2 of his *The Old Testament*) the very sentence which Freud would quote: "Dung is an element of Hades (compare at p. 7 the signification of the beetle in Egypt; for gold as dung of Hades, Mammon=ilu Manman=Nergal. comp. BNT, 96)."

So, back I go to p.7 of his book, to 'compare', but I get confused because, although Jeremias does indeed write about the "dung beetle", the sacred Egyptian Scarab, he does so in a quite different way from the passage just quoted. He acknowledges that the sacred Scarab represents the Sun and its passage through the arc of the day, writing "dung-beetle (Scarabeus) representing the new life in Egyptian mythology," adding straight afterwards, in brackets and with no further explanation, once again, "(dung being the element of the Underworld)."

Two other self-quotations follow, from his *Monotheistische Strömungen innerhalb der babylonischen Religion* of 1905 and BNT, his *Babylonisches im Neuen Testament*, published in the same year, but about the New and not the Old Testament. In the same note 2 to p.7 of his *Old Testament*, he asks readers to "notice also that the Kingdom of Ea corresponds to the Underworld," and refers to another note, on p.14 of the same work, that the goddess "Ishtar corresponds with Ea."

By now, I am getting seriously confused by Jeremias's writing and am far from sure that I understand him.

A little before this, Jeremias cites *Ea*, not *Ishtar*, and considers him the brother of Enki. Enki is the god of deep fresh waters, of wisdom, and also of magic, and expert in the use of spells to drive away demons (see Pettinato), but Jeremias turns him into a deity of the underworld... and then I get completely lost because earlier on (p.7) he had said "the new world proceeds from the phallus of the Deity", an established fact in Sumerian mythology, but here transferred to the *Unterwelt*. Jeremias goes on to add that "the Phallus is present at the gate of the Underworld in various myths," without saying which these are. Then the Phallus gets confused with the "Vapour which arises from Hell" and everything (Underworld, Phallus, Vapours, Gate of Hell) finally merges into the excrement which the Sacred Scarab carries around, but without giving a scrap of evidence for why this should be connected with Gold. (In 1953, in a fine jumble of rings, bulls, seals, cylinders, Egypt, and the history of numismatics, Desmonde retrieves the story of the Sacred Scarab, likewise turning it into proof of the anal origin of Money. A few years later, in 1962, Desmonde wrote an interesting book, hardly ever cited, *Magic, Myth and Money*, which takes as its starting point a consideration of the previously quoted Bernhard Laum, and will also have some relevance to my own book. I think it is inevitable that, starting from the same point, many subsequent topics will take on similar characteristics, which is why many of the themes addressed by Desmonde have also been developed here. Substantial differences remain, both because Desmonde is a follower, however acute, of Fromm or perhaps of Owen, and because he seems in his examination of Laum's theses to attribute a richer significance to food – food consumed in religious practices, food connected to sibling and community relationships; in other words, food that is shared out – than to *obeloi*, which will appear later on in this book, thereby giving a reading of Laum's hypotheses. In any case, it is strange that Desmonde's book is so rarely cited. Is this because he is not an analyst, as he himself says in his introduction, or because on page 157 he writes, unlike in 1953, "We have not discovered any information pertaining to ancient anal practices related to money"?)

In his exposition, Jeremias superimposes Hades, the Land of No-Return, Egypt, Babylon, Nergal, Ea, Enki, Ishtar, the primordial waters both fresh and salt, *Abzu* (the abyss), Hell, the Serpent which eats dust as it creeps along, Greek cosmogony, Babylonian cosmogony, and a pile of other things, but offers nothing that might explain the relationship between Dust and Dung, or between Dung and Gold. The only reference Jeremias brings in support of his idea is a possible interpretation, confined to a note however, based on an Arabic term which might permit a certain, faint overlap between Dust and Dirt (but 'trash', 'dirt' in his text, and not 'dung'), attributing it, what's more, to Winckler, before going on to attempt another overlap between the act of submission which might be adopted either prone or supine by defeated kings bowing down to their conqueror and "kissing" the soil. This ritual of submission was also required by the Pharaohs with or without placing the victor's foot on the neck or throat of the vanquished, which Jeremias superimposes onto the idea of licking the earth, specifically licking the dust – "biting the dust" or "eating dirt", as we might say today. It is an act of submission which certainly underlines the significance of the defeat for the vanquished on the one hand, and of the victor's superiority on the other, but has little to do with Dung, and less still with Gold. Still in note 2 on p. 234, straight after the famous *ilu Manman*, Jeremias attributes to Winckler the (to me mysterious) suggestion that we should read *Isaiah* 1:20 ("If ye be willing and obedient, ye shall eat the good of the land: But if ye refuse and rebel, ye shall be devoured with the sword", *Isaiah* 1:19-20) as הרא, *hàra*, excrement/shit. Literally Jeremias's text reads, "H. Winckler suggests (compare also *F.*, i, 291) reading Isa i. 20 as הרא – that is, as in the Arabic, 'to eat trash, dirt' instead of *hereb*, 'to be devoured with the sword.'" At first glance I cannot understand where Jeremias (or Winckler) can see "to eat trash" in this passage from Isaiah: at most, there is a "will be eaten" referring to "will be devoured". In the Vulgate, this passage is rendered as "*gladius devorabit vos*", "the sword will devour you." In the Septuagint we find μάχαιρα, knife/ dagger/ blade, and also sword, which ὑμᾶς κατέδεται, which "will destroy you", while in the Hebrew text we find the verb *Le'echol*, לאכול, to eat, (not to devour, which would be *Lizlol*, לזלול) in the passive voice and future tense, and, as Dr Avezov points out to me, in an archaic form: that is, *Teuchlu*, תאוכלוּ (today one says *Teachlu*, תאכלו). This "you will be eaten" is understood today and in common parlance in the sense of "be digested", and perhaps this is the link intended by Jeremias (or Winckler), given that the products of digestion become הרא, *hàra*.

On the other hand, however, we do not know the exact sense in which this term was understood when Isaiah was writing, and so – a matter of some importance – if we say "it was a close shave when the plane flew so low over the house", we certainly don't mean that the pilot had turned into a barber with a razor and that the house had a beard. So I think that "being digested by the sword" has no literal "digestive" meaning. As Jeremias himself said, "The literal eating of dust cannot be meant." Ernest Klein, in his substantial, and weighty, etymological dictionary of the Hebrew language gives חרא, *hàra*, as "void the intestines", from the Aramaic חריא, *har'ia*, excrement, shit; cognate with the Arabic *hur*, which means the same. The Arabic word corresponding to *hàra* is indeed *hur*, which I imagine is the term to which Jeremias is alluding, given that he doesn't write it. While the meaning of "excrement", *as in Arabic*, is clear, the transformation of "*hereb*" into "*hàra*" is not clear in the least, at least to me. He cites the term חרא, *hàra*, as the one to be read instead of *herev*, חרב. Jeremias refers to the SWORD, *hereb* (which would be better transliterated as *herev*, since the last letter is *Vet*, not *Bet*, חרב). Is the consequence of this that, if we adopt

Winckler's suggestion, instead of reading, "You will be devoured by the SWORD"/"The SWORD will devour you," we should read "You will be devoured by EXCREMENT"? or "You will devour excrement"? or "You will be devoured and reduced to excrement"? or "Excrement will eat you"? or "You will be digested and reduced to excrement"? The last statement could be valid, but in a figurative, and certainly not a literal, sense. In any case, all this leaves me decidedly disconcerted for the further reason that in the Hebrew text of *Isaiah* 1:20 we actually read חרב, *ḥerev*, sword; not חרא, *ḥàra*, excrement; not *ḥur*. Moreover, if we take up the Jeremias-Winckler "suggestion", how are we to read the passage from *Num.* 22:23 (and 31), which Jeremias himself quotes a few lines later? (While also referring to the *Thousand and One Nights*, *Siegfried*, and *Theseus* on the "gleaming" of the Sword.) In *Numbers* we read, "And the ass saw the angel of the Lord standing in the way, and his sword drawn in his hand: and the ass turned aside out of the way, and went into the field: and Balaam smote the ass, to turn her into the way." And as a further example, how are we to read, *Gen.* 3:24: "So He drove out the man; and He placed at the east of the garden of Eden Cherubims, and a flaming sword which turned every way, to keep the way of the tree of life"? And what about the more than 400 references to the Sword, *ḥerev*, present in the Old Testament? All in all, what with swords, excrement, eating and being eaten, it all looks like quite a muddle to me. According to one rather possibilistic-abstract line of interpretation, my Hebrew teacher Dr. Avezov, bringing water to Jeremias's mill, whispers to me that the roots of Sword and Shit could share a common root – הר – which would indicate something relating to the action of Fire, Flame, Burning, Digesting, Destroying, what remains after a process of consumption. But Klein's text clearly distinguishes between the root חר with a terminal א from the root with a terminal ב, further distinguishing in the second case between 1) Dried out, Dehydrated, Desert, Abandonment, No-further-care, Withered on the one hand, and 2) Ruin, Desolation, War on the other. Both these roots mean "waste", although of different kinds. Food "destroyed" by digestion becomes "waste", חרא, but is still a different "destruction" from that carried out by the Sword, which is and remains חרב, *ḥerev*, Sword, as opposed to *ḥarev*, to dry. I think we can still speak of "destruction", but to replace the second term with the first just seems bizarre. Even so, Jeremias marches on unperturbed and adds (same n. 2 on p. 234), "Then the figure of speech 'to eat dust' would be attested also in its drastic meaning in the Old Testament." As with the Euphratesians, he starts out with this conviction and stays with it. However, at this point, given his insistence on "the figure of speech 'to eat dust'" and "its drastic meaning",

I cannot help thinking about his quoting on the previous page a statement of Luther's (p. 233, n. 3) at the start of this same chapter XXVII. The idea passing through my mind is that this apparently innocuous quotation in which Luther compares the Serpent to a chicken may be the real interpretative key to this whole bizarre charade, highlighting that "Luther", or a certain view of "Luther" (is it a coincidence that the very term *har'ia* appears, albeit in a different context, in Luther's text (p. 178) against the Jews?), and not Winckler, Ovid, Nielsen, Gunkel, Müller, Isaiah, not the Euphratesians, or Zimmern or Delitzsch, not "the eastern Myth which has passed into popular legends and fables", is the true *primum movens* of this whole complicated business for Prof. Alfred Jeremias; the "true" *primum movens* of this whole (strangely not) *vexata quæstio*.

All of which is irrespective of the fact that in *F.*, i, 291 which corresponds to p.291 of the first volume of Winckler's *Altorientalische Forschungen* Isaiah 65:25 is quoted ["The wolf and the lamb shall feed together..."] and not Isaiah 1:20, while Micah 7:17 ["They shall lick the dust..."] is also present. The point is that there is not the slightest sign of the term *ḥereb/ ḥerev*, sword, in Winckler's book, still less *ḥàra*, excrement!

But let's return to Ea/Enki: apart from the fact that Ea is the Semitic version of the Sumerian Enki (see Pettinato) and not his brother but the god himself, earlier on in the Sumerian version which I know Jeremias disliked, and later in the Babylonian-Semitic, Ea-Enki is certainly the deity of the Abyss, the *Abzu* (or *Apzu*, or *Apsu*). But Ea-Enki is not a chthonic deity, since the deep fresh waters are "simply" one of the elements which generate the World in Sumerian and Babylonian mythology, and the *Abzu*, the fresh deep Waters, is not Hades, or Hell, or the "Land of No Return", whatever that may be, but is certainly closer to the primordial Abyss-*Tehôm*, תהום, of Genesis or Chaos, χάος, in its original sense ("crack, crevice, and symbolically *abyss*, which relates to CHAINO, χαίνω, *I am opened* or CHAO, χάω, *I am empty*," than to the Christian Hell. Mircea Eliade (1957, p. 31) somewhat confuses *Apzu, Tehôm, Inferi*, Chaos, Abyss and the World Beyond the Grave, whereas in his 1937 text (p, 53) he seems to adopt a more correct position with regard to Ea-Enki.

Ea-Enki is "the wise, the foresighted, the cunning" and it is he who creates Man out of clay!

Ea-Enki is not an infernal deity. In fact it is Enki who gives Nergal detailed instructions for his journey to the Infernal Regions, and it is the same Enki to whom Enmerkar wants to build the temple I mentioned at the start of this book; the same Enki who, hidden behind a wall, informs Ut(a)napishtim of the gods' intention to destroy the world in the Flood, and suggests that he build a big ship on which to keep himself and the animals safe. All of which had been done before, in Enki's name, by Ziusudra (Sumerian) and later by Atra(m)ḥasis (Babylonian). And Inanna (Sumerian)-Ishtar (Babylonian)-Astarte (Phoenicians and Canaanites) certainly undertakes a perilous journey to the "Infernal Regions", in the attempt to increase her own power, but her attempt fails and not only does she not become their queen, but she gets well beaten and only succeeds in escaping from the clutches of Ereshkigal by the less than elegant stratagem of sacrificing her own husband Tammuz in her place, again thanks to Enki-Ea!

But once again Jeremias reiterates that "Ishtar corresponds with Ea, for the Underworld and Apzu coincide" (*The Old Testament*, vol. I, p. 14): and not content with that, on the same page he repeats that *"The rulers of the Zodiac are Sin, Shamash, Ishtar. According to the law of 'analogy' they become Anu, Bel, Ea."*

The law of analogy, no less!!!

In this connection, Pettinato's text (2003) is of great interest, retelling and commenting on the Assyro-Babylonian (and not Sumerian) stories about the Infernal Regions, which include a strange literary device by which Nergal would continue to live in heaven with the other gods while a kind of simulacrum of him, Erra, would remain happily married to the queen of the "Infernal Regions". It should be added in fairness that there are those who speak of a presumed journey to the underworld by Enki himself, and perhaps it is these to whom Jeremias lent an ear, but in this case too Pettinato's work shows with extraordinarily authoritative clarity that the only true kind of the "Land of No Return" is Nergal. I have a historian's duty to add that in the Chicago Assyrian Dictionary, under *Apsu*, among an immense number of definitions of *Abzu*, or rather *Apsu* in this case, all of which are "canonical" – that is, Ea-Enki-Deep fresh Waters – there is another, but only applicable to the late Babylonian period by which Ereshkigal and Ea (who belongs to the Semitic-Babylonian period, and specifically to late-late-late Babylon and is no longer the Sumerian Enki) are made to coincide in the celebration of a ritual, thereby giving a modicum of support to Jeremias when we read that "Ea is present (in the ritual) as the Apsu, the Apsu is the sea, the sea is Ereshkigal" [CAD], which to me seems no more than a metaphor referring generally to "The World Below/The World Beyond": in other words, the immense sea which Gilgamesh has to cross, rather than a description of the characteristics of the *Unterwelt*, given that Ea-Enki is and remains Lord of the fresh deep waters (see also below, Bottéro, 1989). And in any case, what is still more important is that there is not the slightest trace of Gold. To sum up, in his wish to exclude the *Euphratesians* and believing that everything starts with Marduk, without considering that Marduk is just a late arrival, a parvenu in the Mesopotamian pantheon preceded by an enormously long period of development, Jeremias wants to equate the union between Apzu and Tiamat which creates the world in the Babylonian *Enûma Eliš* and also creates their very noisy sons An and Enlil, with the Biblical Genesis and thus decides that the Apzu, the fresh waters, united with Tiamat, the salt waters, become respectively the Abyss (the waters) and Chaos (p.6), slipping entirely over the fact that the Chaos, the *Tohu Vavohu,* תוהו ובוהו, is something else. Compare this idea with Genesis 1:2:

> The earth was without form and void (the earth was *Tohu Vavohu*, which the Septuagint translates as Chaos); and darkness was upon the face of the deep (that is, *Tehôm*). And the spirit of God moved upon the face of the waters. (King James Version of the Hebrew Bible)

Tiamat "survives" at the base of the myth of Leviathan-Rahab in *Job* 26:12. (It is a curious fact that in Hebrew, Whale is *Liviatan*, לויתן, and *Gen*. 1:21 speaks of "great whales", with various nuances depending on the different translations of the Bible but always referring or alluding in some way to *Leviathan-Liviatan*, starting with κετος in the Septuagint and *Cetus* in the Vulgate. God, in other words, creates the Great Sea Monsters, but in the text of the Bible these are specifically called *Taninim*, תנינים, plural di *Tanin*, תנין, which literally means CROCODILES ! In other words, God created Light, Heaven, Earth, and also crocodiles, which become one of the foundations of the creation. Not to mention the עוף/עופות, *Of/Ofot*, translated in all languages as Birds/Fowls, but which could in fact be translated as Poultry: in other words, God created Light, Heaven, Earth, CROCODILES, and even POULTRY! And not to mention the Red Sea, ים סוף, Yam Suf, that's to say the Reeds Sea (not RED). Which gives us much to ponder.)

The primordial goddess, Tiamat, whose union with Apzu generates *Laḥmu* and Laḥamu in the Babylonian theogony, and not Mammu (for the Sumerians the primordial goddess was Namma, wife of An, mother of Enki); Tiamat connected (as his great-great-grandmother) with Marduk. Tiamat, who is at the base of the mysterious Tohu Vavohu, תוהו ובוהו, Chaos, connected in any case to *Tehôm*, תהום, (see also Zimmern, 1901 p. 5; cf. also Graves and Patai, p. 35, nn. 3 and 4; and Wechsler and Schoffer, pp. 21-23; although personally I believe that the hypothesis formulated by Giovanni Garbini about the origin of this mysterious term, *Tohu Vavohu*, is much the most brilliant: in note 4 on p. 37 of his book *Il Poema di Baal di Ilumilku*, Garbini speculates that the Hebrew term *Tohu Vavohu* is merely a deformation of the archaic Greek *baau*, dark, which is in turn connected to the Semitic *bahaw*: see also his *Note di lessicografia ebraica*, 1998. An additional thought of my own rather than Garbini's, is that through various shifts that I cannot illustrate here, the word remains almost unchanged, under the radar, in the infant game of peek-a-boo which alternates Dark and Light.)

In late Babylon, in the poem *Enûma Elish*, Ea-Enki-Nudimmud kills his great-grandfather Apzu who had wanted to kill his excessively rowdy sons and grandsons ("to be rowdy" is a frequent translation of the Akkadian verb *ḫabāru(m)*, to make noise, clamour – very often associated with sexual activity), having been annoyed by their racket and unable to sleep (the couple Tiamat-Apzu first creates *Laḫmu* and *Laḫamu*, sea-monsters/ serpents [?] and parents of *Anshar* e *Kishar* who are the parents of *Anu* and *Nudimmud-Ea-Enki* who was "controller of his fathers"; Pettinato, 2005, p. 104, lines 10-17); and Mummu (the first error made by Jeremias, see. Bottéro, 1989, above) is not the son of Tiamat, but Tiamat herself, Mummu being an epithet – *Ummu*, Mother – which later serves to indicate Apzu's vizier. Old Tiamat, wife of Apzu, is angry with Ea and Anu for the murder of their great-grandfather and creates eleven frightful beings led by Kingu/ Qingu, her new husband, to punish her great-grandsons. Then the cavalry arrives, captained by the new god Marduk, Enki/Ea's son, who routs them all and, quartering Tiamat, uses the pieces to construct heaven, earth, the mountains, and the rivers, and then mixing the blood of Kingu with clay he gives life to men who, naturally, are Babylonians (see *Gen.* 9:3-4 where, after the Flood, God says, "Every moving thing that liveth shall be meat for you: even as the green herb have I given you all things. But flesh with the life thereof, which is the blood thereof, shall ye not eat." And *Leviticus* 17:11, "For the life of the flesh is in the blood: and I have given it to you upon the altar to make an atonement for your souls: for it is the blood that maketh an atonement for the soul.").

Being more courageous than the gods, Marduk becomes their new leader, to the general applause of the Annunaki and Igigi. Bel, the Chaldean god of whom Berossus was the priest, also carried out an epic struggle, almost entirely equivalent to that of Marduk (see Eusebius of Cesarea for Berossus on *Marcaja-Thalassa*, p. 24). For some Biblical scholars Yahweh also fights with the Dragon, the *Leviathan-Rahab-Tehôm-Tiamat*, splits it in two and uses the body to create the waters-above, *shamaim*, שמים, and the waters (below), *maim*, מים (see *Tiamat-Thamte-Thalatta-Thalassa* in Zimmern, p. 14). Sometimes (in the middle-Babylonian and Chaldean period) Bel is Marduk, Marduk "of the fifty names" – in other words, a step towards monotheism; and Babylon (Tower of *BaBel*) derives from *Ba-Bel*, Door/House (*Ba/Bait*) of the god Bel. House in Akkadian is *Bītu* and beam (over the door?) is *Binītu*. Door in in Hebrew is connected to the letter Dalet, ד, that is, דלת, which Eliade connects to the letter Delta, in turn linked with Vulva; in fact, the Sumerian graphical sign for Woman is actually a Delta, an inverted triangle with a vertical dash at the lower angle.

Still on the subject of Apzu (in this case, only the Abyss and not the Abyss-god-husband of Tiamat), Gilgamesh (in Tabs. XI-XII of his Epic) has lost his *Pukku* and his *Mekku* while trying to find the "plant of youth/ plant of restlessness" (see Pettinato, 2004) in the depths of the Apzu, while in other sections, for example, *Enkidu's Descent into the Infernal Regions*, the Sumerian scribe has used the same literary device and sends Enkidu into "Hell" because Gilgamesh has lost the *Mekku* and the *Pukku*, this time down the drain of the "Land of No Return", which is not the Apzu. So, it may be that this literary superimposition also created some confusion in Prof. Jeremias. Later on, Marduk too will have to present himself to the Annunaki, who in late Babylon will have become Judges of the afterlife (see *Il poema di Erra*, Bottéro, 1989); perhaps an anticipation of the fate awaiting him in History?

So, what exactly were the precious *Mekku* e *Pukku* extracted from the roots and branches of the tree-*ḫuluppu* which Inanna had planted and which was infested by the demons which Gilgamesh defeats (Lilith seems to have her origin here)? We do not know why they were so precious as to dictate such a painful reaction on the part of Gilgamesh who grieves for them and despairs both with his friend Enkidu and with a string of gods, leading to a descent into the infernal regions by Enkidu to try and recover them, losing his life in the process. I will only point out that in the ePSD, *Pukku* is translated as *"Wooden Ball"* but in the ETCSL as *"ring or ball"*, while *Mekku*, is generally translated as *"Tool"* but also *"Mallet"*. Should be perhaps conclude that Gilgamesh played Croquet? And perhaps Cricket too? Baseball maybe? Perhaps a forerunner of Golf? And was he so fanatical about his favourite game that he lapsed into a deep depression when he no longer had the equipment to play it with? Or are we dealing with a classic case of displacement, onto to those "playthings", of his disappointment at not having brought home the Plant of Youth which had been taken from him by a snake during a moment of absent-mindedness after all that effort?

We do not know. But a suspicion has occurred to me: since Bottéro (1992) writes about the *Mekku* and the *Pukku*, but uses the terms *Cerceau* (circle, ring) and *Baguette* (baton); and since in many of the representations of him, the god *Shamash* appears seated with a circle and a staff in his hands; and since in the texts which accompany such representations, including the stele of the Code of Hammurabi, but also in various translations, the terms hoop, ring, *anneau*, circle for the *Pukku* and *rod, bâton* for the *Mekku*, I think we could hypothesise that the *Mekku* and *Pukku* might be the great-great-grandfathers of the Sceptre and Orb (transformation of the Ring) which represents the World, symbols which all emperors enjoy holding in their hand in the classic iconography. Bottéro himself is not far from making this speculation, while confining it to a local scope. And circlet, ring, *cerceau*, and not wooden ball is the symbol which will have some space in this study.

Meanwhile, however, in the work of Jeremias, the Scarab, Money, dust, Shit, and Gold have disappeared. Have they too ended up at the bottom of the *Apzu*, with the *Mekku* and the *Pukku*? Who knows?

Only three or four years earlier, Jeremias had written a small text which I mentioned previously, *The Babylonian Conception of Heaven and Hell* (I quote it in English for the reasons given before), in which, besides attempting to superimpose the Greek deities and divine functions on those of the Babylonians, but not yet Greek astronomy onto that of Babylon, he writes:

> But among the Babylonians, as also among the Hebrews and the Greeks, representations of Hades reflect the melancholy thoughts roused in human souls by mourning for their dead. The soul of the dead sinks into joyless existence, the misery of which has been foreshadowed by the phenomena of mortal sickness. The loss of a corporeal manifestation has already deprived it of all adornment and all exercise of the senses. Where is the soul? Simplicity sought it in the tomb; the shade of the dead man finds it hard to part from the body which gave him form and substance" (p. 48).

According to Jeremias, in the *Unterwelt*, which as we have seen he calls *Hades* (Ἀίδης !), there is melancholy, sadness at one's own death: dust, cold, darkness. Nothing else.

No Gold. No Mammon. No Sacred Scarab. No Devil's excrement.

There is only *Sheol*, שאול, which I could translate, stretching it a little and with a bit of poetic licence (leaving aside the fact that *Sheol* is the place of questions, of anxiety and not knowing: that is, of those who do not have Faith, Sheol from "ask", *Lish'ol*, לשאול, in turn connected to *Sha'al*, to consult an oracle; connected to the Assyrian *Shu'alu*, tomb, tombstone, keeping this meaning in the *Jewish Encyclopedia*; and in E. Klein) as, "There shall be weeping and gnashing of teeth" (*Matt.* 22:13): in other

words, a tremendous, solitary, desolate, frozen "Nothing".
In Pettinato's translation (p. 99, 2003) of the poem *Nergal and Ereshkigal* (Sultantepe and Uruk version; cf. also Bottéro, *L'Épopée de Gilgameš*, tab. VII, Col. IV, lines 32 and 38, pp. 150ff) the *Unterwelt* is described thus:

[towards the house from which whoever enters never departs]

[on a journey whose path] is without return

[through the house where the inhabitants have no light]

[where the dust is their nourishment, the clay their food]

[they are clothed like birds, covered in feathers]

[they see no light], they grope in the dark

[where terror reigns] and lamentation

[they weep and groan] like doves

The Other World of Mortals, both for the Sumerians and for the Babylonians, is not Hades or the Infernal Regions, Erebus, Tartarus, Avernus, or Hell. It is not a place of reward or punishment, and has nothing of the Hellenistic, and still less the Christian conception in which one participates in the joy of communion with God, or suffers in remoteness from him, undergoes Pains and Griefs or enjoys Pleasures and Happiness determined by one's conduct *ante mortem*. Bottéro is categorical about this: "There is no prevailing belief in any life *post mortem* in which accounts would be settled with one's life on earth." In this Sumerian Other World there is not the slightest trace of Gehenna, *Gehenom*, גיהנום, to warm the cold, the dark, the sadness (and here I dispense with the complex Rabbinical elaborations, mainly Midrashic and Mediaeval, from long after the Scriptures, of the concepts of *Sheol* and *Gehenom*, and refer once again to the fine book by Graves and Patai, *The Hebrew Myths*, as well as to Ginzberg and Buccellati). And Ascalone (2005, p. 267) specifies:

> Only in the first millennium BCE does the underworld, divided from the human world by the abzu, the primordial ocean in which Enki and his consort Damglanunna dwell, become populated by demonic beings who trouble the life of the *gadim*.

(The *gadim/gidim* are the spirits/ghosts of the dead, those whom Bottéro calls, in Hebrew, *nefesh*, נפש; although a better term might be *neshamah*, נשמה since *nefesh* renders soul in the sense of "a village of thirty souls":

that is, person-body-soul, while *neshamah* has a more metaphysical meaning, and spirit, πνεῦμα, *pneuma*, breath, divine wind is rendered by *ruaḥ*, רוח. These matters, and the dual nature of Nergal/Erra are very well clarified in the excellent book on Heracles by Silvia Maria Chiodi, 2004): not forgetting the fundamental work, *L'epopea di Erra* [The epic of Erra] by Luigi Cagni (1969) which concerns the Babylonian evolution of the god Erra, and the fine work 'The Poem of Erra and Ishum: a Babylonian Poet's View of War' by A. R. George (2013).

Damgalnunna is Enki's wife in the myth of Enki and Ninḫursag, where she is given the epithets Nintu and Ninḫursag (Pettinato, 2001, p. 529); *Nin*, Lady; *ḫursag*, *Ki*/Earth, She who fixes Destinies, Good Fate, depending on the period.

How touching in their sadness, grief, and drama are the words addressed by Gilgamesh to the spirit of Enkidu, who has been allowed for a moment to return to the world of the living (Pettinato, *La Saga di Gilgameš*, pp. 376ff, or in the Introduction, pag. LXVIII).

I will also pass over the fact that in the second volume of *The Old Testament in the Light of the Ancient East*, Jeremias, perhaps taking on board Jensen's suggestion that *Sheol* is derived from the Assyrian *Shillan*, 'West' (*Jewish Encyclopedia*), he lets himself go and fills twenty pages (pp. 32-51) with references endorsing his "passion/ obsession" for the Underworld-*Unterwelt*. So, Sodom becomes the Underworld, as do "the Sun and the Moon of the desert", and "the desert itself, which is the West", and even "Egypt" being in the South (?) is the Underworld, finally even calling "Esau" the Underworld. Jeremias tends to pass over the fact that it is the *entrance* to the *Unterwelt* which is in the West, not the *Unterwelt* itself. In the Sumerian conception, the *Unterwelt* is beneath the Apzu, as Bottéro (1987, p. 306) stresses: "With the allusion to the main Entrance [not the location, but the Entrance (my specification)] to the infernal realm, we again find the term KI.U.SÙ, for KI.UTU.ŠÚ.A, *The Place where the Sun* (Utu) *sets*."

Jeremias's aim is to demonstrate that the West (oh God, Europe is the *Unterwelt* too!) has taken everything from Babylon. All right. In his *Old Testament* he starts with Greco-Roman astronomy and reveals its many similarities to that of Babylon. And that's all right too, since there are plenty of stars that can be seen from both Babylon and Athens.

But forcing the superimposition of different deities onto each other in an attempt to endow them with a perfect one-to-one correspondence of attributes, playing a game of matching pairs or doubles, a splendid and ever-available all-purpose skeleton key, or calling on the Law of Analogy, cheerfully straddling the millennia, becomes a decidedly dubious operation. And in any case, I don't see the Dioscuri as a good reason for turning the Babylonians into the jolly Bavarians of the ancient world. It is true that the *Euphratesians* invented beer, but it's hard to see the young ladies who serve it at the Oktoberfest as having their origin in Babylon. But Jeremias wants to exclude Ur and the *Euphratesians* entirely.

I have already made this observation in connection with Fromm, and I repeat it here: with all due respect and admiration, completely sincere in this case, for Professor Jeremias, for his love of research and the vast scale of his learning, I think that instead of calling his book *The Old Testament in the Light of the Ancient East*, he really should have called it *The Ancient East in the Light of* (not *the Old*, actually, but) *the New Testament*, since the observational viewpoint he adopts is by no means that of the archaeologist, but that of a man of faith: cultivated, intelligent, engaging, but substantially and profoundly committed to attempting a historico-archaeological demonstration of his own faith.

However, at this point I have no choice but to take a look at his *Babylonisches im Neuen Testament*, already quoted above, and especially at p. 96, hoping to find a key to guide me in the muddle I've got into.

I don't know German, but in a first, stuttering attempt to read it just with the aid of a dictionary, I am left dumbfounded by what I think I have read, but not knowing German, I don't rely on what I think I have understood and ask for enlightenment (not of the French variety) from a learned and kindly teacher of German born and raised in Austria: Frau Evelyn Rachbauer. My amateur, approximate and bizarre translation of p. 96 of *Babylonisches im Neuen Testament* is in fact confirmed *in toto*.

Running through the whole of *Babylonisches im Neuen Testament*, in a long and intense study (and again I am grateful to Frau Rachbauer for her enormous care and intelligence) we find no trace of Sacred Scarabs, Infernal Regions, Mammons, or anything to do with Dung, Dünger, Gold, Money, Shit, Scheisse, Dreck, Geld or Münze, or even a bit of Silber.

The only reference is on the same p. 96 where Jeremias once again states, for the only time, still without quoting his source, that "Gold, according to the eastern Myth which has passed into legend....," a quotation which I now know very well. He is repeating statements he has made before but here too he brings nothing to support his claim. From my love of truth, I must admit that there is one other reference to Gold in *Babylonisches im Neuen Testament*. Frau Evelyn and I have found one on p. 54 where he says, "to the Sun [god] is owed [the offering of] Gold, Incense and Myrrh." The interesting parallel with the Adoration of the Magi (who, it should be noted, come from the East and travel to the West! Are they following the comet in order to go to Hell?) is obvious, but a little less obvious is the fact that from here we pass to the Shit of the Devil (offered to the Divine Child as it was to the pagan divinities and a legacy of the *Euphratesian* tradition!).

And yet Jeremias, on p. 96, establishes that astonishing connection, still without any explanation. That is what I seemed to have understood, and it left me completely dumbstruck: "Mamîtu is the wife of Nergal"; which obviously confirms that gold is *"Dreck der Hölle"*! (From *Babylonisches im Neuen Testament*, by A. Jeremias, p. 96 – my emphasis):

"*In der Matthäusstelle ist Mamon als Götze gedacht und Gott gegenüberstellt wie Targum Prov. 3,9 : ‚Ehre deinen Gott mehr als deinen Mamon;' Berachoth 61 b : ‚Es gibt manchen, dem sein Geld lieber ist, als seine Seele.' Mamon (Mammon) ist babylonisch man-man, ein Beiname Nergals, des Gottes der Unterwelt (kakkab ilu man-man ist Saturn, der Planet Nergals, Mamîtu Nergals Gemahlin). Das Gold ist nach orientalischem Mythus, der in die Sagen und Märchen der Völker übergegangen ist, Dreck der Hölle*".

("In the passage from Matthew, Mammon is thought of as an Idol placed in opposition to God, as in Targum *Prov.* 3,9: "Honour thy God more than thy Mammon"; Berakhot 61 b: "There is one who prefers his money to his soul." Mamon (Mammon) is *man-man* in Babylonian, an epithet of Nergal, god of the underworld (*kakkab ilu man-man* is Saturn, Nergal's planet, Mamîtu is the consort of Nergal). According to the Eastern myth which has passed into popular legends and fables, gold is the dirt of Hell."

In the text we find the term *Dreck* and not the term *Dünger*, indicating dirt, not excrement.

In German, the Sacred Scarab which Jeremias cites so often is *Mistkäfer*, but he uses the term *Dreck* and not *Mist*, or *Kot*, or *Dünger*, nor even *Scheisse* or *Kacke* to identify the "excrements" of Hell.

Now, the Targum, תרגום, is the Translation (TARGUM = TRANSLATED -TRANSLATION) into Aramaic (the common tongue of the period) of the Hebrew Bible (written in a hieratic, and therefore sacred, language). In the Introduction to the Edizioni Paoline Bible of 1964, p. XVII, it is pointed out that TARGUM also means Explanation and Interpretation, since after the Babylonian Exile, in order to prevent the sacred language dying out, "Two competent persons were chosen from the community to read the text during sacred gatherings, one in Hebrew while the other translated it verse by verse." A little later we read that "The translator was called *me-turgheman*, from which is derived the term 'dragoman', translator."

And thus far we are in agreement. But the point is that in Proverbs 3:9 we find no trace of the sentence Jeremias quotes: in fact, the ninth and tenth verses of Proverbs 9 say, "Honour the Lord with thy substance, and with the firstfruits of all thine increase./ So shall thy barns be filled with plenty, and thy presses shall burst out with new wine." In the Targum of Proverbs 3:9, "thy substance" is called *Hon*, הון, [literally, *Mehonkha*, מהונך, 'the goods which come from thee (to the male)']; not *Matmon*, מטמון, 'treasure/ hidden treasure' (connected by E. Klein with *taman*, טמן, 'hidden'), and not *Mamon*, ממון, 'capital committed to a project, finance' (in E. Klein connected to *Amen* אמן, 'believe in, have faith in, place trust in, rely on'). This meaning is modern: perhaps a contamination from two millennia of (distorting) Christian interpretation, as A. Toaff shows in the case of some Easter myths in his *The Bloody Satanic Sacrifice Rituals of the Jewish Race: Blood Passover*. *Hon* is closer to 'fortune', not in the sense of a (good) fate, which would be *Mazal*, מזל, but in the sense of substance, a sound inheritance – worldly goods, in fact. In other words, Jeremias, *pro domo sua*, stretches the concept to authenticate a god, Mamon, who must obviously be ugly, dirty, and wicked, but who isn't in the text. Here I take issue with the fanciful conjectures of Jacques Attali about the etymon of Mamon on p. 41 of his book, which involve an improbable and far-fetched translation of the Hebrew *Moné*, 'to count/enumerate', which is directly connected to the Akkadian *mīnu(m)*, number, payment, and *minûtu(m)*, "to count/counting/numbering", and translates the Sumerian UTTU. In his more recent *Dictionnaire amoureux du judaïsme* the term Mamon "strangely" does not appear.

Mammona, μαμωνᾶ, does not exist in antiquity. Mammona, Mamonà, appears only after 70CE in *Matthew* 6:24 and then in *Luke* 16: 9, 11, 13. My fantasy is that Matthew, as a state functionary, in other words a tax-collector, a *publicanus* (being accustomed to "recovering" money for the Roman Treasury, money which the Israelites were not happy to be robbed of) had to look for the dough in hidden/ subterranean places, and imaginatively drew on his own experience as a Treasury official to create a neologism – that is, if we are referring to Money, a word which had never existed before. In all seriousness, however, Matthew was a child of his time, the first and second centuries CE, and it is in this period that the Mishnah, a summary of the Halakhah, the Law, was being written. Immediately after the destruction of the Temple, the Rabbinical "academies" came into being, writing down and collating the Law, and creating commentaries on it. In the Mishnah we find the term Mamon for the first time. We specifically find it in the treatise called *Neziqin*, 'damages, injuries' (its first chapter is called First Door/ Bava Kama, the very one cited by Viderman) where a distinction is made between "matters of mamon" and "matters not of mamon", which are also called *Isur* (see the *Jewish Encyclopedia*). Greatly simplifying, a distinction is established between the Civil Code, Mamon, which governs damages/ offences/ injuries to the person, and the Religious Code, Isur (prohibition/ ban) which obviously deals with religious matters. It is from this perspective that we need to read the metaphor in *Matthew* 6:24. A. Toaff (2008) writes about a ritualised, strictly codified law, *Halakhah*, and it is to this that Matthew is referring, to the rigid application of codified compensations. Both Matthew and Luke write about this: a blow received, a coat or cloak taken away, considering, among many other things such as time, objects, physical labour as a means of restitution and also, but not only, money as a medium of exchange but after a rigid evaluation of many aspects of the blow received; see also Viderman who writes about five basic elements of evaluation: 1- the overall material damage, 2- the value of the thing damaged, 3- any possible medical treatments that may be necessary, 4- inactivity caused by the damage, 5- the shame aroused by the damage (see also Exodus 21:1, 22: 23, Talmud and the commentary by Jacob Neusner). It has nothing to do directly with money, only with the rigidity of the ritual compensation. These elements exactly reprise *Isaiah* 50:6: "I gave my back to the smiters, and my cheeks to them that plucked off the hair: I hid not my face from shame and spitting."

Thus, the opposition between God and Mammon is not between God and Money, but between the pure Heart which loves its "enemy" and goes beyond the damage/ offence/ injury and the Heart which demands compensations prescribed by law for the blow, which demands the tunic as compensation for damage or an offence!

And it is in this sense that we must read Matthew 5:40-41 and Luke 6:29.

But let's return to the Sumero-Akkadians. Leaving aside the goddess Mam(m)a/i, also called Mam(m)i.tu, which is an epithet of Ninlil, who in other literary compositions, as a young lady – ie. before marrying Enlil – is called Sud, but after her marriage also Nin.tu, the lady who brings into the world, the lady who spreads the thighs/knees in relation to lovemaking, coitus, and childbirth; and also Ashnan (equivalent to Ceres, cf. Bottéro, 1989, p. 120); Sud-Ninlil mother of Nin-urta (Lady/lord of the ear of barley); Sud-Ninlil wife of Enlil (En, Lord, and Lil, Wind-Air-Atmosphere, god "of the Air", of all that is found between Earth, Ki, and Heaven, An) present from the before the birth of Nergal, given that she is his mother; Sud-Ninlil-Nammu-Ninmaḫ who collaborates in the creation of Man at the request/ suggestion/ challenge of Ea-Enki. In other passages the birth of Man happens at the prompting and protest of some other, lesser gods, fed up with working for the great gods. But other, different versions exist. Most interesting are the observations of Bottéro (1989) both about the god Wê who is sacrificed to give life to men with his own blood, and about the term *awīlu*, man, p. 621. Ea-Enki creates U.MU.UL (a term which designates a "wretched/sickly creature" in Pettinato and ePSD, but which Bottéro translates as "My time has not yet come," alluding instead to a feature of immaturity-incompleteness-being unfinished) after seven "failed" attempts. If we rule out a later god Mamu, god of dreams (cf. Saporetti, 1996) and "mist" [MA.MU, dream, in Sumerian; *mamud*, in Babylonian; *šuttu* in Akkadian; in ePSD] the only remaining possibility is a (male) Mamîtu whom Jeremias nevertheless labels as the *wife* of Nergal. So if we accept that Mamon is descended from this Mamîtu, given that once again Jeremias does not think it necessary to quote his source (a very bad habit indeed!). This Mamîtu appears in the poem, "An Assyrian Prince's Vision of Hell", a Babylonian literary composition in which there are fifteen hideous, ferocious and deeply wicked judges at the court of Nergal, one of whom is called Mamîtu and has the head of a goat and the hands and feet of a man. And he is the only Mamîtu connected, either by the Euphratesians or the Babylonians, to Nergal. If this is the Man-Man to whom Jeremias is referring, then the other fourteen infernal judges present in the poem should also be taken as facets-epithets-"dioscuri" of Nergal. But in any case, this Mamîtu does no more than increase, by his mere silent presence, the terror of the poor Assyrian prince, one Kummā, the dreamer. Besides, this Mamîtu in the poem, "An Assyrian Prince's Vision of Hell", the only *Mam(Man)-something* present in the Land of No Return, has no connection to riches, gold, or anything else of the kind; as if Nergal neither has them, nor ever did. Unperturbed, and without batting an eye, Jeremias decides nevertheless that Nergal and Mamon are one and the same and lends Freud his support for the claim that according to Oriental mythology, which has passed over into popular legends and fairy tales, gold is the excrement of Hell. (Freud 1908, 'Character and Anal Erotism', p. 174, n. 4)

Another explanation, not from the Old Testament, but from the Akkadian, of man-man/mammān would risk insulting Jeremias's intelligence, given that the correct translation of the Akkadian "man-man/mammān" is "someone, anyone" (ePSD) or "somebody, whosoever" (CDA). This term renders the Sumerian LU.NA.ME with the identical meaning and is represented by a graphic sign which has nothing to do with divinity. Hence *"kakkab ilu man-man"* would be "a certain shining/celestial god" and definitely not "the shining god Mammon"!

Now, for Professor Alfred Jeremias, Academic, Intellectual, Orientalist, Bible Scholar, Theologian, Lutheran, that's that: if *Mammon* is the god of Hell and gold is the product/ excreta of Hell, the excreta of Hell can be no other than the excreta of Mammon of Hell, the excreta of Hell can be no other than the excreta of Mammon who represents it, *ergo* Gold is the excreta of Mammon! I stress again than that the definitions of the chthonic deities and the world of the Dead in the work of Jeremias and, to some extent, of Freud, there is a fusion and confusion not only between Babylon, Greece, Egypt, and Rome, the Middle Ages and the *Malleus Maleficarum*, but also Pluto, Plutus, Hades, Hell, Avernus, *les Champs Elysées*, Tartarus, the Land of the Dead, *"The Land of No Return"* and so on. This is the basis of the "oriental myth", as I noted a little while ago, that was borrowed by Freud. At least as far as I have been able to reconstruct it. In any case, with the greatest respect to the Master, the thesis reported above is a bit feeble, and the Folklore with which Freud wants to demonstrate his "evidence" is even more so. In other words, the examples from folk tales which he adduces all seem very "recent" and certainly do not have the density of Myths and ancestral Symbols: see *Dreams in Folklore*, 1911. Without undue effort, I could find in the folklore of the small, out of the way valley I grew up in at least four traditional stories connected to Gold, which have nothing to do with shit, but with a very dangerous and terrible relative of the Basilisk, with a little ingot of gold on the tongue, a creature you risk encountering at night on the *Bià Antìga*, the "old road" built by the Romans which cuts the valley in half, in other words relating to castration and ingratitude; and the "quaintly attractive" valley in the foothills of the Alps, where I was born, is not renowned for the quality of the literature produced there.

A final detail: for Professor Jeremias it is a matter of no importance where the gold comes from, and he seems to consider it obvious that it is dug from under the ground; so the ground becomes the only place from which gold appears, which is clearly absurd and historically untrue.

Pliny (23 BCE – 79 CE, *Naturalis Historia*, Book XXXIII, 66-79) and Diodorus Siculus (90 – 27 BCE, *Bibliotheca Historica*, Book III, chap. XII, p. 166) are the first to write about gold-bearing deposits, Las Medullas in Spain (Pliny) and gold mines in Nubia (Diodorus) where gold is excavated on an industrial scale employing masses of slaves at great financial cost and moving enormous quantities of earth and water; whereas Strabo (64 BCE – 25 CE) writes of deposits in rivers, and Ovid (*Metamorphoses* I, 137), writing about the Iron Age which follows the Age of Bronze, says (although the Bronze Age wasn't such a great time of peace):

Nor was it only corn and their due nourishment that men demanded of the rich earth: they explored its very bowels and dug out the wealth which it had hidden away, close to the Stygian shades; and this wealth was a further incitement to wickedness. By this time, iron had been discovered, to the hurt of mankind, and gold, more hurtful still than iron. War made its appearance, using both those metals in its conflict, and shaking clashing weapons in bloodstained hands.

At least until the end of the Babylonian era, VI century BCE, and beginning in the time of Gilgamesh (cf. G. Pettinato, *The Epic of Gilgamesh*, tab IX, lines 172-197), precious stones "grew" on trees in the garden of the Sun god, by no means an infernal deity, at least during the day, since at night he passes through (passes through and does not rule over) tunnels found under the mountains of Kur (which sometimes indicates the "*Land of No Return*" and sometimes, further back in antiquity, the Forest of Cedars – in other words, the Lebanon – where Gilgamesh kills the monster Humbaba with the help of Enkidu. Sometimes Kur is the West in general, at others it is an unspecified dangerous land, and at others the source of the principal rivers, those of Genesis) but the tunnels through which Gilgamesh passes, astounded by the sight of the Garden of Shamash in search of eternal life, are described only as dark and cold, and serve only as an obligatory passage enabling the Sun to be reborn in the east at the dawn of the next day:

> [After he has marched for the twelfth double-hour] the light shines!
> He is astonished to see every kind of tree of precious stones:
> the carnelian bears its fruit,
> a vine hangs from it, beautiful to gaze upon...
> the lapis lazuli bears leaves,
> and it too bears fruits pleasant to see
>   [Lacuna of 7 lines]
>
> [ ] cedars [ ]
> their fronds are [full] of white stones
> wood of the sea [ ] chalcedony
> as if on shrub[s and bu]shes carnelian blooms...
>   the carob he takes in his hand and behold it is chalcedony, gems, haematite
> [ ]
> Abundance and wealth he can [admire] as [ ] turquoise:
> the reed bed [on the shore of the]sea has [mu]ch abundance.
> Gilgamesh strolling through [this glade]
>   lift[s his eyes towards] her,
> Sidur[i, the innkeeper] who lives (far off) on the shore of the sea.

Traces of this "belief" are found later in the Greek story of the Garden of the Hesperides (Hesiod, *Theogony*, 215, but also in Pliny, *Nat. Hist.*, V, 46; Ovid, *Met*, IV, 638; see Graves, *The Greek Myths*, § 133) with the golden apples which grow on the trees, and their fine guardian dragon, Ladon. All these elements, depending on how we choose to date them, derived from or strongly influenced Hebraic and, later on, Christian iconography. Or precious stones which have come from the sky: Siduri, in this case, the innkeeper, is also an epithet of Inanna-Ishtar, which is also the name given to the planet Venus. And clearly the name Siduri is not so remote from the Greek σιδηρός, Sideros, which leaves traces in the Roman Iron Age with the term Sideritis, Magnet, in other words, magnetised iron.

Sideros gives us the term Siderurgy, metalworking, making clear in this way that precious things like iron, which until the end of the Bronze Age was only meteoric iron, come from the Stars and certainly not from the Underworld.

Put yet another way: until the VII century BCE, at least through the entire Bronze Age, and at least in Lydia and the Fertile Crescent, gold – the shit of the gods, or of hell, or of heaven – was sought only in rivers and other watercourses, and in gold-bearing sands, or gathered from gold-bearing veins on the surface. See also *Deut.* 33:19, "for they shall suck of the abundance of the seas, and of treasures hid in the sand." Only much, much later, as I hinted earlier, does the concept of a mine appear, understood as a costly and systematic method of deep excavation, and thus – at least hypothetically – open to alignment with the concept of the Underworld/Hell which "hides" precious things. So, at least until the Lydia of Midas and Croesus, or on the island of Aegina under Pheidon of Argos in 895 BCE, gold was "collected" in a basket like mushrooms, or with a sheepskin immersed in the river Pactolus, Πακτωλός, anciently called Χρυσορρόας, near Sardis, with its source on Mount Tmolus (King Midas transfers his curse onto the river), and also the Golden Fleece and the Argonauts. In this connection, I think that at least in the ancient world we should revise the dogma of Socially Necessary Work as the constitutive element of an item's value.

Returning for a moment to the innkeeper, Siduri, from the Berlin/London fragment of one version of the *Epic of Gilgamesh*, Bottéro (1992, p. 265) reports the words which she addresses to Gilgamesh when he asks her for directions to that he can continue his journey in search of eternal life:

> Why do you wander, O Gilgamesh? The eternal life you seek, you will never find! When the gods created men, they assigned death to them, reserving immortality only for themselves!

Instead, you should fill your belly; spend day and night in leisure; make every day a holiday; dance and enjoy yourself day and night; dress yourself in clean garments; wash and bathe yourself; look tenderly upon your little son who is holding your hand; and make your wife happy, keeping her close beside you! This is indeed the only prospect for men.

These words seem very similar to *Qoheleth/Ecclesiastes* 5:17; *Eccl.* 10:9; or *Eccl.* 8:15, "then I commended mirth, because a man hath no better thing under the sun than to et and to drink, and be merry: for that shall abide with him of his labour the days of his life, which God giveth him under the sun." And shortly after this, 9:7-9 continues, "Go thy way, eat thy bread with joy, and drink thy wine with a merry heart: for God now accepteth thy works.

Let my garments be always white; and let thy head lack no ointment.

Live joyfully with the wife whom thou lovest..." and then in 9:10, "Whatsoever thy hand findeth to do, do it with thy might."

But it is unfortunate that in the ancient world, Mammona, Mammon, ilu man-man, Mamîtu, Pūdu ilu (who is not a Nuragic/Sardinian deity but another "discovery" (!) by Jeremias, another epithet of Mammon – that is, "The god of the Shoulders" (?), since *pūdu* in Akkadian is "shoulder, back", whereas in Sumerian *murgu* is "back" (ePSD), perhaps believing – still following Luther, but that is another story – that the "Back" is the "Unterwelt" of the face) and the now famous "oriental mythology, which has passed over into popular legends and fairy tales [that] gold is the excrement of Hell" (BNT, 96) *simply never existed.*

In the Old Testament I have found no trace of Mammon, whereas I have instead found traces of a Ba'al Hammon or Ba'al Hamon, who is a Canaanite, Phoenician, and Philistine god, Ba'al whose epithet, Hammon, means "of the multitudes", a divinity who demands human sacrifices, and specifically sacrifices of the firstborn, as does the god Moloch. These sacrifices of the firstborn hint at cannibalism (*Deut.* 28:53, "And thou shalt eat the fruit of thine own body, the flesh of thy sons and of they daughters, which the Lord thy God hath given thee, in the siege, and in the straitness, wherewith thine enemies shall distress thee;" and Baruch 2:1-3 where "the Lord hath made good his word, which he pronounced against us, and against our judges that judged Israel, and against our kings, and against our princes, and against the men of Israel and Juda, To bring upon us great plagues, such as never happened under the whole heaven, as it came to pass in Jerusalem, according to the things that were written in the law of Moses; That a man should eat the flesh of his own son, and the flesh of his own daughter") and go so far as to foretell the intervention of Nebuchadnezzar

II as the saviour from this horror, and the occupation of Jerusalem as a just punishment for it. These words seem to echo a passage from *The Great Sage*, a late Babylonian poem which Bottéro dates to a thousand years before the Biblical account and which tells of the "plagues" sent by the gods before the last one (the Flood). Here it is written of the "plague of famine", "When the sixth [year, the sixth plague] comes, your daughters will be your food, and your sons your meals." These are the human sacrifices and those of the firstborn against which Yahweh rages, demanding that the firstborn be dedicated to Him and not sacrificed, "ransomed" – that is replaced – by an animal sacrifice, often itself the firstborn, who can take his place and "compensate" the deity for the lack of a sacrifice. (The connection with the sacrifice of Isaac, first and only son of Sarah, seems obvious, including the substituting of the son with a Goat; perhaps also Ishmael's expulsion into the desert, which would seem to foretell his inevitable death, like that of the scapegoat abandoned in the desert, had it not been for the arrival of an angel.) We also find echoes of this tradition, the sacrifice of children, long after the assertion of official Monotheism, as is testified by the story of Jephthah's daughter who is sacrificed for a vow made by her father before a battle in *Judges* 11:30-31; and appears again in the *Iliad*, with the sacrifice of Iphigenia and the "replacement" of the human victim with an animal, a hind. The tradition is also present in the episode of Athamas and Phrixus, almost identical to that of Abraham and Isaac; and I think that the death of the firstborn of Egypt, a dramatic prologue to the Exodus, can also be connected to this "pagan" custom, a custom which, once it has been condemned by Yahweh, is transformed into an instrument for his own use – His personally, *motu proprio*, not that of the Angels – to punish the enemies of His people (Exodus 12:12, "For I will pass through the land of Egypt this night, and will smite all the firstborn in the land of Egypt... I am the Lord!" Exodus 12:29, "at midnight the Lord smote all the firstborn in the land of Egypt, from the firstborn of Pharaoh unto the firstborn of the captive that was in the dungeon; and all the firstborn of cattle.") Hence, as is often the case, the tradition is maintained in transition, but begins to be "transposed" outwards. And it often remains, below the surface. A whole book would need to be written on the concept of "first fruits" since they are directly connected to the first-born, to their sacrifice to the god Ba'al (Hammon, Moloch, and Ba'al Pe'or, "forefather" of the medieval witches, and "grandfather" of Belfagor) and the sacrifice which Yahweh requires of the first fruits both from the fields and of livestock (literally, offering/ sacrifice of the first born). The episode of Ishmael and Isaac is a further example but this is not our aim, here and today.

At the end of his splendid book, Bottéro (1989) makes some further observations about certain (very) late Babylonian texts, writing of "dry and quite obscure works of which a dozen fragments remain. So far, in their difficulty and esoteric character, they would seem to have discouraged Assyriologists" (1989, p. 775); and in note 3 he adds "Only H. Zimmern and Fr. Thureau-Dangin had begun to study them almost a century ago..." Bottéro calls these texts "mythological exegeses of the liturgy": that is, explanatory material about the liturgical actions of the officiating priest based on and connected to mythological elements. Beyond the scant relation of these obscure liturgical fragments to classical Babylonian mythology and the equally obscure rituals about whose development and real symbology and function we know nothing, in these texts appears the equation Apzu= Underworld/Unterwelt="Land of No Return" so dear to Jeremias (see Bottéro's text for a more detailed investigation). I am not in a position to say if these studies by Zimmern to which Bottéro refers, are those which underpin the reading applied by Jeremias. However, it would be like a joke by a mocking fate that what Jeremias took as his starting point, the architrave, the cornerstone I might say, of his whole enterprise, turned out over time to be merely a marginal, decadent, indeed long dead esoteric element, the confused expression of a society and a cult with no more life to it. What was the beginning for Jeremias, could be no more than the dying breath of a now exhausted god. And Bottéro, underlining this fact poetically, concludes with a quotation from the poet Carducci: "Jupiter dies and the Poet's hymn remains" (Rime Nuove, Dante, 1906); *Marduk* is dead, but the liturgy goes on twisting around itself, an empty and repetitive liturgy which outlives divinity itself, almost through inertia, a confused expression of a now worn-out faith dragging itself towards its inevitable epilogue which will bear the name of Cyrus the Great.

6.
NAÏVE BARTER AND CHESS

I now return to the Sumerians and to the Poem of *Enmerkar and the Lord of Aratta*. I think that in the few lines written by/for Enmerkar there are a great many elements which concern us as psychoanalysts, and I think that psychoanalysis is capable of using its own tools to investigate both the Poem and the topic of Money, if necessary renovating crystallized conceptions when these show themselves to be inadequate for a more detailed examination. If this process subsequently requires us to tolerate lacunae and unsaturated spaces in a fabric which we might wish were beautifully smooth, homogeneous and fully coherent – well, all I can ask is your patience. I will give my reasons, and will leave wishing to the world of fairy tales.

The Poem both reveals and conceals. Some Things are implicit and other Things are explicit. Some are conscious and some are unconscious. In the Poem of Enmerkar there is a King, a deity of which the King himself is the representative, and a "request" for exchange, if not for a sacrifice and/or submission.

This sacrifice should perhaps be seen as an apparently disadvantageous but necessary exchange, one that is owed, or considered to be so, and which in this case takes the form of submission (or of furnishing precious materials for the Temple of Enki). If we think of chess, much of the game is based on this Exchange-Sacrifice. Moreover, the term "checkmate" derives from *Shah-Mat*, *Shah* meaning King, and *Mat*, Fallen, Powerless, from the Hebrew *Mat*, in turn from *Lehitmotet* (see Pianigiani; other sources place the invention of chess even further back, with the Persians and a late Sanskrit origin in India during the 5th century BC). Here too there is a King and a "fall" at the end of the game, a game made of sacrifices-exchanges in pursuit of a King's defeat (the letter t of *Mat* is a Tet and not a Tav: hence it means 'fall', 'collapse', or 'die', *Lamut*. However, if we prefer an Akkadian derivation, we would instead have *mātum*, to die). This is a King who collapses, falls, decays in a ceaseless conflict of Doubles which is no sooner ended than it is renewed in another game, a new battle with new sacrifices, new players, new Kings, new Doubles, new victories, new defeats, new falls, and new Oedipuses.

And how many of these elements, so profoundly interwoven, and in such complicated ways, are "played out" on the chessboard of the psychoanalytic setting, as we all know only too well.

And in that chess-playing, conflictual context around 3,000 BC (or 4,000 if you prefer) Writing was invented to demand goods (Goods not Evils) which, being considered precious (from the Latin *pretium*) for the temple of Enki, take on an unmistakably religious and sacred dimension. These Goods must have a recognized and shared value, otherwise they become the beads of coloured glass which Columbus offers to the savages. (Though it must be remembered that the beads certainly had great value for the "savages", which the tomato, potato, vanilla, cocoa, and tobacco did not, whereas for the "savage" Columbus they were worth a lot. It seems that, together with such delicacies, this discovery also gifted syphilis to the western world, but given that today the French call it *mal de Naples* and the Italians call it *mal francese*, the discussion is best left unresolved.)

Or else these goods are only the object of theft and robbery, which some writers consider "pre-monetary" features. *Ergo*, as we consider the acknowledged and shared concept of the Value of those Goods, we find both Enmerkar and the Lord of Aratta, as it were, around a chessboard, knowing and respecting (even as they infringe them) its strategies and rules, and the value of the pieces, and understanding what is at stake.

The Value of a Good/Product which may perhaps become a Commodity is dependent on a number of elements, including demand, availability, quantity and quality of raw material and of "intrinsic" work, the energy needed to produce it, warehousing, distribution, accessory materials, the technology employed, the immediate possibility of a sale (Say's Law does not always work as a clock) and creativity/ inventiveness ("a Stradivarius is not only 400 grams of wood," as an advertisement used to say, something Marx seems to have "forgotten"); and I find it hard to believe that the "time" employed by Enmerkar in the invention of writing is equivalent to the "time" need to produce two fathoms of silk or a suit, and that Enmerkar's work can be classified under the concept of *abstract work*.

Here two interesting lines of discussion could open up, on the one hand about *abstract work* which, for Marx, is not intellectual, non-material labour but a species of "quintessence" which would constitute the common basis of "all" work, and which I do not want to go into, since it speaks for itself. But on the other hand, Marx makes an observation which cannot be passed over in silence, beside the fact that what I have said about Weber and Keynes also applies to him, as it does to many others: "Yes, it's true: that's what you said, but you didn't really mean it." Marx did not, in fact, forget the Stradivarius, and in *Theories of Surplus Value* (in *Marxism and Art: Essays Classic and Contemporary* (ed. Solomon, M. Detroit, Wayne State University Press, 1979) he wrote:

> The same kind of labour may be productive or unproductive. For example, Milton who wrote Paradise Lost for five pounds, was an unproductive labourer. On the other hand, the writer who turns out stuff for his publisher in factory style is a productive labourer.

Turning up my nose a bit, while showing due respect and deference to my elders, given that Marx is older than me, I pretend to nod but, with some puzzlement, commit the claim to memory. Marx goes on to claim that:

> Milton produced Paradise Lost for the same reason that a silk worm produces silk. It was an activity of his nature. Later he sold the product for £5. But the literary proletarian of Leipzig, who fabricates books (for example, *Compendia of Economics*) under the direction of his publisher, is a productive labourer (my italics).

And immediately afterwards:

> A singer who sells her song for her own account is an unproductive labourer. But the same singer commissioned by an entrepreneur to sing in order to make money for him is a productive labourer, for she produces capital.

Besides the obvious projective element associated with his relationship with Engels, which I don't feel like emphasizing because I find it a bit disrespectful to do so, even when we are dealing with Marx, who is far from respectful (but you've got to do what you've got to do!), I think it is difficult to maintain that Marx is "only" analysing an economic concept and simply differentiating capital from non-capital. If that were the case, why does autonomous labour not create capital while employed labour does? Since when? And why should a proletarian produce "compendia of economics", and why specifically in Leipzig?

The point is, there is nothing but scorn in these assertions, scorn and acrimony which have nothing to do with economic analysis. I certainly do not adore Milton and do not believe he is a genius, but to reduce an author who writes a literary work which is nevertheless important for a whole nation and, during the period of which he is a representative for the whole of Europe, to "a silkworm which by its nature cannot help producing silk" seems merely.... Well, I will make no further comment.

Anyway, in order to be exchanged and perhaps come a Commodity, such a Value must find a certain "object", a Price which must in turn be accepted, acknowledged, and shared by the parties to the exchange (and many giants have struggled with the concepts of Value-Price-Labour). In this respect, Semerano's etymology of the term Price[1] is very interesting:

> Pretium,-ī price, money paid for a service or acquisition; penalty... in the sense of means *to pay*: "fulfil"/"discharge" also a component of *interpres, interpretis*. *Interpres, -pretis*, one who resolves a situation, who acts as an intermediary, interpreter.... The popular Latin *interpretis* etc. is the form truest to its origins: it shows that the second component petr- derives from the Sem. ptr, corresponding to Akk. Paša-ru (Aram., Hebr. ptr: interpret; originally "loosen", "resolve", "clarify" also in the sense of the Lat. "solvere", from "Sé + Luĕre", as "to sell" (...) pretium derives from the allotropic form pašaru: Akk. paṭa -ru: formerly Akk. piṭrum in the sense of "exchange", "redemption".

The sense of the Latin "*solvere*" is still linked to money in the terms "solvency" and "(in)solvent". The term "resolved" from *resolvere* is *patur* in Hebrew, with the root *ptr*, which I believe is from Akk. *paṭārum* loosen, discharge an obligation.

All of this – Value, Price, and Chess understood as a defined and stable environment in which to "play" – is already present in Barter, and does not come into being with Money, Commodity, or Capital. Two camels for twenty sheep is already Value and Price. And even in a raid, operations are carried out based on the Value/Price of the "Commodities" being seized: women have a different value from cooking pots or fish-bait.

Hence, all the "vileness" attributed to money is already present in barter.

---

1 "Prèzzo : Rom. pret; Prov. pretz; Fr. prix, formerly pris [whence Engl. price, prize]; Sp. precio; Port. preço ; Cat. preu – Lat. PRETIUM [not PRECIUM] – Lith. prek- is, which combines with perku I buy, with Cz. práce, negotiation, and Gr. prì-sthai [or perhaps pri-amai, πρίαμαι; *my insertion*] I buy, pipràsko I sell. Bergham derives it from the Sanskrit root PRAKH- ask, request, see also the entry for Prece [prayer in early Italian] while Curtius connects it to the root PR- / PAR- expanded into PARK from the Sanskrit. para-yâmi I make transactions, oya- pâras, negotiation/ trading, a-prtas employed, to which he connects the Gr. prattô / prassô (for pragtô, pragsô) negotiate, conduct business and, more restrictedly, redeem, collect money (cf. Appraise, Practise, Barter). However, most agree in ascribing the term to a root PAR- change, which in Sanskrit is found in the form PAN- : panate [= par-nate, = Gr. parnatai] barter, trade, to which we can certainly also annex the Gr. per- àô [different from peraô, I penetrate], pernêmi I sell (cf. Emporium).

We need only read the Poem of Enmerkar; it's all there: fraud, value, exchange price, wealth, property, accumulation, added value, quantities, even some quasi-standards (especially measurements of weight, which recur constantly) and all the rest.

There is nothing virginal and paradisal about barter. Because they are unmediated, we can encounter in barter all the force, brutality, and bullying connected with immediate Necessity, with *ur*-gent needs, with offers "you can't refuse!"

But barter/bar[a]ter/baratto (the term probably derives from πράττειν, in Latin *permutatio*; but in my opinion, we cannot by any means rule out a connection with the Akkadian term *Paḥaru(m)*, "put together", "close", "bring a conflict to an end") already contains within it culture, bureaucracy, hierarchy, the administration of justice, value systems, rules/ codes of conduct, procedures, rituals of exchange and negotiations standardised according to times and customs previously established, shared, and respected by the interested parties (as they fundamentally were in the time of Predation too, judging by the familiar example of the Romans abducting the Sabine women).

Likewise, in barter and in "the age of barter" there were sacrifice, divinity, conflict, the so-called "common ground" in which request and offer coincide, a common ground which can never be achieved without a subjective evaluation which is always involved when determining the "necessity", mainly interior-internal, in which the participants in the exchange find themselves. I am referring to the familiar example that in a desert water can be more precious than a bag full of diamonds.

"Water for diamonds", whether applied to "internal objects" or to the rat cited by Galiani, opens up highly complex scenarios, a "necessity" which constitutes a simply explosive mix involving drives, fantasies, conscious and unconscious desires, childhood experiences, traumas, reparations (real or presumed), claims, hatred, pride, narcissism, romance, gallantry, baseness, etc. on the part of those contracting the exchange.

Looking closely into so-called "silent barter", a splendid example of transfer-compromise between Predation and Reciprocity, we can see that it contained all these elements since, if the natives had lacked building sites-ships-internal market-necessity – in short, the whole supply chain underpinning the chessboard, the Carthaginians would not have travelled beyond the Pillars of Hercules to trade with the peoples of the Atlantic, a region which was then called Libya (Herodotus, *Histories*, IV, p. 336. [Harmondsworth, Penguin. 1954]).

The Carthaginians also tell us that they trade with a race of men who live in a part of Libya beyond the Pillars of Hercules. On reaching this country, they unload their goods, arrange them tidily along the beach, and then, returning to their boats, raise a smoke. Seeing the smoke, the natives come down to the beach, place on the ground a certain quantity of gold in exchange for the goods, and go off again to a distance.

The Carthaginians then come ashore and take a look at the gold; and if they think it represents a fair price for their wares, they collect it and go away; if, on the other hand, it seems too little, they go back aboard and wait, and the natives come and add to the gold until they are satisfied. There is perfect honesty on both sides; the Carthaginians never touch the gold until it equals in value what they have offered for sale, and the natives never touch the goods until the gold has been taken away.

Truly a splendid description.

And in this virginal barter all the following continue to coexist in parallel and to mingle: fear of loss, dependency, autarchy, predation, symbiosis, parasitism, envy, gratitude, due and undue appropriation, search for security, interdependence, avarice, rage, frustration, desire, and all the rest. The story of the most famous barter in history, that between Jacob and Esau, who bartered his birthright for a pottage of lentils, testifies abundantly to this, the Biblical account endorsing Georg Simmel's thesis that the exchange/ barter which tries to standardise the values being exchanged is the attempt to overcome a state of constant and repeated predation, which runs in cycles, each side "mimicking" the other in their adoption of the roles of predator and prey. In the example from *Genesis*, it will take a few years for Esau to overcome his anger at being "swindled" by Jacob and thus escape the spiral of predation and counter-predation. To say nothing of the fact that in the everyday life of the *Euphratesians*, almost nothing, apart perhaps from salt or sugar between neighbours, as I have suggested, was bartered *sic et sempliciter*. I will clarify this later. It is also interesting that in Spanish *barato* means good value, inexpensive.

It's not all *Hey presto!* like Copperfield the magician, or like Gilgamesh, the Postman who Always Rings Twice. Nor does it arrive whole, *ex novo*, with the Reformation, with Capital, with the Taylorian "division of labour" or the alienation of the product or of labour.

There is nothing virginal about barter: underlying it, there is already an immense, underestimated, highly complicated and subtle game of chess.

I will not expand here – though I will later – on the other fundamental element present in the so-called epic of barter: the diffusion of systems of weights and measures and the presence of scales across the whole of Mesopotamia and the south-eastern Mediterranean during the Bronze Age.

7.
MERX, -CIS

I referred to Merchandise little earlier, referring to the goods traded by *merchants* in *markets*. Open any newspaper and you will find the term "Market" everywhere, in various forms all derived from the Latin *Merx,-cis*:[1] the global *market*, a fall in *market* values, *merch*ant banks, the com*merc*ialisation of emotional life. In the English-speaking world, the goods which become Merchandise are also called Commodities, again from Latin: *commoditas*, "fitness, adaptation, convenience, advantage," from *commodus* "proper, fit, appropriate, convenient, satisfactory," from com-, here probably an intensive prefix (see com-), + modus "measure, manner". Which gives rise to expressions such as the commodification of women, or of labour. If Money is absolute Evil, the Market and the Commodities

---

1 MARKET: early 12c., "a meeting at a fixed time for buying and selling livestock and provisions, an occasion on which goods are publicly exposed for sale and buyers assemble to purchase," from Old North French market "marketplace, trade, commerce" (Old French marchiet, Modern French marché), from Latin mercatus "trading, buying and selling; trade; market" (source of Italian mercato, Spanish mercado, Dutch markt, German Markt), from past participle of mercari "to trade, deal in, buy," from merx (genitive mercis) "wares, merchandise." This is from an Italic root *merk-, possibly from Etruscan, referring to various aspects of economics.
The god Mercurius was probably the god of exchange. According to [Walde-Hoffmann], the god's name was borrowed from Etruscan; in principle, the same is possible for the stem *merk-altogether. [de Vaan] Meaning "public building or space where markets are held" is attested from late 13c. Meaning "a city, country or region considered as a place where things are bought or sold" is from 1610s. Sense of "sale as controlled by supply and demand" is from 1680s. Market-garden "plot of land on which vegetables are grown for market" is by 1789. Market-basket "large basket used to carry marketing" is by 1798. Market price "price a commodity will bring when sold in open market" is from mid-15c.; market value "value established or shown by sales" (1690s) is first attested in the writings of John Locke. Market economy is from 1948; market research is from 1921. (from Online Etymological Dictionary)

it deals in are not much better. In order to avoid confusion in the pages which follow, I intend to use the term Merx in a broad range of senses to indicate the Market, Commodity, Goods (and not Bads!), Merchandise, and so on. But what is the "Merx"? I won't get involved with Sraffa's book, *Production of Commodities by Means of Commodities* (1960) since, with all due respect, I think it would be hard to find anything more profoundly distant, Enki forgive me, from what I am writing. Until the first half of the nineteenth century, "merchandise" simply indicated goods/ products which, not being utilised by their producer, were taken to the market in order to realize their value in another form which was in turn profitable, useful, exploitable by the producer of the initial goods. This trade consists in "giving that which we do not need in exchange for what we do need." In other words, the "relatively superfluous" is exchanged for the "relatively necessary", as Genovesi put it very clearly in 1765.

The term "capitalism" which, as far as I have been able to discover, is inseparable today from "market", appears for the first time in its "modern" sense around 1744 in an Italian economic text, *Dell'impiego del Denaro*, by the previously cited Scipione Maffei, who in some ways anticipates by a decade the great and, even more often cited, Abbot Ferdinando Galiani and, by fifteen years, Quesnay. Before then, the terms in use were Yield, Profitability, Substance, Fortune, Wealth, Stocks, Livestock, Riches, Estate, and so on... Capital was very often a "simple" and "obvious" attribute of rank/ nobility. 'Stock' derives from a wooden rod cut in half lengthwise, on which the details of the economic transaction were noted. The creditor kept one half, called the *stock*, while the debtor kept the other, called the *tally*, from the Latin talea (Martin used the word *foil*). This system for keeping accounts, which could in turn involve third parties, was in use in England for about six hundred years from the twelfth to the seventeenth centuries, and seems to have been something between the Roman *rausdusculum* and the Greek *symbolon*.

Capital is from the Latin *Caput, -itis*, Head, including Head of cattle: it does not seem to be a very modern term.

And it is even harder to call Quesnay's concept of Capital "modern".

The distinction introduced by Aristotle (Use Value and Exchange Value) had not, until then, had a great effect, in the sense that it was obvious and universally accepted.

We do not know how (?), but around the middle of the nineteenth century, "market/ merchandise" begins to be a term which still indicates Goods to be Traded, but where the emphasis has shifted slightly, "imperceptibly", from Goods to Trade.

And it also starts to smell a bit, and not because the Merchandise is going off: it is because if I cultivate something and eat it, this is Use Value, but if I cultivate two things and eat one, the second has Exchange Value. The first is "good", but the second is anything but.

But if my cousin, using a tractor, is able to make his land yield twice as many onions as my grandfather was able to grow using oxen, there is no reason why my cousin has to eat twice as many onions as my grandfather did.

Do certain *ummannu* perhaps need reminding that money, unlike onions, does not rot?

And if, as has happened thousands of times in history, seven fat years are followed by seven lean, what do we eat in those years if we can't make use of money? (And while on the subject of dreams, whether interpreted by Joseph or Daniel, it is to the Sumerians and their Lists of Professions that we owe the certification of the acknowledged and honourable profession of Interpreter of Dreams)

What are we going to use to buy grain, oil, or wine in a region, neighbouring or distant, which has not suffered famine, locusts, phylloxera, or the attentions of Landsknechte? The answer we are generally given – though not by Thomas Aquinas who is often more modern than many moderns; and let's pretend we've forgotten *Ecclesiastes* 7:12, "for wisdom is a defence, and money is a defence" – is as follows: there would be nothing wrong with the Market and Exchange Value (in theory!) if it were not for the fact that as soon as we start selling Products (a term, from *pro ducere*, which slightly predates Calvin) we risk creating a Market Fetishism which rapidly deteriorates still further into a Fetishism of Money. This is such a terrible pathology that if it is not harshly and speedily dealt with by some "wholesome and vigorous property tax" (this is always the solution), it inevitably becomes a Bulimia of Money which renders the subject afflicted with this disease completely autistic and shut off from the world, exclusively absorbed in a frantic activity of accumulation, hoarding, putting at risk even his children's saleable organs at risk, not to mention the honour of his wife and the safety of those around him. So, setting aside the clear and obvious moralistic root of such apocalyptic assertions, I want to ask, what does Money Fetishism mean?

In my work I come across quite a few Fetishes of many different types.

The psychoanalytic process and psychoanalysis itself can become a fetish.

Any thing, any idea, every single element of the universe around and inside me can take on the role of a Fetish (in their clinical practice and in their theories many giants from Freud to Greenacre, from Glover to Payne, have considered Fetishisms and Perversions, concepts which have not always been clearly distinguished from each other, and many others have followed them, going all the way to Masud Khan, who certainly "understood" a thing or two about the subject! Here, however, I am concerned with Money Fetishism and not the Fetish in clinical practice, although some interesting observations could be made about it: but not today. It is curious that even the Foreskin can become a Fetish-Fetish, as Masud Khan illustrates in his 1979 work, and to the moralistic way of thinking, Fetishism is a sin and sins must be thrown on the fire. But I do not think it helpful to ban thongs or stiletto heels because they sometimes become the object of fetishistic attention. I do not think a rabbit's foot should be thrown on the bonfire, naturally with its owner, because particular talismanic qualities are attributed to it. I do not think it is "moral" to hide all little girls' faces, because in that way their companions will not detect morbidity. (Will not detect them, or have they already been detected by the imposition of the veil itself?)

Nor do I think that *The Little Prince* should be burned in Berlin, just as I do not think that Haiti should be razed to the ground because of its Voodoo dolls.

(Naturally, all these "negations" will mean something their opposite.)

Rather, the important point for me, the thing that needs to be understood, is the partiality of the other person's desire; it is the magical thinking with which that specific person addresses life; their respect for it and ability or inability to move in it with love; the reasons for the emphasis on Death. Not the bonfire. The bonfire, all bonfires, are always lit by "moralists" like Savonarola who started lighting fires in Florence long before he ever went shopping, or like Frollo singing "Hellfire" to Esmeralda in *The Hunchback of Notre Dame*, a cinematographic and theatrical parody of the much more tragic Heinrich I. Kramer, the terrifying author of the *Malleus Maleficarum*. And these morally superior people very rarely wear dark priestly habits, but camouflage themselves in transparent, luminous, attractive costumes of *liberté* and *tolérance*.

But whatever the case, Merx, so some sages tell us, is responsible for the start of the process because, to begin with, the Merx is exchanged for Money in order to get the merx we need and then – here is the first fetishism – the merx weakens the human spirit and we start to exchange merx for its own sake and then we exchange merx not so much *with* money as *for* money, and money completely corrodes the human spirit (again!) and we sell merx in order to have more money and again and again and again, and we end up in Gomorrah. Money itself becomes merx, and so begins a turbulent and frenetic waltz in which we exchange Money for Money with the sole aim of accumulating it. Harpagon *docet*. Aristotle himself falls into this moralising trap.

And the moral of all this: essentially, you only need to touch Merx and it instantly strikes you blind. (In this context, there is much interest, curiosity, "amusement", and certainly "instruction" in some observations by Calcagno on Circumcision as an instrument for controlling-preventing-curing that horrifying disease called masturbation.)

At this point, however, let's be clear: when those of a secular persuasion have finished sniggering at the transubstantiation which happens during the celebration of the Holy Mass, do they cheerfully accept the fact that "money becomes merx"? (How many times in history have there been disputes over transubstantiations, consubstantiations, dual nature, Monophysitism, the Trinity!)

I have conducted an experiment which will make Popper embrace me with tears of joy: I cleared my desk, cleaned and polished it.

I put a hundred euros on the left and a cigarette lighter on the right. It seemed like a good deal.

After two weeks the money had not been transformed into a Merx-Lighter, and this was understandable since it would be a shameful loss of value on the part of a one hundred euro note to turn into an object worth one euro: but still more worrying, since it still seemed like a good deal to me, the Merx-Lighter had not yet turned into Money.

This was very upsetting. But I told myself, "At least transubstantiation is an act of faith." No priest has ever tried to sell it to me as science.

At the end of the experiment, I went to the tobacconist and used the hundred euros to buy cigarettes.

Magic! The Merx *turned into* money, for the tobacconist.

And the money *turned into* merx, for me!

The serious conclusions of the experiment, which will make Popper proud of me: the merx *did not change* into money. The money *did not change* into merx. They were *ex*changed for each other! The merx were not transubstantiated into money and the money was not transubstantiated into merx.

I wanted cigarettes and the tobacconist was willing to sell them.

It is the desires of men which move merx and money, not the money or merx which transform or transubstantiate "themselves", or behave like a mob or like stalkers. Let's remember the words of Motterlini, as well as Freud's from 1913, which I quoted earlier.

Merx, mercis: commerce, trade; *mercor*: I conduct business; *merces*: price paid, salary, income; *mercimonium*: merchandise; *mercatus*: traffic, business; *Mercurius*: Mercury.

Merx is cognate with *mereo, -es, -ui, -itum, -ĕre*: to earn; and *meretrix*: one who earns, who receives pay, and also *meretrix* as a particular case of exchange "Pounds for Cigarettes", but with sex as the commodity.

From the Akkadian *Makurru* or *Makkuru*: Goods, Possessions, Merchandise; a term clearly related to the god of trades and business dealings, i.e. Mercury (Semerano, 2007). Besides being the god of trades and business dealings, Mercury is also the god of thieves. Not only mere pickpockets but proper stealing, robbery.

(Hobbes' observations about this are superficial and frankly ridiculous: *Leviathan*, chap. 10. If it were not for his general, caustic, irreverent, and corrosive polemical cynicism, out of which some curious pearls occasionally emerge, we would have to wonder if when Hobbes was writing *Leviathan* there was perhaps an overproduction of paper which had to be used up by royal decree. Or maybe Hobbes, as an acute polemicist, was provocatively trying to resuscitate a deceased Marduk who might, at last, cut his Leviathan to pieces?)

And the first example of theft as something to be prohibited and execrated, seems to appear in the Seven Laws of Noah, where theft is closely linked to the use of deliberately false weights in the various transactions which, in an era before coinage (?), were carried out by weighing the goods themselves on Scales or Balances, and by a complex, codified system of measures managed by the scribes and the "weigher of silver", as we will see later.

Here I will only observe that the Italian verb *"comprare/comperare"* to buy (I will ignore *emo, emĕre*, hence "emporium", which has a different root; see also Semerano) derives directly from the Latin *comparāre* (hence the English, "compare"): to put two different elements in relation to one another in order to find a precise correspondence between them. And nothing serves better than a *bi-lanx* – "two plates" – to render the idea of a correspondence that is absolutely equal, so as to become over the ages, and to remain for many more ages, the very icon of Justice, Right, Equity, in all its forms. From Merx in one's own front yard to the formulation of Laws, from the *Psychostasia*, weighing the heart against a feather in Egyptian mythology in order to receive the judgement of the gods, specifically Ma'at, the goddess of the Feather, on the soul's destiny, to the weighing of a heavy goods vehicle. I have already mentioned the concept of Ma'at in Freud's *Moses*.

And while *vendĕre* (sell: which gives us the English noun "vendor") derives from *venum dare*, put on sale – that is, give in exchange for money (hence the adjective "venal") – the Italian verb *Pagare* (Eng. pay, Fr. *payer*)[2] derives from *Placare*. And here I can explain Enmerkar's yell, which I declined to translate at the start of this book: *"ana* KUG.BABBAR/ kaspim nadānum" literally means "give in exchange for money", i.e. Sell, where KUG.BABBAR is a Sumerian term while *kaspim* (genitive of *kaspum*) is the corresponding Akkadian term.

---

2   On p. 24, n21 of her book Dr Schimmel notes, *"En Hébreu la monnaie était désignée autrefois par le mot 'sangs' (au pluriel)"* – "In Hebrew the coin used to be designated by the term 'bloods' (in the plural)." While it is indeed true that in Hebrew money is associated with Blood – that is, *Dam*, or rather *Damin* (*au pluriel*) – this word refers specifically to coins in one's pocket, and not to money in general. In other words, it refers to the small change one might give a child as their weekly pocket money. In this sense, I do not think that Blood refers to money in connection with blood-blood but, ironically, to the disbursement-exsanguination-fainting of the parent faced with their children's pressing demands for money to spend on "trifles", so that this pocket money is called (*au pluriel*) "bleeding pockets", *Dmei kis* (*Dmei* being the contracted form of Damin). I have the fantasy that, in making this assertion, Dr Schimmel was influenced by, or herself influenced, an almost identical assertion by Jacques Attali (or they both drew on a common source, or else came to the same conclusion). On p.31 of *Les Juifs, le Monde et l'Argent* Attali writes, "Ainsi l'une des plaies, le sang, est-elle nommée par le même mot, dam, qui désignera plus tard l'argent (damin) " – "Thus one of the wounds, blood, is called by the same name, *dam*, which will later designate money (*damin*)."
Dr Schimmel's observation is aimed at demonstrating that Money is a pacifier and the example she produces does not seem very apposite to me, whereas the next seems more relevant and appropriate: *"En Hébreu, le mot paix e payer ont la même racine"* – "In Hebrew, the words peace and pay have the same root". And indeed, *Shalom*, peace, has the same root as *Shilem*, 'he has paid'. However, I will add that the etymology of the Italian verb "pagare" (to pay) is also connected to a term which is to do with Peace, in the sense that *Pagare* derives from *Placare/Pacare*, to pacify. This echoes the need to restore calm, peace to a situation which has been altered, heated, and in which it is necessary to bring back peace, *pax*: i.e. paci-fy, *pacem facěre*. 'To pay' is also rendered in Latin as *Solvěre, Dare, Luěre* (*Luěre* as in *luo*, λύω, Greek, to pay rather than to loosen). Therefore, the connection between Peace and Pay is also present in the Latin etymon, and clearly also in French, Italian, and Spanish. In English too, 'to pay' is connected with 'to pacify'. It is a little different in German, although here too there is a link between Equity and Criterion of Value, the term Wertmesser being connected to the Knife (Messer) which guarantees peace both in relation to Sacrifice as equal distribution and/or equal compensation (Entgeld) and to Reparation (Wergeld, blood money) (in M. Catarzi, Bernhard Laum, 1997, p. 38). Hence there seems to be an obvious and direct connection between Pacifying-Appeasing-Peace and Money. It is equally clear that blood flows in situations where there is no Pacifying-Appeasing-Peace.

According to Semerano, the *venum* (accusative: dative, *veno*) derives from *Winum*, a Semitic transformation of the Akkadian *Minum* (count, amount to, number), in turn derived from *Manū(m)* (compute, sell) which gives the Latin *Manus* (hand), the first tool for calculation, and I will talk at more length about this in connection with Rome.

Thus, we have a product which initially was not consumed directly by the producer – i.e. the Merx – which is a Good (from *Banû*, beautiful, good, precious). Such a Good is subsequently put on sale (*vendere, venum dare*), and then *comparatus* (compared, i.e. bought) and thus the vendor is placated/ pacified in his need for "compensation", either with another merx or with money/ coin as a stable and established correlative of the merx on offer (*placare, pagare,* and also *pacĕre* and *pangĕre*: fix, establish a fixed point).

Where Money Fetishism is in all this I frankly do not see; I really cannot see it.

Nor do I see around me, or in history, endless crowds of people afflicted

---

It is less clear that Money itself is Blood. However, in avoiding bloodshed by restoring, or potentially restoring Peace and Pacification, Money is certainly connected to the element of Aggression-Sacrifice-Equity-Knife-Blood, mediating and replacing it. It could be said that the Coin has symbolic features strongly connected to Blood. I believe it is the way the French language has evolved which stops us easily detecting this nexus between *Payer* and *Placare-Pax*, which is obvious in Latin, even though the conjugation of *Payer* often involves us losing the Y and returning the original I. In any case, the French etymology of *payer* recorded by the Nouveau Petit Robert is identical to that of the Italian verb. Peace in German is *Friede*, from the old German *Fridu*, Cold, Calm, which gives *Zufrieden*, Content. No money, blood, merx, knives for Ειρήνη which instead remains Pact, Rite, in other words Exchange/Barter with the deity: a closed Pact, preserved in the ăron, the Ark of the Covenant, of sacred Cedar wood (Sum. *Erin*, Akk. *Erĕnu*: see Semerano).

To sum up: the English verb "to pay" derives from the Latin *"Placare"* and *"Pacare"*: Pacify, Calm, Render calm. That is, the act of payment means re-establishing a state of peace with someone else. We have the same situation in French (*Payer* deriving from *Faire la Paix*) and Spanish (*Pagar* deriving *Apaciguar*) and in Italian. As I mentioned above, in the Hebrew language, *Shalom* (Peace) has the same root as *Leshalem* (to pay) and *Shilem* (he has paid). *Shalom* descends from the Akkadian *šalāmum* (peace/ to be in peace) and *salīmu*, (peace/ friendship) and is connected with *šamu* (to pay) and *šimu* (barter). In the Akkadian language we can also find another word for Peace: *pa-ha(r)um*, peace-calm-closure of a conflict/dispute, which is in turn derived from the Sumerian term psychopathology-har, the accumulation of goods/ savings (specifically silver), and also barter/exchange, which in Akkadian becomes *pa-hu* (barter/ exchange); hence the Latin term Pa-x (see also Semerano).

with a pandemic of bulimia or blindness, unless I start seeing the world through Savonarola's spectacles, or H.I. Kramer's with his *Malleus Maleficarum*, or those of some raving millenarian fantasist in the many bizarre forms in which they conceal themselves.

All I can see is the perfect concretisation of a piece of Justice, Equity, Respect.

That such Respect may or may not be pleasing to legions of individuals who boast of being maintained free of charge in the most varied ways is quite another matter; a matter which, even if "only" from an oedipal perspective, psychoanalysis ought to address a bit deeply, at least along the lines followed by Freud in his *Mass Psychology*.

Another small thing, very small, absolutely marginal and entirely absent from the discussion of Merx, is given by the "thanks" which all participants in an exchange, at all latitudes, give each other reciprocally after concluding the transaction.

Sometimes the response is "You're welcome," or "Not at all", or even "Think nothing of it", but they are always polite and more or less elaborate formulae whose meaning is always related to the "thanks".

Why "thanks"? It has never occurred to me to thank the automatic cigarette dispenser, even the one owned by the very tobacconist with whom the magic of transubstantiation was achieved.

And yet when I buy cigarettes from him personally, I say "Thank you" and he replies, "You're welcome," or "Thank *you*." But to the automatic dispenser I say nothing, and wouldn't think of doing so. Sometimes, when it's failed to give me any cigarettes, I'd be lying if I said I don't think anything at all. The same must go for the machine itself, since it likewise says nothing to me.

I don't know: we must feel a mutual dislike, or maybe there's an element of projective counter-identification going on.

Joking apart, I think this little phrase, "Thank you," is a very important element which represents, or should represent, the end of an enormously long journey which begins when we are children, one on which all parents expend, invest, and waste floods of time and energy.

Depending on the child's age, it is more than obvious that the urgent "need" and the dramatic and almost deadly dehydration which the child experiences when he or she demands water – a dramatic dehydration which superficial parents do not even notice – conceals a claim by the child to control its servitude on the one hand, and on the other a denial of its dependence on its parents by reversing the situation.

The "magic word", the requirement to "Say please" and the insistent

"Now what do you say?" are an attempt to make a breach in a complicated system of fears and arrogance which characterizes the innocent, tender, uncontaminated, pure and, by now, terribly dehydrated child (see Rousseau).

This endlessly repeated demand tries to make sweet little Émile aware of the fact that what he has been given until that moment is also an act of love, given freely and selflessly, but that it must be acknowledged.

Indeed *because* it gives Pleasure to do the giving and Pleasure to do the receiving. Because it is a meeting of two desires.

And both must be satisfied. Both the offer and the request. And so we give "thanks" to those who are the authors of this satisfaction of our desire. I think it is very clear what I am saying. When a prayer (see *"Prego"* and *"Je vous en prie,"* the Italian and French responses to "thank you") is offered up to Heaven in the hope of obtaining the favour of the celestial, or chthonic, divinities, we are requesting a "grace" (in the Romance languages) that will satisfy a certain desire, or asking the other to "think" of us and our needs (in the Teutonic languages, where "thank" and "think" share a common Indo-European root related with "I have good thought for you/ think of gratitude") although the exchange in question most often concerns a "need" that is not material but psychic/ emotional.

At that precise moment we find ourselves in total dependency – that is, at the *mercy* – of Him or Her who can satisfy, or whom we presume can satisfy, our request. And we offer sacrifices, vows, promises, gifts if the request for Grace is granted, or in order that it may be.

It makes no real difference whether the requested benefits are material or spiritual.

I don't think there is any need to cite the endless quantities of literature, poetry, songs, plays, iconography, traditions, and sayings which address this subject and requests for "grace" of all kinds, from Love, to Peace of Mind, to Revenge, Fortune, Health, Money, and so on.

Etymologically, "grace" seems to derive from the Greek χάρις, -ιτος, *Charis, -itos*: that is, Favour, Gift, Benefit, represented by the Graces of ancient Greece, daughters of Zeus and Aphrodite, dispensers of Radiance (Aglaia), Youth (Thalia), and Joy (Euphrosyne). *Charity* has the same root.

Emotionally, it all derives from a desire that wants to be satisfied.

From showing gratitude, being grateful to him or her who has been its author/ means/ possibility.

Even if I have already paid, i.e. pacified, the "Creditor".

"Gratitude" for the benefit, for a Benefit, received or which one desires to receive.

"Gratitude" for the benefit given to me or which I hope and desire will be given to me.

In Italian, *Grazie*; in Spanish, *Gracias*; in German, *Danke*; in English, Thanks: from the Akkadian *Danqu*, grateful; and I believe that *Dunqu*, grace (Semerano) is constructed from *damāqum*. Not forgetting the Greek *eucharistó*, εύχαριστός, grace (connected to Eucharist) to which the response is *parakaló*, παρακαλῶ, from παρακαλέω: I invoke, pray.

Which brings us to the French *merci* which perfectly, even phonetically, sums up what I have been saying about Merx, and from which the English 'mercy' is directly descended.

So, far from generating fetishes, the concept of Merx represents the actualisation and concretisation of a state of grace, completeness, and satisfaction of one's Desires-Needs precisely by means of what one receives having desired it.

And so we give 'grace', gratitude, to the "Giver" who allows us the satisfaction of our desire.

I thank the tobacconist and the tobacconist thanks me. We thank each other for the Merx. For the transaction. For the mutual satisfaction.

I am "grateful" to the tobacconist for the *merc*handise and he is "grateful" to me for the com*merc*ial transaction.

"Grateful" means thankful and, once again referring to Semerano, must be connected to the Akkadian *karšu*, internal organ, belly: by extension heart, feeling; and I would add, to *karāsu*, bind, unite.

A strange coincidence: the resemblance between the Greek *Charis*, χαρις, and the Sumero-Akkadian *Karšu*.

A really strange "coincidence", since the Sumerian and Akkadian *Karšu* precedes the Greek *Charis* by at least 2,000 years. But maybe it isn't a coincidence.

As we close this discussion, I could say that the Merx is a Gift which must be Paid for. And the Gift is a Merx (Benefit, Product, Service) which must be given for free/gratis...

(I have always "much enjoyed" little word plays like "the Meaning of the Void and the Void of Meaning.")

From a relational perspective, the difference between Merx and Gift lies in the fact that, whereas the Gift – or rather, the act of giving – creates an element of dependency in the receiver deriving from the selfless attitude of the parental function in our species and in the care of offspring which every living being undertakes towards its little ones, the Merx, without denying this aspect of gratitude, but providing a corresponding value to what is received, loosens the bond which would be created by the Gift. While maintaining Gratitude for the Grace received, and by the payment to the "donor", Merx restores independence and autonomy to the beneficiary: or rather to both the person who 'benefits' intransitively by receiving a benefit, and the person who 'benefits' transitively by giving it.

In other words, we replace a dependency-relation with one of equal and fair independence or inter-dependence which leaves both subjects free to be actors in and authors of the exchange, thereby emerging from an Adult-Child logic in favour of a logic of autonomy which makes them free whether to propose and choose an interpersonal relationship or not.

The Gift may constitute an element which "binds", which does not loosen, which generates dependency but could also be generated by the position of being dependent, whereas the Merx (Com*merc*e – i.e. Money) restores and generates freedom, but could also be generated by the demands of freedom itself, as we can find stated with great clarity in a passage by Xenophon on the subject of Agesilaus, King of Sparta:

> If, indeed he had *sold* his favours or handed out benefits *in exchange for money*, no one would ever have believed they owed him anything; however, there are so many who received an *unconditional benefit* that they took him as their benefactor, both because they were the first to obtain the favour and because he considered them worthy to have the obligation of gratitude imposed upon them (Xenophon, *Agesilaus* 4.4; my italics)

This is a very beautiful and subtle description of what passes between a king and his subjects. An exquisite description: "he considered them worthy to have the obligation of gratitude imposed upon them"! A bond that will never be broken. In the end, even Mauss agrees with this. In this case, it is a bond which unites King and Subjects as Symbolic Father and Children, and not as Master and Servant, at least in the currently accepted sense. And, for Agesilaus at least, it is a responsibility he accepts, but at the same time one requested and chosen by warriors, adults and Spartans. And the Spartans are not exactly a querulous chorus of defenceless babies, always prone to whining and recriminations. Given the implicitly hierarchical nature of this relationship and the responsibility which derives from freely electing a King, and from the acceptance of the responsibility which such a role entails, it seems obvious that, at least in this case, there is no diminution in the element of respect, despite the distinction between the parties' allotted functions.

From what has been said so far, I think some reflections need to be made on the dialectical relationship between Servant and Master, well known from Hegel.

If for a moment we replace the Master-Servant dyad, in the sense that has been asserted over the past two centuries, with the Adult-Baby or Parent-Child, we get a picture which moves a long way from Hegel's own reading of it, and which Marx uses to radicalise both the element of conflict and the element of collusion with a single aspect of the dyad, a picture which is open to symbolic, psychological, and economic readings very different from those that are commonly accepted today. The current interpretation of the Hegelian dyad serves as a theoretical and ideological basis for maintaining a continual protest and complaint by the Children-Servants: that is, individuals who are essentially adolescents paralysed in a position of counter-dependency, demanding to be acknowledged, *gratis* (!) by the Grace of the Adults-Masters.

The "Children" demand water, want the Gift, but do not want to hear of words such as "please" and "thank you" because that would require them to accept the idea and the state of dependency in which they find themselves; a state of physiological dependency in which one then needs to be immersed in order to free oneself laboriously from it later: if one can.

(Davide Lopez wrote beautifully about this in his most recent book, taking a close look at some aspects of how psychoanalysts are trained and their relationship with their patients.)

It is much simpler for many "Children" to try to take over in fantasy, and attempt to do so in reality, the "means of production": that is, the wallets and wealth created by others, by the Parents-Masters, trying in this way to maintain in perpetuity a situation of total infantile dependency which tyrannizes and drains what has been generated by countless previous generations without ever achieving the necessary and responsible evaluation embodied by the free, mature concept of Merx.

If we observe him in an oedipal light, the Baby-Servant should naturally be destined to become an Adult-Master with no need to kill the Laius-"persecutor". The *Moirai* within each of us will come vividly to mind.

Instead it is necessary that the Babies make full use of the dependency-protection in which they find themselves and face up to it as best they can in order to become healthy adults once Fate has done its work.

And at this point to see what they are really capable of, if they are up to the task of taking responsibility in the way that maturity, and also real independence entail.

I would just like to emphasize once again that a position of freedom and independence is highly demanding and costly to achieve, and cannot be given by anyone to anyone else (and cannot be "stolen" from anyone by anyone else; someone can only be deprived of it, but no one else can acquire it, as I observed earlier). In this respect, *et cum grano salis*, Hegel is not entirely wrong when he claims that:

> Those who remain slaves do not suffer an absolute injustice; indeed, he who does not have the courage to risk his life in order to attain freedom deserves to be a slave (*Phenomenology of the Spirit*, (b). Self-consciousness, § 435).

And if, in the question of Public vs Private to which I referred earlier in connection with Fromm, we wished to see a further permutation of the Hegelian dyad, with the bureaucracy and machinery of public administration (someone calls them 'tax consumers') on the one hand and the producers of real wealth (someone calls them 'tax producers') on the other, a phenomenon which has affected the whole of the contemporary West and goes back to Bismarck (or perhaps, to a very partial extent, to some English Charity laws from "a few decades" earlier), but which expanded massively and abnormally from the 1970s onwards, we could have the photograph of a Child-Slave, or rather of Children-Slaves multiplying like locusts, dependant on the activities and people who actually produce wealth – the Parents-Masters – even as they try to overturn their dependency. I am speaking of the monstrous increase in public spending, using and expanding both the debt and the deficit of the State. But isn't that what Keynes said, that it is good to expand the debt? Or is that not what he said? Or maybe he said a bit yes and a bit no? Who knows?

Whatever the case, today and not only in Weimar, we have some splendid cats to skin.

Long before Bismarck, we could have found situations like this in ancient Rome (and also in Athens) where the *clientes* are essentially in a position of total dependence on the *patronus*, who extended the role of *pater familias* to his own *clientes*, thereby generating much more complex and elaborate relations, codified in Roman Law, than a bad modern publicist would credit.

Obviously, straddling and expanding the elements of Conflict which permeate all "Families" is of great help to the power games played by false prophets and agitators, giving them space to put into practice the famous principle of *divide et impera* with which they can parasitically find a way to make a living as peacemakers, "resolving" the conflicts they themselves created by fanning the flames.

We are still a long way from fully understanding Benedetto Croce's profound reflection which shows us how to overcome the old Hegelian scheme in the dialectic of the Distincts. In psychoanalytic language, I think this Crocean observation has much in common with the Winnicottian notion that "there is no such thing as a baby": in other words, there is no baby without a mother/there is no mother without a baby; in the sense that Mother and Baby are not monads which overlap so that one must prevail/ predominate over the other, but a vital unity precisely because they are an indivisible couple, but not therefore in-distinct. Indeed, the distinction between the two components is absolutely vital in ensuring that the whole process of care and development can take place. Without this distinction many of us know only too well what terrible consequences can befall not only the baby but also the mother herself. I also think that the acute observations made by Lopez about the concept of Distinctness are not too remote from the spirit which moved Croce.

However, I think it is entirely legitimate for the Children to try to hold their own theses together with whatever scraps they deem necessary, such as the idea that Marx's theories are fundamentally no different from Girard's, for example, or Gargamella's, or those of Nostradamus; and everyone is free to make the connections they want, but they should bear in mind what Keynes wrote (did he say it? did he not say it? Yes, he wrote it!) in a work of 1926, *The End of Laissez-Faire*:

> But Marxian socialism must always remain a portent to the historians of opinion – how a doctrine so illogical and so dull can have exercised so powerful and enduring an influence over the minds of men and, through them, the events of history.

And this must mean something. If, however, you wish to clutch at straws, seeking improbable syntheses between completely different authors, political climates, cultures and values, playing with improbable exegeses, vertiginous interpretations of terms, moralism camouflaged by theoretical and collusive indignation, go ahead, be my guest. Besides, Hayek's 1949 work could give an answer to Keynes.

At this point, it would be interesting to attempt an examination of how to quantify accurately the real "cost" of a psychoanalyst's units of work (costs of training, moving house, supervisions, working time, conferences, rental/ mortgage on the consulting room, living expenses, and everything else that gets written off by work because becoming a psychoanalyst is not just an economic investment but an enormous emotional one as well): i.e. how to determine the minimum fee below which we can no longer speak of Work but philanthropy, since Work is also a merx and must be quantified. But I will not address that here. On that subject, and provisionally, I think the 20 Gold Reichsmarks proposed by Lou Andreas Salomé for her own psychoanalytic work are a sufficient starting point for those "psych-whatevers" who feel like reflecting upon it.

Nor will I address the problem posed by the Apple which the Witch offers as a gift to Snow White, or Virgil's *timeo Danaos et dona ferentes* (*Æneid*, II, 49): in other words, the topic of the persecutory, aggressive, manipulative, and destructive valence which the Gift can take on when it slips towards "the dark side of the Force." This will be the business of Luke Skywalker and Obi-Wan Kenobi in the next episode of their Saga. On the subject of the Gift, I merely note a passage from *Exodus* (23:8) which reprises some considerations in the Laws of Noah about what we could today call the Impartiality of the Judge, or of whoever is invested with a similar function:

> And thou shalt take no gift: for the gift blindeth the wise, and perverteth the words of the righteous.

Or *Ecclesiastes* 7:7: "a gift destroyeth the heart."

And, albeit in a slightly different context from that of the Judge, I think the terms *pro bono* (*cui pro-id-est?/cui bono erit?*), 'free of charge', 'reduced fees' and 'low fees' should be handled with great care and attention when applied to psychoanalysis (and although it is not an exhaustive treatment, the previously cited, somewhat stigmatized, paper by Theodor Jacobs provides numerous points for reflection of great interest in connection with this).

8.
NUMBERS AND THE KNIFE

In any case, everything I have written about so far – i.e. merx-barter-grocery scales-gift – needs NUMBERS in order to function. Numbers: 1, 2, 3, 4, 5, 6... And it must be possible to add and subtract these numbers, as Pettinato indicates explicitly (*I Sumeri*, p. 33) in recording the blessing of the god Enlil on his bride, the goddess Sud:

> The art of writing, the tablets adorned with writing, the stylus
> The surface of the tablet,
>   for carrying out calculations, addition and subtraction, and the dark blue mensorial rope...
> are rightly in your hands.

Numbers, at least as the expression of positive natural integers, are at least 4,000 years older than the first letter written by Enmerkar.

Numbers, weights and measures, are already present in the story of the building of the Ark; money is not. Whether positive natural integers are to be considered in turn as the expression of aggregations or as a simple sequence 1+ 1+ 1, is not important for the time being.

What I do think is important, however, is to bear in mind that, in various parts of the world, batons and bones have been found, dating from 15,000 to 8,000 BCE with notches carved into them, SIGNS suggesting Calculations (from *calculus*, stone) of flocks and various produce, signs that are the forerunner of the Abacus (which was indeed sometimes made using stones), from the Hebrew *Avaq*, dust, in turn from *Afar*, soil. (On the abacus, which is often positional, "101" sheep is "written" differently from "11" sheep and "1001" sheep, and so, in a way, the "zero" is already present. For the "notches" "used in most of the world and by almost all societies" in relation to the birth of writing, see Gaur, 1984.)

That is: one sheep one sign, two sheep two signs, and so on. Zero signs, zero sheep.

Everywhere. All over the place. Among the Sumerians, the Egyptians, and then, much later, the Latins, the Chinese, the Indians. In different periods, but everywhere.

At least among peoples who have adopted the Written Word.

A separate discussion would be appropriate for the "Incas", or rather the Quechua, who had numbers, as is testified by the *yupana* and the *quipu*, but there is an ongoing debate, based on various hypotheses about the *quipu*, about whether they had or had not invented true writing.

I'll pass over Mayan numbering and writing, its "cartoon strips"; another time. (For a detailed study of Mayan Glyphs, I refer to the work of M. D. Coe and M. Van Stone, who recount the troubled – see Knorozov – and very recent history of studies of Mayan writing; I also indicate Gaur's (1984) authoritative opinion about Mayan writing, opposed to Knorozov's.)

In any case, Mesoamerican and South American cultures did not influence the beginnings of writing in Mesopotamia, at least as far as we know today, suspending judgement for now on Atlantis (*pace* Plato) and on a series of hypotheses by which the Sumerians were the first to "discover" South America (see the elaborations by the Brazilian anthropologist Schwennhagen, who hypothesises a Phoenician landing in Brazil; and many curious stories can be found on the website of the Ancient Hebrew Research Center). Should these last hypotheses ever be proved correct, we would in any case have to conclude that such cultures were influenced by the Sumerians or the Phoenicians, and not the other way around. On the other hand, many scholars believe that the origin of writing in these cultures was in fact very recent, historically speaking, and besides, any possible development of it was blocked by the Spanish and European conquest which decreed its end.

As far as Numbers and the Knife are concerned, in these cultures too – that is, in both the Mayan-Toltec culture and culture of the Aztecs, speaking the same language, Nahuatl – recent attention to the end of the world according to the Mayan calendar indicate those people's great interest in Measurement (i.e., Numbers) along with expressions of sacrificial practices in which the Knife overwhelmingly appears: the offering of the beating heart of a human victim is one of the best-known sacrifices. Likewise present and widespread, anticipating a concept I will develop later on, is the practice of Circumcision, as is clearly evinced by the decorations of certain temples in Chichen Itza; and Quetzalcoatl, the Plumed Serpent, offered blood taken from a cut in his own penis to give life to humanity, a practice to which the Emperor, the Inca, returns in times of drought or famine. (In this case, could we speculate that there is a relationship between writing, knife, culture, sacrifice, and penis? We would need to add another ingredient: "added value", which is ubiquitous since, without a sufficient surplus to support/maintain specialist but not directly productive individuals, it is not possible to create any "culture", whether astronomical, religious, mathematical, agricultural or literary, since all individuals have to struggle merely to survive; and this is the serious problem which the Sustainable Degrowth movement does not wish to face because it is merely a kind of suicide. Obviously, for Penny Black theorists, added value, surplus, and profit – everything that is 'plus' – are subtly different concepts which cannot be crudely bundled together as I am doing.)

I cannot comment on the Tartarian tables rediscovered in Romania in the early twentieth century, which seem to be connected to the culture of Vinča, in Serbia; I will only point out that Pettinato closely relates them to the type of writing used in Uruk III (*I Sumeri*, p. 53); it is also interesting to note Haarman's observations which challenge the hypothesis of a Sumerian origin for writing, re-assessing both an Egyptian origin and one that was European in character preceding the Mesopotamian. As some authors, Pettinato among them, I think, have rightly observed, we now know that the Sumerians were able to write, but at the moment we do not know if they first learned the ability from others, and we cannot for the moment rule out this theoretical possibility. Haarman's investigation seems to go in this direction.

However, the Sumerians have left us the Songs of Gilgamesh, whereas from the Vinča culture we have only some signs carved on a few scattered fragments, the subject of which is still a matter for debate. So, for the moment, I think we can leave things as they stand.

I will also pass over the fact that Numbers are the basis for playing cards (in which the seeds of Money and the Knife are often present, even though sometimes in the form of Diamonds and Spades), dice, and a great many other games, both of chance and otherwise. These games have, for many years, taken up much of the time and mental space of many humans, shift a ton of money, and arouse much emotion, besides stirring up and perpetuating, in the phoney and threadbare guise of competition, aspects of Predation and Robbery. For this very reason I am not much in agreement with Löwenkopf (Klebanow and Löwenkopf, 1991), who believes that the true gambler is the businessman, because the gamblers I have known both socially and professionally, even if they consider gambling a profession, make constant use of sharp practice and *Fur*-tiveness which links these people – albeit in more elaborate forms – to the ancient practice of Predation and Robbery rather than to wealth-creation/"adding value"; thus, Arsène Lupin and Moriarty are certainly true professionals, but they are definitely not businessmen.

It was the Indians, those of Asia and not the Sioux, long after Enmerkar – around 628 BCE – who added a further abstraction in the form of the Zero understood as a number, having already invented the modern pictogram of the number (the so-called Arabic numerals), and the history of numbers was further enriched.

But the number 1, *1*, *I*, -, *l*, is still there; that original hyphen, that number that is not thought to be first, but is surely the Prime number, that Sign which, carved with a fingernail, or more likely a blade of flint – certainly not iron – carves a concept, a "simple" concept, for the first time: one sign one sheep, two signs two sheep, three signs three sheep. Another thing that no one had ever done before.

At least *one*, *two*, and *three* have been maintained over the millennia without resulting in a "Babel".

And that Sign or Incision endorses the newly arrived abstraction, its "sedimentation" and hence the symbolization of the one-to-one correspondence between a pebble and a certain Thing, between a dash carved on a rod and a sheep: correspondence between the Thought of a thing and that Thing.

If I write XLVII, and with a knife I can both carve and write it, I thus indicate a quantity, a class of objects. And in doing so, I say and write a word. It does not seem like a word in the familiar sense by we understand them, given that numbers seem to have taken a different path from "words", but in fact it is: because if I say *apple*, I mean all the fruits that are produced by the *Malus*, and I mean only the fruits with those shared characteristics, excluding pears – and, obviously, bananas. In the same way, if I say 47, I am referring to different objects that are nevertheless comparable, in that they have their number as a shared characteristic, that certain quantity which becomes the common feature of different objects: 47 sheep, 47 men, 47 pebbles… and I am excluding all the classes of objects which have 46 elements, or 48, or 50. And one, 1, /, I is a Sign.

And the Sign is the basis of writing; pictographic in the beginning, so the giants tell us, then ideographic, then phonetic (the work of Albertine Gaur, 1984, cited earlier, is still interesting, although showing its age).

But where does the number start from? From a sign, of course, as we Lilliputians say.

Well, from a smear in the sand, an incision made perhaps with a

fingernail, or with a splinter of flint on the back of a rod, or on a rock, maybe by an unknown, perhaps semi-nomadic, shepherd. Maybe the great great-great-great-great grandfather of a ḫāwiru.
A sign for signifying something, maybe sheep. Sermonti[1] would probably not agree with this claim since, in his opinion, the "sign" – in this case, the letters of the alphabet – descends from the heavenly constellations which human beings borrowed and used for constructing the elements of writing. Though I dissent from his general thesis since, in order to call a certain group of stars "The Bull", I must first call something earthly a "Bull" and not the other way round, I have nevertheless been interested by his hypothesis that it was the moon, as it wandered through the pentagram of the sky, which gave the alphabet its rhythm and sequence which we know today (at least in Greek and Hebrew, but in the Sanskrit and runic alphabets the sequence seems slightly different; and obviously all the non-alphabetic scripts are excluded). As it journeys, the moon first encounters Taurus (which is A), then Gemini (which is B), then Cancer (Gamma), Leo (Delta) and so on, with certain adjustments (Sermonti, p. 41). This would give us A, B, Γ, Δ, E, Z…

Sermonti also makes interesting reference to more ancient writers who speculated that the alphabet "descends from the stars"; and there is also a very interesting 1978 work by Bausani entitled *The Alphabet as an Archaic Calendar*, which comes to the conclusion that the birth of the Ugaritic-Phoenician alphabet should be dated to around 2,300-2,000 BCE. And (quoting "*La scrittura celeste: nell'alfabeto un'antica testimonianza archeoastronomica?*" by Dr Stefano Serafini, 2004) it was not only Sermonti who tried to investigate this subject: "as far back as the seventeenth century P. Athanasius Kircher had juxtaposed alphabetic signs and asterisks in search of an original language, which for him would have been Hebrew. In the modern era, other learned men (in truth much less brilliant than he was) have been interested in the same juxtaposition,

---

1 Very briefly: for Sermonti the alphabet is drawn on the night sky with the outlines of the northern constellations. The first half of the Greek alphabet (from α to υ) retraces whole constellations, marking one lunar station for each sign. The second half (from ξ to ω) reproduces whole constellations laid out in the Milky Way. The circle of the zodiac makes a line of animals (from Taurus to Aries) while in the galactic arc there is a line of heroes, birds, and the infernal dog (from Orion to Cygnus to Canis Major). The astral reference point fixes the order and the permanence of the alphabet, which was first a calendar and only later a glossary, and its initial letters (acronyms, or rather its pictographic representations) gave the alphabet its starting point. Winckler's (1907) observations about this are also very interesting.

always moving through glottological comparison, like the Egyptologist Gustavus Seyffart (1796-1885) or J. Broome who in 1872 published *Astral Origins of the Emblems and Hebrew Alphabet.*"

Having said this, I still find it curious that humanity's first Written Word is in fact Carved. In cuneiform. That the method is that of leaving a sign with a knife, or more exactly the point of a knife. Enmerkar's poem goes on:

> From the herald, the Lord of Aratta
> took the artistically worked tablet;
> the Lord of Aratta studied the tablet –
> the spoken word has the form of a nail, its structure pierces.

Its structure pierces.

The reality of its execution is the piercing of clay. By a knife.

But at the same time, the reality for the reader is the piercing the mind of the Lord of Aratta. As if by a knife. Which will later become a stylus made for the purpose.

A knife always has, and has always had, a triangular and more or less accentuated orthogonal section, and when inserted into clay it leaves the imprint of a nail, of a dash.

It pierces clay. It pierces the mind.

A dash tending to the shape of a segment which varies but is always, more or less, a segment. The wedge (*cuneus*), cuneiform.

The Sumerians who, long before the Bible maintained that there was a time when everyone spoke Sumerian (a claim made also by the Assyrians about Assyrian, and then by the Chaldeans, the Babylonians, and naturally by the Hebrews), carved numbers (and only later began to carve the first words) with dashes; exactly like Egyptian numbers, like Latin numbers, and like Chinese and Indian numbers with some variations such as a horizontal position instead of a vertical one, a more than legitimate variation, I would say, after at least 15,000 – 8,000 years. I do not think the fact that in the ancient Ionian and Hebrew numbering systems letters are used to indicate numbers (Aleph/Alpha assumes the value 1, Bet/Beta 2, Ghimel /Gamma 3, Dalet/Delta 4, and so on) changes the substance of what I am arguing, since they are much later representations.

With no "Babel", those signs which indicated quantities of things have stayed unchanged for millennia. And all thanks to two "simple" elements: the Sign, graphic or conceptual, which is transformed into a Symbol; and the knife. And even today, if we think about it, a pen, a stylus, and a fountain pen/graphic stylus are not very different from a knife.

I find Denise Schmandt-Besserat's (*Before Writing.* Austin, University of Texas. 1992) summary of the evolution of the number illuminating, precise, and concise, for the time being:

> Before the invention of agriculture, hunter-gatherers had little need for keeping accounts. Counting was probably only used for recording the passage of time: they engraved bones and used the method of one-to-one correspondence. With the advent of cereal cultivation and the establishment of an economy of redistribution, accounting assumed an important role. In order to record primary goods, Neolithic farmers invented the system of tokens, using concrete calculation. With the birth of cities, productive growth increased the need for accounting and widened the possibilities of the token system. The development of States was a severe challenge to accounting systems and the token system disappeared following the invention of abstract numbers. Pictographic writing was a consequence of the invention of abstract numbers.

While I disagree with Dr Schmandt's concept of an "economy of redistribution" for the reasons which I will illustrate in the chapter on the City, I think these few lines are a well-made summing up.

Though I do not accept the concept of redistribution as an economic category when used in a religious-sacrificial context, the term acquires remarkable significance, as Catarzi observes on pp. 38 of his fine contribution, which I mentioned a little earlier, to a study session held in 1995 on Bernhard Laum, an author I also mentioned earlier and will speak about at more length later on. Summarising Laum's thinking, Catarzi writes:

> In Old Germanic the term *Geld*, present in *Geltung*, *Gültigkeit* (Validity), *es gilt für* (it is worth, it stands for), means sacrifice as compensation (*Entgelt*) and blood-money (*Wer-gelt*) [*Wer* from *Wider*-, against] in relation to the deity and the community. Furthermore, in the German language, as in English and French, the same term (*Preis*, Price and *Prix*) is equivalent to the Italian "*prezzo*" and "*premio*", and the estimation of value (*Wert-Messer*) [from *der Messer*, masc.] shows a singular correspondence with the sacrificial knife (*Messer*) [from *das Messer*, neut.], the distributor of value. Something has value if it can be used as a means of compensation and the value is the power to acquire a compensation. In this dynamic, the original element is not guilt, but its convertibility into something else. The *primum* is the possibility of thinking about a discomfiture or injury, which then prompts a request for help, i.e. revenge or the wrath of the deity – thus, thinking about it in the form of substitution – and the original valuation, which for Laum is the subjective choice of the right animal to sacrifice, presupposes just such a substitution and the setting – in this specific case, taurinity (taurus/bull instead of man) – in which the dynamics of substitution occur.

Equity, Reparation, Value, "Taurinity", Sacrifice, Knives; these are all elements which will fully emerge later on. They are elements directly connected to the concept of the City.

I will speak about the City later. But I am going to speak about abstraction now.

9.
THE SYMBOL AND SOME TRANSFORMATIONS

So, we have a Sign, the number, which in the absence of the object becomes an abstract expression of the object itself. That is, a Symbol.

I think a lot of Giants have killed each other over the question of the Symbol, more even than those created, no one knows how, by God and cited in *Genesis*, the *Nephilim*: but what I mean by Symbol here is a form of abstract representation of something which, starting from an experiential datum and in the absence of that datum, consciously or otherwise, renews the experience and the memory of the experience which originated it and therefore entails its continual re-signification (hence it is a similar process to Après-coup, which should not be seen as an occasional episode of resignification, but as a kind of Operating System or Anti-Virus program, which runs in the background and intervenes when necessary).

In other words, I am thinking of the Symbol (which is not a part-object but the object itself, albeit abstracted) as an experiential, emotional, affective, and cognitive "memory Sign" of a certain context, a certain experience, in its emotional, cognitive, and sensory totality; that is, I find myself in perfect agreement with what Seminara (2006, p. 43) has written on the subject, or as Searle (p. 98: J.R. Searle, *Mind: a Brief Introduction*, New York, Oxford University Press, 2004) has said, a symbol is a sign characterized by "first-person ontology".

And the question of how Abstraction comes to be formed has been and continues to be abundantly debated (but it is a topic I do not wish to address here, just as I do not wish to consider whether it is Condensation or Displacement which generates Symbols: there is too much confusion between Symbol, Symbolized, Content, Containing, Container, or even Saucepan.

In any case, I think it is important to emphasize the origin of the term Symbol – *Sym-Bolon*, σύμ-βολον – given its precision and distinctiveness: as everybody knows, the Symbol, also known as *Tessera hospitalitatis*, is a ring/token which is broken in two, creating a specific "key" which allows the mutual recognition of the two contractors to that specific operation/ contract/ agreement/ pact. Besides emphasizing the ring, which we will return to as the element which seals a pact, the other thing which seems fundamental to me is that without a "re-joining", without one part coming back into relationship with the other, the Symbol itself has no meaning. It is merely a broken ring, a split coin, a useless token, just like an abstraction, an abstract thought which has been detached from its origin and is flying solo. No longer an abstraction, it becomes a delusion.

Returning to the Number/ Symbol/ Sign: this "Memory sign" is not necessarily conscious; but considering for a moment just its conscious aspect (which is the tip of an iceberg), when I see and touch a wound, a bruise, a cut, and talk about it to someone else, neither he nor I make any effort to "feel" what he and I see, having experienced it, although in different ways (but in my case definitely on my skin), by means of an immediate identification. Thus, the Sign (in this case, the Wound, Bruise, Swelling, Scar) represents in the most direct way possible an experience which takes on the characteristics of sharing, understanding, and universality within a certain group of people who are sharing that experience.

Obviously, all of this does not only apply to bruises.

The greater the number of people who share a given experience, and the more profound the experience, the more the Sign acquires shared, one-to-one characteristics for the group, the kind of language it uses, its jokes, its core values, its dietary customs, its medications, and the stories that are told to its children, becoming truly universal with elements such as Birth, Death, Disease, Love... despite being experienced in ways, times, and situations that are in themselves entirely different.

Each of us, without necessarily knowing the language, rites, and specific culture which underlies or sustains a certain manifestation, recognizes the difference between a kiss and a spit, a blow and a caress, a cemetery and a maternity unit, between a souk and a factory, between town and country, between desert and forest, between health and illness, between living and dead. Joy is joy and fear is fear. A stone is a stone, while a flower or a snowflake are a flower or a snowflake, whether they be pink or blue. And the more deeply the Sign concerns the individual's Flesh, or rather each individual's Flesh, the more deeply and literally the Sign is carved. Flesh-Sacrifice-sacrificial Victim-Scapegoat. Mirror neurons permitting, obviously.

Nicola Parise (1997, p. 29) gives a very appropriate summary of the concept of Sacrifice-Symbol:

The Sacrifice-Symbol represents something which it is not, for which it fulfils a vicarious and equivalent function. The animal sacrifice represents the human sacrifice and is in turn represented by one of the tools designated for carrying it out.

The Symbol is *not* the thing it represents since it performs a deputing function, but at the same time the Symbol *is* the thing it represents since it acts as an equivalent.

In any case, some of the observations made by Freud, again in his *Moses*, remain very valid, in my opinion.

As I said earlier about the concept of the "Pre-Monetary", which is moreover closely connected to the concept of Symbol/Sign, whatever these may be – I wrote a little while ago about the meaning of Sign and its relation to Symbol – I do not feel much inclined to enter into a full investigation of the texts by Cassirer, or Kant, or Lévi-Strauss, Saussure, Chomsky or Benveniste. Giants all. On the other hand, if a half-comprehensible piece of Heidegger happens to turn up, I think the next chapter won't mind waiting a bit:

> The characteristic in the primitive relationship with signs, in all fetishes, magic, and the like, is this: For primitive man, the sign coincides with what is indicated. The sign can itself stand for what is indicated, not only in the sense of replacing it but such that the sign-tool itself always is what is indicated. This remarkable coincidence of the being of the sign and of what is indicated does not imply, as it has been interpreted, that the sign-thing has already undergone a certain 'objectification', so that the sign is taken as a thing and is thus displaced into the same region of being as the thing signified. But this 'coincidence' is basically not a coinciding of two previously isolated things. It is just that the sign-thing has not yet become free from what it signifies; and this is because such a preoccupation and such an elementary life with signs and in signs is still totally absorbed in what is indicated, so that the sign-tool itself to some extent cannot be taken separately....
> So that the sign can now fulfil its function as purely as possible: that is, so that it may acquire the character of the handy and the element of conspicuousness, the sign is produced from what is always already on hand. This 'materialization' of the sign, if we may put it that way, has however nothing to do with any sort of materialism or materialistic point of view, as if the indicating and the sense of the sign were tied to 'matter'. Instead, 'matter' here really does not have a material function but a specifically 'spiritual' one, which is to guarantee the universally constant accessibility (Heidegger, *History of the Concept of Time: Prolegomena.* Bloomington, Indiana University Press. 1992: p. 208).

How far from men must Zeus "of the wise counsels" have felt as he looked down from Olympus and saw all those primitive, stupid, useless, feeble flames! But given that I find myself in the midst of these "primitive men" struggling to symbolize, and since I am one of them, I will take Heidegger's loftiness as a kind concession and his Olympian detachment as a sign of generous "benevolence" towards me and the rude, primitive human race.

I also believe, however, that neither Ernesto De Martino, nor Mircea Eliade, nor Clifford Geertz, nor my friend Roberto Malighetti, nor many other scholars share such high-handedness towards "primitives". And, as I mentioned earlier, I think neither Nietzsche nor Ferdinando Galiani would much appreciate Heidegger's self-assurance.

In any case, even "primitive" men, those like me who struggle to symbolize and do not have access to the pure and adamantine abstract intelligence which Heidegger has available to him, little by little and sometimes by accident – maybe because of a pharaoh sneezing at the battle of Kadesh or an unknown Sumerian dream-interpreter having a coughing fit while smoking a Marlboro and writing a strange book on money – sometimes intentionally and sometimes inadvertently, introduce small variations into symbols, which in turn, without ever entirely losing their original form, take on ever more aspects, sometimes inverting their meaning without losing it; or else they become associated with new terms and circumstances. Sometimes they become so far removed from their origin that they cease to be recognizable to a first glance or to common sense, although, as Eliade (1957, p. 88) says:

> History cannot radically modify the structure of archaic symbolism. The symbol's structure cannot be destroyed by the continual addition of new historical meanings.

And this is how Babylon, KÁ.DINGIR.RA ^{ki}, *Babilim*, BA-BEL, i.e. BA-BE'EL, i.e. "The Gate of the god Bel/Be'el", becomes "Confusion", "The House of Bel/Be'el" (translating House rather than Gate takes into consideration the fact that the Gate is the entrance to the House and that Bet, the second letter of the Hebrew alphabet, represents the House, Bait, and is characterized by an opening on the left side which indicates the entrance: i.e. the Gate) becomes "The Great Whore". (Where have I heard this term before? Oh, of course: *Meretrix* for Merx! The strange association between Babylon and Merx.)

It is in this way that *ilu*, god, not *mammon*, becomes *il* and then *el* which, merging with *e* (House in Sumerian), could become *e'el*, house of the god El (*Eanna* is the word for Temple, House of AN, father of the gods but here understood as Heaven and as Uruk dedicated to the goddess Inanna) and in the meantime there is a gap in which Ba'al is generated, since the vowels are not written in Hebrew/Aramaic, and house has become *Bait* or *Bet*. (The same happens with Yahweh or Yeho(w)ah.) And *el*, the bull god El, perhaps linking up with, superimposing itself upon, *ilu*, becomes *eloah*, and then *elohim*, *elohai*, and so on. (These transitions obviously do not take much into account the correct application of the much more complex Akkadian grammar and the yet more complex Sumerian. They are not far removed, however, from the Hebraic conceptions which Bible scholars have identified as Elohist and Yahwist).

In a carefully reasoned argument, Leone Caetani (1914, p. 75) suggests that Allah should also be linked to *ilu* and *el*, being a phonetic deformation of the ancient roots; or that he may be the male/spouse of the more ancient goddess *al-Lat*.

Every "primitive" man adds a little to the story himself, and the outcome of the parallelogram of forces, conscious and unconscious, individual and collective which are applied to the Symbol over the centuries by millions of individuals, gives us the outcome of the entire process: the outcome which that term/ symbol/ vowel/ word/ image presents to our eyes today. First, a multitude of divinities, maybe one for each "primitive", and the Penates, Lares, and Manes would fall into that category; then a multitude of gods shared by a multitude of small communities; then one of these gods emerges into prominence over the others without entirely overshadowing them; then that god becomes the Lord of the gods; then he becomes almost the only god, taking on almost all characteristics as Father or Son, Victor or Saviour of men and/or of the gods themselves; and then he grows even greater, further emancipates himself, and now becomes the Father of all, on whom all the other gods and men now depend on or descend from.

This is naturally associated with new theologies/ theogonies/ cosmogonies in the service of new cultural and political expressions, which are themselves the expressions of new faiths, in an interweaving with no real beginning or end; it has no cause and effect but a process in which the whole participates in the whole.

From another point of view, we could hypothesise a god, El, the great bull, and let's suppose him to be a first god, although this supposition implies a pre-existing tendency towards monotheism and so the god would hardly be "first"; then El finds a house, E, and so we have E'El. Then the house is transformed and we have Bait-El, so it is a short step from here to Ba-El and Ba-Bel, and thence from Ba-Bel to its contraction into Be-El, i.e. Be'el, who represents power, calling on the Lord of the house as the divine Lord. Another short step and we have Ba'al, a term which indicates the Master, present also in the Akk. *bēlum*. From a religious viewpoint, Be'el slowly becomes the Double of Ba'al and replaces him (Liverani's observations about this are of great interest); and then, even more slowly, as I mentioned, we pass from the god El to Be'el, who is transformed, keeping the El, into Eloah and then Elohim, which is the plural of El and Eloah, so that Elohim as the name of God is simultaneously singular and plural (Gen. 1: 26, "Let us make man in *our* image, after *our* likeness"; Gen. 3: 22, "Behold, the man is become as one of *us*." Here we find Make, *asah*, and not Create, *barà*) which, it seems to me, is a very good summary of the concept of synthesis and fusion operating in the direction of monotheism, towards El Elyon, the Most High.

And at the end of the process of transformation we find ourselves beginning all over again with a first god, Ba'al, who is still present but has become negative, and a god "generated" by the first god whose place, or places, he has oedipally taken: that is Elohim, or Marduk (or Yehoshua). (In any case, on the subject of the polytheistic Hebraic fragmentation preceding the one, "final" monotheism, see the still highly interesting work of Robert Graves and Raphael Patai, 1980, which is much more precise than my brief exposition; Liverani, 2007, also makes very interesting observations on the subject.)

The ancient gods were often attacked, proscribed, and vituperated, and we witness bloody battles between the adherents of different cults. The Bible is full of episodes of this kind.

And yet none of these things prevent Ba'al and El sometimes being closely tied by friendship and affection, as is shown by a short Syrio-Palestinian poem of the sixteenth century BCE quoted by Saporetti (1996, p. 162), in which El is concerned about the fate of Ba'al and in a dream intuits that his friend is alive somewhere and organizes an immediate search for him. I must add a small personal observation on Saporetti's fine work: it is probably the result of a professional distortion on my part, but what a strange impression it gives one to read a book about dreams, admittedly in this case the dreams of the Ancients, and not to see Freud quoted once; or at least Jung; not "even" De Sanctis"! A very strange sensation indeed.

On one side the Devil, on the other side, God. Good and Evil. Creation and destruction.

On one side Ba'al Zvuv, the Lord of the Flies, on the other, Yahweh, the Lord of Creation.

Yah-weh has his roots in the rites of the god Sin, the Moon, known anciently as YAH, and later called YAW, and note that the letters which form "I am that I am", that is His tetragrammaton, are Y, H, and W with the addition of a Hei, G-d's letter par excellence: that is, if we use Latin characters. But in Hebrew, the tetragrammaton is slightly different, consisting of a Yud, a Hei, a Vav (not a W), and another Hei. However, it seems that Vav used to have a more obscure sound, closer to W. And in Hebrew even today, the Moon is called not only Levanah, but also Yareaḫ, the god Sin, masculine.

Now I'm going to play for a moment like Attali, who will reappear very shortly. And what if the previously cited god Wê comes onto the scene, as Bottéro (1989, p. 620) reminds us? We would have a god, Ya-reah/YAW, more important than the Sun in ancient times, so much so as to determine the Lunar calendar, who merges with a god, Wê, who is killed in order to give life and spirit to man, made of clay, who would in this way also have the Spirit of the deities. (It seems that Wê was chosen precisely because he was endowed with *têmu* – or rather perhaps *eṭemmu*: i.e. spirit. The Divine Spirit? Distilled Spirit? Alcohol? Maybe the High Spirits of Cheerfulness, Euphoria? Spirit as *esprit*, barbs of wit, vivacity, perspicacity, intelligence?) The fusing of the two gods would create a brand-new god; Ya- Wê, or even YAW-WÊ![1]

Abram's wife, Sarai, will change her name to Sarah just as, in the same way, Abram will become Abraham – by the addition of a Hei, to underline the presence in them of G-d.

For Jacques Attali, however, the name Sarah should be linked to *sahar*, the moon; but my Hebrew teacher, Dr Avezov, cited earlier, answered my direct question as follows: "The moon, *yareaḫ*, or its synonym *sahar*, the quarter moon, often used as an Islamic symbol [here again are the Viennese Croissant and the Islamic Crescent] can obviously only be confused with the name Sarah by using a transliteration into Latin characters because in Hebrew it is clear that they have two different roots: 1- the quarter moon, and 2 - a root which indicates presence, importance, control, and power."

Setting aside the Croissant, we will see later how Sarah is connected

---

[1] I am obviously playing a little with words, but shortly before finishing this book, reading a work by Mander (2005), I came across an example which he presents on p. 43, curiously relevant to *WE*. Mander writes about the fact that the Sumerians/Akkadians, like all people andin all languages, enjoyed word games, whether phonetic or semantic, and he notes that in the word *awīlum*, syllabised as a.wi.lum, man, links can be traced between the god WE, sacrificed in order to give rise to man (read in this context as WI) and the term divinity, *ilum*. The word Man, *awīlum*, would thus contain the seeds of its own origin/creation.

with Queen/ Princess (Akk. *šarratum*) and not with Moon.

Ba'al, however, lives on in the same way: in modern Hebrew, Ba'al still means Lord, Master of Property. Perhaps *Ba('al)* connected to *Ab*, i.e. *Aba*, Father. Certainly connected to Akk. *bēlum*, lord, master.

But Ba'al continues to have yet another everyday, human existence as "He who is the material author of the act of plucking the Flower:" that is, the "privileged" male responsible for deflowering, *Ba'al beilah*. In modern Hebrew, and losing a little of its sacredness, *Ba'al beilà* is more simply "he who penetrates" and no longer the one who makes that first important, sacred penetration. Incidentally, it is curious that the entry for Ba'al in *The Jewish Encyclopedia* has this, among other things:

> According to another conception, *Ba'al-berith* was an obscene article of idolatrous worship, possibly a *simulacrum priapi*.

The term *Ba'al-berith* is interesting since *berith* is in fact *Brit*, Covenant. Thus we have a Ba'al, the precursor of Yahweh, whose prerogatives include being Lord of the Covenant; and, moreover, this covenant is directly connected to an image of Priapus!

In the end, a great many ancient divinities live on, hidden away, as is well represented by the episode of Rachel who steals the images of Laban's gods from him, hides them (*Gen*. 31:34), and takes them away with her.

Ba'al-berit must also be cognate with Ba'al Peor (*Num*. 25:1-18; *Joshua* 22:17; *Jer*. 3:6).

What's more, Ba'al, smiling slyly, is still Lord, Master, Proprietor, Possessor of Awareness, he who is conscious/ aware. A prudent person, someone of sound judgement is *Ba'al hakarah*, literally Lord of Awareness.

And while on the subject of personal gods, ignoring Cicero and more or less primitivist approaches, I am also much interested and amused by Dumézil's observations, quoting St Augustine (*De Civitate Dei*), on the ancient Roman deities, clearly mocked by Augustine like something out of a comic play or Disney's *Sleeping Beauty* (again!), where the little fairies hold a snoring competition around the infant princess's crib, thereby giving each of them a specific character. I will mention only a few, referring to Dumézil's book, and obviously to St Augustine who describes many others, only to hint at a further hypothesis:

> The group of divinities who govern a child's birth, nutrition, and schooling [who are addressed for their grace and protection] from *De Civitate Dei* IV: 11 and VII: 3.1: "After Vitumnus and Sentinus have given the child life and feeling, Opis gives him vigour [Ops, also Opis, goddess of the earth, like

Cybele], Vaticanus opens his mouth for his first wailings, Levana lifts him up from the floor, Cunina cares for him in his cradle, Potina and Educa give him things to drink and eat, Paventinus manages his fears; and when the child goes to school and comes home again, Abeona and Adeona look after him, watched over by Juno Iterduca and Domiduca."

And it's the same for agriculture, where we find Veruactor (to aerate fallow land); Reparator (to reinvigorate fallow land); Imporcitor (for ploughing deep furrows); Insitor (for sowing); Obarator (for shallow ploughing); Occator (for harrowing); Sarritor (for weeding); Messor (for reaping; and I wonder if Messor has anything to do with the German *Messer*, knife, whether *das* or *der*: Pianigiani believes instead that the Mediaeval term Messere is connected to Mio Sire [My Lord], and not to knife. However, it is also true that by no means everyone had the right to carry a knife, just as owning a horse in Mediaeval Japan was reserved for nobles); Conditor (for storing produce).

And so on: a swarm of divinities for transportation, and another for the sequence in which trees were to be planted, for pruning, for the series of actions associated with weddings and the first sexual encounter between spouses. Pixies fluttering about everywhere. Every human action, every sphere of life seems to have had its own dedicated divinity. In a different way from Augustine, Cicero is also determined to desacralize the nature of the gods and divination. In a more scholarly manner, De Gubernatis (1899), cited by Jung, does not mock Roman religion practice like Augustine, and shows genuine interest in the "small" Roman divinities, relating them to those of India in an attempt to show that the Latins belonged to the Aryan group.

But in Latin and Italian grammar, an action is called a Verb[2], *Verbum*, whereas the object is called *Vocabulum*. And the *Verbum*, closely conected to Verb, turns out inevitably to be connected to the Logos as well as to action, to doing.

Therefore, the Verb, divine or otherwise, can be seen as the action over which a single divinity presides: thus there are as many gods as there are actions; one god for every Verb, and every Verb represents a god!

Read in this light, the words of John 1:1-3 acquire a yet more curious meaning:

---

2   Vocabolario Etimologico Pianigiani: "Verb – this is what grammarians call the Word which denotes Action." Semerano (19XX: "In grammar, *Verbum, -i* is opposite to *vocabulum*.... *Verbum* has the ancient meaning of *date, relation, work*."

In the beginning was the Word, and the Word was with God and the Word was God.
The same was in the beginning with God.
All things were made by him: and without him was not any thing made that was made.

A kind of primordial DNA, endowed with its own force, perhaps even pre-existing the god himself? (see also below p.215-216, note 1)

And if this were the case, we could speculate that it is the gods who are given birth by the *Verbum* instead of the *Verbum* being produced by the gods. A curious hypothesis, even by my standards; one to be developed.

Another interesting feature of the *Verbum*, Logos, Word is that, from the Sumerians to the Hebrews, it is absolutely inherent in the thing itself: that is, "the thing designated, the form of the sign, and the sound of the word are the three interchangeable sides of a triangle: the sign does not represent the thing, but is the thing," writes Seminara (2006, p. 63).

The Sign is; the Sign is not; perhaps the Sign is the thing; but perhaps it isn't; maybe the Thing is only a Sign; there is a symbolic equivalence; no, it's a symbolic equation.... I don't believe there is an Absolute. I think every writer tends to emphasize an important aspect which is present in all research and in all aspects of life, generating contradictions that are more apparent than real.

Now, however, I will turn to someone who knows a lot about (divine) Signs: Noach (or Noaḥ; or Noè; or Noah).

10.
A HALF-SERIOUS HALF-FACETIOUS ANTE-DILUVIAN DIGRESSION

At this point, before I go on, having touched on some important facts about certain cornerstones of the *Oikos*, I think it may be useful to take a "small" step back in time and take a closer look at some aspects which I've only hinted at, or which may not even have come into view yet. Mentioning naïve Barter, the Merx, and Mercury, I also introduced Noah. Noah goes into the Ark at the age of 600, says the Bible.[1] And when the Flood ends he will live for another 350 years. He must therefore have met Enmerkar. We know nothing of their relationship, but they must have written, visited, seen, each other; they will certainly have been members of the same club and the same gym, and dined at the same restaurant, perhaps "Chez Deucalion", where Pyrrha was a waitress.

Mind you, right after the Flood there won't have been many eateries left open. The importance of these unverifiable meetings is that they echo the parallels which we will encounter later on between other characters in Sumero-Hebrew history: characters who, at a certain point in the narratives, take on the natures and characteristics of authentic Doubles in the terms described by Girard, or at least in the sense of *hoi bioi paralleloi*, Οἱ Βίοι Παράλληλοι of Plutarch. If Enmerkar invents Writing, Noah is no less impressive in inventing the Ten Commandments.

Two of the greatest inventions, Writing and (the Writing of the) Laws,

---

1   On the subject of the Flood and the Bible, and the scientific, philosophical, archaeological, and perhaps also religious implications which the discovery of the Sumero-Akkadian tablets are having and will have, I recall the respectful, but also ironic, words with which Bottéro begins his 1986 book (*The Birth of God*, University Park, University of Pennsylvania Press. 2000):"On December 3, 1872 the Bible forever lost its immemorial prerogative of being 'the oldest book known,' 'a book unlike others,' the book dictated, written, by God Himself....on this particular day, before the Society of Biblical Archaeology in London... G. Smith [one of "the first Assyriologists"] announced his extraordinary discovery: a history that was strikingly close to the biblical narrative of the Flood, even in details, but that preceded it and had obviously inspired the story in the Bible" (pp. 3-4).

which humanity has ever seen; and all of this in the same timeframe and the same geographical area! Magnificent. Writing and Law. An Axial Age indeed! I am only partly joking. The fact is, I wanted to give full vent to my enthusiasm and have let myself get a bit carried away.

Noah's Commandments aren't actually ten in number. Beside the fact that the Commandments weren't even Commands for the Hebrews, Noah's number "only" seven. And furthermore, they aren't even "written" on stones.

And he and Enmerkar probably didn't know each other at all.

And maybe they weren't even Doubles; but I couldn't resist a narrative construction that would allow me to joke a little, to put on a little show and introduce these concepts.

In any case, however you look at it, the law of Noah consists of seven instructions/ precepts, of which six are negative (Thou shalt not do something) and one positive (Thou shalt do something). This is another subject on which there is a boundless literature concerning adherence or otherwise to the divine plan, or the birth of the individual's fundamental freedoms, or the foundation of a national State, plus a heap of other things. Professor Davide Astori of Parma University has made a very fine investigation of all this, as has Bindman – a bit bizarre but intriguing – and Dallen – somewhat generic and didactic. However, it seems to me that we cannot do without the highly serious and detailed work of Aaron Lichtenstein, who analyses the seven laws of Noah in parallel to the 613 Mitzvot.

My specific reason for bringing Noah and his story into the picture is one of his seven Commandments. The same commandment was later taken up by Moses – pardon – by G-d himself. And among the Commandments of Noah or Moses, or however you want to translate the term (*Mitzvot Noaḥ*: in other words, the Precepts of Noah; or *Brit'Olam*, the Pact/Law for the World; or *Aseret Hadibrot*, the Seven Sayings of Moses; or *Luḥot Habrit*, the Tablets of the Law/Alliance/Pact/Covenant, or perhaps instead *Brit* understood as a mutually binding Contract) the one I am talking about is "Thou shalt not steal" – more familiar in the words of Moses, but already present in Noah: that is, before the Flood. (Scout's honour, I won't say a word about Fromm; even though I've never been a boy scout.)

Probably out of carelessness and habit, I have always thought of this commandment as a generic "You shouldn't appropriate other people's things/ You should respect other people's things"; but there's more to it than that.

Bindman's reading of this is of great interest – he's not alone in addressing the subject – observing that the act of rebellion/ appropriation for which Adam and Ḥavah were punished can be seen as a Theft, an undue appropriation of something found on another's property; the Apple from the Garden (and the other tree, the Tree of Life) would belong to God, being on the land of God and not in the part of the Garden which was placed at their disposal. Hence, man's first sin, the Original Sin, would be the Theft and violation of God's private property (not stolen property, since it was God who created the world and the Garden, and they are specifically His: He did not steal them from anyone; private property and not de-prived/stolen from someone else). Besides this observation, there is also the human and concrete, not exegetical and speculative, sense of this affirmation-exhortation-prohibition-commandment "Do not steal" implicit in quite a few Biblical quotations. Here is a sample of three:

> Thou shalt not have in thy bag divers weights, a great and a small (*Deut.* 25:13).
> ... thou shalt have a perfect and just weight (*Deut.* 25:15).
> ... and the sabbath, that we may set forth wheat, making the ephah small, and the shekel great, and falsifying the balances by deceit (*Amos* 8:5).)

A brief aside to make clear that here too we find ourselves in a condition where the Semitic world again shows deep roots in the Sumerian world: the patron goddess of Lagash is not Utu but Nanshe, goddess of Justice; a hymn dedicated to her around 2,500 BCE describes the types of criminal who will suffer her wrath; among the various kinds of criminality which constitute an *abomination* in the eyes of the gods, Nanshe, goddess of Justice, assisted by her husband Haia and by Nidaba, goddess of Writing and Accounting, indicates as criminals those:

> who substituted a small weight for a large weight,
> who substituted a small measure for a large measure.

And she reiterates that her rules of justice will serve:

> to comfort the orphan, to make disappear the widow (Kramer S.N., *The Sumerians: their History, Culture and Character*. Chicago and London, University of Chicago Press, 1958, p. 125).

Almost identical passages will be found 300-400 years later in the *Prologue* of Ur-Nammu.

Weights that must be equal and fair. Scales that must not be false. This is what God is instructing Noah with the words "Do not steal," and more correctly than Fromm. (Oh, look at that, would you believe it? I've broken my promise!)

Perhaps the statement "two weights and two measures" comes from this.

If, when I'm at home, God tells me to keep only "a perfect and just weight" in my bag or my pocket, and failing to do so literally means performing an *abomination* in the eyes of the Lord, this must be a matter of real importance.

I've always got a horse chestnut in my pocket. I don't believe it will keep colds at bay, as my grandmother Gina used to say, but every year, when the conkers break free from their husks I bend down and pick one up to give as a present; and another to keep for myself. Maybe this is a fetish, but it is my way of remembering with affection someone I loved, my grandmother Gina. But I am also convinced that when – as has sometimes happened – I have forgotten my Granny, I haven't considered it an *abomination* perpetrated against her.

Therefore, I think that the weights to which God is referring must have a very different meaning from my conker; and even if it's highly likely that the Sumerians and the descendants of Noah loved their Grannies, this has nothing to do with grandmothers.

And the operative feature of Noah's commandment, its concrete and human significance, can be found at Ebla, a Sumerian city whose archives were rediscovered by Italian archaeologists (see also Matthiae, 1977; for more on this see also Gelb I.J.):

> Weights come from public palatine buildings, military complexes, private residences or houses, and from one of the temples, testifying for their use in several areas of the city and for different activities. In this paper we will briefly discuss the evidence from each category of buildings, considering the weights linked to a broad and general weighing "function" (Ascalone and Peyronel, 2006).

At Ebla and all over Sumer, the inhabitants all "carried weights in their pockets" because they WEIGHED things. They WEIGHED nearly everything.

Given a standard fixed by the authorities, a standard which was the result of calculations about the production of all kinds of goods, obtained from experiences calculated, archived, and developed over centuries and often "averaged", the value of an item came to be transformed into a weight, and then exchanged for another item likewise expressed as a weight.

In other words, they weighed the quantity of a certain desired piece of merchandise which, by means of definite, agreed units of weight, matched another quantity of another type of merchandise.

Weighing, related on the one hand to standard weights and on the other to commercial standards, was the way in which exchanges and transactions were carried out. This was the system: a highly refined system. Still with no Coin, perhaps, but it was a far more complex, sophisticated, and subtle mode than the naïve Barter I referred to earlier; see Peyronel, 2008).

Thus, before Ebla, before Gilgamesh (as we will see later on), before the Flood (it must mean something if Noah and also Utanapishtim are able to build Arks with precise dimensions which accord with the dictates of the gods), before Money and before the Coin, there may be evidence of a complex and finely worked out system of weights and measures widespread through all levels of the population. So well worked out and extensive that it enabled the free distribution of onions. But that's not all. There is likewise evidence of the very refined technical skill needed, not only to be able to adulterate and modify weights in order to "steal", but even more to be capable of producing, reproducing and recognizing what comes to be called the Just Weight, creating, making concrete and tangible, the basis for the very concept of Justice.

And it is not only necessary to have a competence distributed throughout society in order not to be robbed and to be able to rob, but the knowledge, acceptance, and dissemination of the established standards must be accessible, widespread, and shared.

And, last but not least, it must be possible to build and use the device for which weights are intended in their function of providing Equity and Justice: the Scales. It is a still more astonishing fact that the scales are connected to Justice, Law, Legislation, and that this derives from the development of a frenetic weighing activity which encompasses the entire Bronze Age, though I will only be giving marginal attention to this, returning to the concept later on in relation to certain aspects of the ancient Roman world.

Just as I have no great qualms, at this point in the discussion, about going back to before the Flood, I see no particular problem about leaping ahead a few millennia. All I wanted to do was to establish this point: the importance of Weights.

They will appear in various forms, with artistic features, delightful representations of animals or delicate human heads; they will appear in abstract shapes, polyhedral, spherical, ovoid. But always precise, multiples or fractions of what was ordained by Utu-Šamaš, the divinity who oversees Justice. They are the *Abnu(m)* (from Ascalone and Peyronel, 2004):

> If we look in the epigraphic documentation, it is possible to find other hints for this kind of research, confirming the link between Šamaš, the concept of justice and the symbolism used to express rectitude, correctness and equality; the only deity associated unequivocally to the Akkadian term abnu = weight, is the sun god Utu/Šamaš: in this case the most appropriate translation of "weight of Šamaš" would be "standard of Šamaš", indicating a weight used as a "correct" standard of measure, which certainly represents a strong evidence for the above mentioned concept of justice.

And in this process, we cannot forget that Enmerkar is the son/grandson of UTU/Šamaš with all that this entails: the responsibility, above all, to be Just. And we cannot forget that, at the same time and in the same place, the ancient Middle East, Noah is chosen by God to generate a new humanity, precisely because he is Just.

Just, from *jus, juris*: Legislation, Justice, and hence, Law.

And it is precisely in terms of what is "just" that the fates of Sodom and Gomorrah will be decided.

Moreover, Enmerkar is a predecessor in the dynastic line, a dynastic ancestor of Hammurabi, while Noah would be a progenitor, a biological ancestor, of Hammurabi.

Now, what does Hammurabi have to do with this? At the moment this may seem a piece of redundant or useless information, but that is not the case since, as we shall see, Hammurabi seems to be a direct descendent of Noah, or rather, of Ham. He will lead us to the name, among others, of Nimrod.

I'll dust off the lineage of Noah:

> Now these are the generations of the sons of Noah, Shem, Ham, and Japheth: and unto them were sons born after the flood.
> The sons of Japheth...
> And the sons of Ham: Cush, and Mizraim, and Phut, and Canaan....
> And Kush begat Nimrod: he began to be a mighty one in the earth.
> He was a mighty hunter before the Lord; wherefore it is said, Even as Nimrod the mighty hunter before the Lord.
> And the beginning of his kingdom was Babel, and Erech, and Accad, and Calneh, in the land of Shinar.
> Out of the land went forth Asshur, and builded Nineveh, and the city Rehoboth, and Calah... out of whom came Philistim (*Gen.* 10:1-14)

And Shem? We'll leave him in peace for now.

Instead, later on I'll write about that Nimrod, son of Kush, whom Rashi, the very great mediaeval exegete of the Bible, and many others with him, identify with Amraphel: that is, Hammurabi.

Ginzberg's observations about Nimrod are very interesting, bringing to light a story which can be almost completely superimposed onto the one which sees Herod in the role of Nimrod, with the Comet Star which announces the birth of Abram, and with his splendid Massacre of the Innocents, 70,000 male infants it seems (see Ginzburg, 1925, vol. 2), in order to avoid being overthrown.

I am reminded of something.... Something very similar happened to Moses (*Exodus* 1:22).

If Jesus and Abram have such a similar beginning, who knows if, deep down, Noaḫ and Enmerkar aren't, if not true Doubles, at least somewhat related.

11.
THE *UR*-BS, 𒌷
NOT THE POLIS, ΠΟΛΙΣ

After that semi-serious digression, it's back to Uruk and Ur; but please let's leave behind the Asiatic Mode of Production and Despotism, whether of the Oriental or Hydraulic kind. If only people had been a bit more patient and taken the trouble not to generalise, they would have created fewer Babels-Confusions; if only they'd paid attention when they read the story of Joseph.

S. N. Kramer (ed. Bottéro, L' Oriente antico, 1992, p. 11) even writes about a bicameral system which is challenged by Gilgamesh.

But in this case too there are "ifs" and "maybes"...

Why the wisecrack about Oriental Despotism? Because of a whole supposedly "scientific" politico-journalistico-scandalistico-ideological line of thought which operated between the second half of the nineteenth century and the first half of the twentieth (C.E.), although it was Hegel himself who initiated it. This ideological set-up, beginning with Marx or Engels, or maybe Morgan, aims to show, alla Rouseau and Voltaire, that the first humans (perhaps Lucy, Australopithecus afarensis, and her cousins?) lived in a kind of primeval Eden (surrounded by lions, hyenas, vultures, snakes, scorpions, diphtheria, parasites, in all extremes of weather, naked, without a cigarette lighter, or even a cigarette); this "scientific" line of thought goes on reiterating that, in various but largely equal forms, the "birth of civilisation" (established ope legis after a tree-dwelling consciousness-raising group discussed it?) brought no end of chains to enslave even the most distant members of that post-arboreal society.

To some extent, Freud indulges in this too and lets himself be seduced by the "primitivist" angle, but not too much, fortunately, or at least not too often.

With a dizzying logical somersault, this primitivist approach first comes (like little epigones of Rousseau) to condemn the brutal system which enslaves individuals who had previously been free and happy; a brutal system in which Oriental Despotism and Civilisation are to all intents and purposes synonyms and both are initially considered in a negative way, meaning that they would subjugate the beautiful free and spontaneous child who has only just come down from the trees; but immediately after that, such a brutal and castrating imagined (or, in the psychoanalytic sense, projected) system comes to be lauded in Antiquity as a system of "collective" cohabitation which would represent the closest thing to a "state of nature", a kind of second-rate primitive communitarianism; and indeed praise is heaped on that system which would have "deprived" the individual of his personal freedom, his naivety and primordial purity, given that such a system would, in the end, be the ultimate in "equality" since there would be no individuals beside the Sovereign (Lucy again?) who would have the freedom to make decisions. Obviously no "primitivist" would agree with this summary of mine.

As with so many human affairs, however bizarre they may seem at first glance, even this position does in fact have some truth in it; in the striving towards a monotheistic religion, with all its possible and imaginable stages, twists and turns, we find exactly this feature: the only true power is God's, and the rest either doesn't count or is subordinate to God. Somewhat paradoxically, all men become equal before the deity, and even more equal before a single God, something which does not happen and cannot happen if the divinities are fragmented, given that there will be first-division and second-division followers of first-division and second-division gods.

And so is born the demand to live in Maat, the demand for Equity and Justice which now characterizes the entire late Bronze Age and embraces ethical, personal, political, economic, and scientific aspects of society; it passes through the Mos and generates the Jus.

And it says that all men are equal. (And it says that Weights must all be equal. But it could also be the case that from the Weights we arrive at God.)

However, it is one thing to attribute egalitarianism (or rather, the striving towards an egalitarian idea behind which an enormous range of monsters may be hiding) to a phenomenon which takes a religious form, which is a profound feeling and becomes rooted in the spirit of every individual, constituting an urge, an afflatus towards emancipation from submission by calling on a notion of true equality; and it is quite another thing to superimpose the King (today we say the State) onto God (Hegel docet), which poses a little problem or two. Therefore, if the argument is about God, this seems like a serious matter, but if instead this process is applied to very human categories, essentially in a logic of domination, we get a (not very serious) something else. Although my summary of "primitive communitarianism" and "hydraulic despotism" is a bit over-the-top, the position it describes, well summed up by Sédillot,[1] leaves me absolutely lost for words.

If it makes sense to go back to the Asiatic System of Production, Bloch's examination of the subject in an article entitled 'The Symbolism of Money in Imerina' (in Parry and Bloch, 1996) seems very interesting and precise. And deep down, but really deep down, re-read a long time later, it can be a curious experience to read Wittfogel (1962).

The absurdity of the confusion between Right and Sovereign, Law and Power, Absolutism and Responsibility, God and State has generated a lot of monsters; unfortunately, this weirdness hasn't been confined to handful of adventure stories, but has become a praxis which has cost the lives of millions of people and even today can count on many followers who claim that the Jus derives from the authority which legislates instead of accepting that it is the Mos, to which even the Sovereign's power is subordinate, which generates the Jus, and hence its transformation into Lex, which in turn can only be accepted as an expression and shared evolution of the Mos itself.

Von Hayek's observations on this seem exemplary and illuminating: he didn't like psychoanalysis, but this shouldn't surprise us since he was a

---

[1] Summarising the concept and endorsing it, Sédillot writes, "It is a regime of constriction, in the wake of which the administrative machine leaves no initiative to man. Products are identified and stored in gigantic State depositories. Harvests, fabric, weapons are warehoused. Distribution and allocation are overseen by officials. There is no place in this system for coinage or barter (1989, pp. 34ff; also quoted in Schimmel, 1993, my translation). Which is like saying that from Aratta, on foot or on the back of a mule, or in an ox cart, my cousin's onions are carried to the capital Uruk, a journey of a thousand kilometres, to be stored in gigantic silos. Naturally, having added his onions to the mass, my cousin goes home again. Then, when he wants to sauté an onion, back he goes to Uruk, another thousand kilometres, to get his share of onions; going back home again, another thousand kilometres, with his allocation of onions. I can't see any point in commenting on this conspicuously ideological thesis. And even supposing that every single village made use of "gigantic" silos, I cannot understand how salads, dates, melons, watermelons, jujubes, and peaches, but also meat, fish, milk, butter, beer, and wine can be first allocated, then stocked and preserved in warehouses, and only distributed after that. To say nothing of the technical items needed by the various professions: from clay to lathes, from bows to fishing nets, from hoes to knives, etc. Moreover, according to this view, we'd have to suppose that the Sumerians pretty much had to make do with a diet of dried fish, smoked meat, and rotten onions, and that their ration books entitled them to one arrow a month. So, no wonder they declined and fell. P. Einzig (1948, p. 24), writing about the fallacious modernist approach to ancient civilisations, wryly quotes Keynes, according to whom, "the fall of the Egyptian and Sumerian empires could have been due to a lack of metal for making into coins." (!)

friend of Popper, and Popper understood little about psychoanalysis.

It is painful to observe how many Giants, Lilliputians, warlords, political leaders, kings, emperors, prime ministers and presidents, baristas and domestics under all the world's skies still hold the delusional belief, that you have only to Say something for it to become Reality.

It's like a parody of Genesis, in which God simply names a thing and it immediately comes into being, is immediately created. It seems a parody but it's a tragedy.

In this stance of apparent free expression but disconnected from deep respect, a stance that is apparently free to take only itself into account, there is no more Mos, no Jus, not even Lex, but only the childish and megalomaniac parody of Fas (and Fas was also the ancient Latin goddess of Justice), the will of the Gods, which becomes mere Doxa, δόξα – blah-blah-blah, in other words, with which too many tribunes, prophets, governors, kings, and scoundrels identify, believing that their mere word can literally create a whole new order, naturally at their service and "in their image and likeness", where everyone can lay claim to what Bottéro calls the "Effective Divine Word," with which – let's be clear about this – with which Marduk made a whole constellation disappear by simply "speaking", the same word with which God *Barà* ("created") the world; a Doxa which holds, with all its blah-blah, that it can create a new world, a new man, a new civilisation, completely new and disconnected from all that went before it; entirely "new", taking no account of the Mos but only its own Utopia – or rather, its own delusion – and with no respect for an authentic Ethos which is praxis – concrete, sequential, and consolidated action, and not the abstract Ethics of the moralists and philosophers.

Antigone is still compelled, every day and in every part of the world, to reckon painfully with the arrogance of Creon.

And yet Locke's observations[2] still have an astonishing clarity. Just as do the painfully clear and moving observations, unfortunately unheard and repressed, of a great Italian scholar, Bruno Leoni.

Anyway, journalists-ideologues-agitators apart, the Sumerians didn't have to ask the Sovereign for permission to take onions to market, and they weighed them carefully before selling/buying them to ensure that they

---

2   "Because men would not be thought to talk barely of their own imagination, but of things as they are; therefore they often suppose the *words to stand also for the reality of things*." Locke goes on, "give me leave here to say, that it is a perverting the use of words, and brings unavoidable obscurity and confusion into their signification, whenever we make them stand for anything but those ideas we have in our own minds." (*Essay Concerning Human Understanding*, Book III, chap. ii, § 5)

weren't robbed. And my cousin did the same. And they didn't even have to justify the use of their own money to some Temple functionary, as some claim to do today. Bernhard Laum, in *Heiliges Geld* (p. 159), arguing along the lines of Babylon and the Euphratesians, and what's more, in agreement with Jeremias, writes:

> The internal market [in onions; my addition] has always been free and individual.

So it is evident that, while foreign and military policy – great public contracts, and the organization of the territory – could only be decided by the "Sovereign" – by a central power – as is the case today, the "*Non-Œconomia* of the Great Man Mountain" was the prerogative of individuals. Otherwise, what would have been the point of building up a body of civil law like that of Ur-Nammu, of Lipit-Ishtar, and of Hammurabi, or proclaiming the laws of Noah and then of Moses, for individuals to draw on? And above all, what would have been the point of equipping oneself with such sophisticated, precise, and widespread weighing systems? Setting aside the perennial dispute about Monetary and Pre-Monetary, I have found it very interesting, though in relation to the Late Bronze Age and the Assyrians, to read the scrupulous observations by Peyronel (2008) on archaeological digs and what they have unearthed, by Kültepe, and also Bulgarelli's (2009) detailed examination of the subject; not to mention the splendid words which Bottéro (1992, p. 150) devotes to private enterprise, including that of women.

The Sumerians, not slaves of a central authority but subjects of the power of a King who was bound to respect their *Mos*, and deriving from it the *Jus* which was expressed in *Lex*, did not pay "taxes", unless they were "tenants" of the Temple or the Palace: in other words, they paid "rent" (*mutuum* in Kant); the free citizens, however, lent or offered a service; service in the temple or palace, but not taxes.

These services were *corvées*. I'll come back to them later. Even then, this did not apply to VIPs, although Ur-Nanshe, founder of the first Sumerian dynasty of Lagash (and also Gudea, I think) had himself portrayed with the basket used for the building and maintaining of canals, as if had done the manual labour himself, but we know that the Istituto Luce's pictures of the Battle of Grain from a few short years ago do not actually mean that Mussolini, if he were alive today, would devote himself principally and practically to agriculture; it's not just "capitalism" that's a bit older than Calvin, but propaganda too.

Here's a brief aside to recall that for the Absolute Sovereign, submission to the *Mos* was such an imperative that respect for the *Mos* rather than the Sovereign became an extremely powerful political weapon used by Assyrians and Babylonians in their campaigns of expansion and annexation in neighbouring realms, which were thus easily accused of "heresy", to the extent that they lost the support of their own subjects when they lapsed from the *Mos maiorum* or the *Jus gentium*. The problem of respect for the *Mos* was also faced by the Sumerians, when the Royal Dynasty changed "house" (see Pettinato).

And the *Mos* was so important that, among his other obligations, the so-called Absolute Sovereign had to submit to the annual ritual of a tremendous slap from the High Priest.

Dumézil reports a similar ritual surviving in India and officiated by the Brahmin; I think the Catholic Bishop's "slap" in the rite of Chrism may have the same meaning. And one of the most important titles of the absolute sovereign was Shepherd[3] of his people. The slap was so hard that if the Absolute Sovereign did not weep with pain, having in this way to remember that he was responsible for his people, their shepherd, not their master, unfavourable auguries were drawn for the year to come, throwing the whole kingdom into dismay.

A tremendous slap, definitely a symbol, but with a certain nuance of concreteness, I would say. And despite being the interpreter of the word of the gods, the Sovereign could not in any way use the *Fas* in opposition to the *Mos*, since this would have been interpreted as a blatant contradiction and therefore heresy, or an obvious abuse, since *Fas*, the divine will, was what directly generated the *Mos maiorum*. An example: the divine will expresses itself in the Ten Commandments which found a custom, a tradition, a way of life and a method for addressing problems, a "style" of the Fathers which conforms to the will of God. Therefore, a Custom/Law/Instruction/Royal Will which challenges the divine word, or a divine word which challenges the custom derived from it and does not comply with it is

---

3   Psalm of David, 23: "The Lord is my shepherd; I shall not want. He maketh me to lie down in green pastures; he leadeth me beside the still waters. He restoreth my soul: he leadeth me in the paths of righteousness for his name's sake. Yea, though I walk through the valley of the shadow of death, I will fear no evil, for thou art with me; thy rod and thy staff they comfort me. Thou preparest a table before me in the presence of mine enemies: thou anointest my head with oil; my cup runneth over. Surely goodness and mercy shall follow me all the days of my life: and I will dwell in the house of the Lord for ever." Very interesting the beginning of Psalm 22, with the misunderstood words of Jesus: "Elì, Elì, lama asavstàni"!

an impossibility and creates problems, just as it has done whenever it has occurred; we need only recall the Kings of Israel and their dealings with various Prophets. But back to the Sumerians.

At this point, after all I have illustrated and after a gestation which precedes the reign of Enmerkar by at least 4,000 years, the Number would be ready to generate Writing all by itself, but many other elements need to come together in order to make possible the birth of the Engraved Word: and that is a Priest-King; a divinity; the sacrifice-exchange of something held in high regard; a sign which has become a symbol; something to engrave on – i.e. a base that "remembers"; and a knife.

In more descriptive terms, as implicit, underlying, elements taken for granted, perhaps even unconscious, but nevertheless present for the "citizen" of *Ur*, we could list:

An administrative system which makes laws and is able to propose them and make them respected, resorting to force if necessary and, at least in the beginning, superimposing itself on an apparatus for the performance of rituals addressed to the deity and almost always with the King himself taking the lead. Here too it would be interesting to note that it is not the army and the religious apparatus that are "eternal" but also, and especially, the Bureaucracy (I am referring to the Freud of *Group Psychology and the Analysis of the Ego*, in which he talks about two "lasting" structures, the Army and the Church, but forgets Bureaucracy which in its various forms is the keystone, in my opinion, of many of the questions I am raising here. Perhaps we also owe Bureaucracy to *Ur*? I believe Giorgio Buccellati, 2013, shares this opinion).

A deity shared and accepted by an entire population which identifies with that cult, accepting its rituals, its calendar, responsibilities, and sacrifices – i.e. its customs and traditions: in other words, the *Mos* – sharing and understanding its symbols.

A widespread and certified quantitative system of measures which also includes the qualities of the materials to be measured and the rates of exchange between different goods.

A system of signs and meanings which are immediately comprehensible and shared by the entire population.

An instrument which enables the practical execution of all that needs to be transmitted so that all the entities described above may communicate among themselves: Writing, in other words, achieved by means of resources that are easy to obtain and use, such as clay and a knife.

All these elements as a whole are called the City, which is not the *Polis*, which will only appear two thousand years later; and we can easily intuit that the City, *Ur*uk, the First City, the city of Enmerkar with whom I began this survey, is a highly organized and very complex *Ur*-ban agglomeration which had already brought about the division of labour. The *Ur*-bs is not only a market, as Max Weber claims, because in order to have a market, in the temple or in a district, first you must produce the goods which will be sold by the merchants.

An *Ur*-ban concentration and not scattered villages, or lone sages, stylites or anchorites.

And although many Giants won't agree, I am greatly fascinated by the prefix *Ur*, which even today, after millennia, continues to indicate the initial concept, not only as a toponym – the city of *Ur* and Uruk – but also its original meaning and its sense of 'origin', *ur*-bs, *ur*-ban, *ur*-banise and through the Latin *urĕre* it also maintains the sensory reference to something that burns: *ur-tica* (nettle), *ur*-ine, *ur*-gent. Semerano cites the common etymon which derives *ur*-be from the term "*aratrum*" (*urbō /urvō, - ās, -āre*, to plough), as in the case of Romulus ploughing the future boundaries of his city, and then adds:

> "But the formation urbs was shaped long ago and corresponds to the Akkadian *urbu* (to enter, to go home) and to the Sumero-Akkadian *ūru* (city, *Stadt*)."

The Sumerian cuneiform sign after *Ur*-bs in the title of this chapter designates the Sumerian word for "city", which is also pronounced ŪRU. The city of *Ur*, and before it *Uruk*, was an enormous furnace, *fUrnus* in Latin, tan*Ur* in Hebrew, in Akkadian tin*Uru*, which constantly "burned" with the baking of bricks, hence a kind of synecdoche/ metonymy which discloses, still contains, the seeds of its own creation. A fire which indicates work, activity, energy, goods, products, ingenuity. All the stuff you need before there is a market. In *Genesis* 11:2-9:

> And it came to pass, as they journeyed from the east, that they found a plain in the land of Shinar; and they dwelt there. And they said one to another, Go to, let us make brick, and burn them thoroughly. And they had brick for stone, and slime had they for mortar. And they said, Go to, let us build us a city and a tower, whose top may reach unto heaven; and let us make us a name, lest we be scattered abroad upon the face of the whole earth. And the Lord came down to see the city and the tower, which the children of men builded. And the Lord said, Behold, the people is one, and they have all one language; and this they begin to do: and now nothing will be restrained from them, which they have imagined to do. Go to, let us go down, and there confound their language, that they may not understand one another's speech. So the Lord scattered them

abroad from thence upon the face of all the earth: and they left off to build the city. Therefore is the name of it called Babel; because the Lord did there confound the language of all the earth: and from thence did the Lord scatter them abroad upon the face of all the earth.

This passage recalls various Sumerian and Akkadian accounts in which a god, sometimes Enlil, sometimes An, perhaps a little envious or a bit oversensitive, unable to sleep because mankind have become too noisy (noise often refers to the busy enterprise of the human race, but even more often to their sexual and reproductive activity), gets angry and sends plagues almost identical to those of Egypt: floods, destruction and death, to get humans out from under his feet. "Thank God" there was still Ea-Enki, the famous infernal deity of Jeremias, ready to patch things up (see Bottéro, the Poem of the Great Sage).

Sennaar is Shine'ar, is Ur, that is, Sumer; and the episode is about the Tower of *Babel*, which shouldn't be Babel's, but Ur's or Uruk's or Lagash's or Kish's, or some other city belonging to the Sumerians (G. Steiner, p. 41, believes, though I don't understand why, that "Tower" is a term of convenience since the Hebrew text uses the term Migdal, referring not to a "Tower" but to "Large, enormous object": a gigantic idol, in his opinion. In E. Klein's Etymological Dictionary, *Migdal* is a term derived from the Ugaritic *mgdl* and the Aramaic *Migdalà*, also present in Arabic as *Mijdal*, and means Tower, Turrett, Pulpit and, sometimes in post-Biblical Hebrew a hanging object. Going back to E. Klein, *gdal* would be a contraction of Much/ Large/ Extended, from which comes the Hebrew *gadol*, large, at least partly justifying Steiner, although erecting a Large Construction which "*ve roshò bashamaim*," "whose top may reach unto heaven," is not like building an underground car park. And I really don't see where idols come into it. Indeed, a Ziggurat is not strictly speaking a Tower-Tower, even though it's made of many-many-many bricks, and is a large, enormous object trying to "reach unto heaven". Furthermore, it's not clear to me why he explains the blasphemy connected to the building of the Tower as being indicated by the verb "to make", used for the Tower, and the expression "divine creation" given that in the Biblical text the root of Create, the action of God is *Barà*, when the verb for constructing the Tower is *Banah*, Build, from the Akkadian, *banum*. Both these verbs, irrespective of any connection to God, take their origin from 'Generating a child'. This element remains perfectly preserved in the roots of *Bar*, e.g. in *bar mitzvah*, *Ben*, and *Ibn*).

It is also interesting that in German, besides being present in the term

Urin (urine), for example, the prefix *Ur-* still indicates something ancestral, ancient, primordial: e.g. *Ursprung*, origin, and *Ursprache*, primordial language. Consulting the Pennsylvania Sumerian Dictionary we read that, while dejection in the sense of defecate is SUH in Sumerian, dejection in general is U.RA; and specifically 'to urinate' is rendered in Sumerian by A.SUR, literally "the movement of water", and also by KAŠ.SUR. Beer is KAŠ, and we know the effect produced by that drink (which is a Sumerian pun since, despite being pronounced the same way, *Kaš*, beer and urine are written with different cuneiform characters). I won't venture into the concept of Dejection proposed by Heidegger, since I think his position has been well enough illustrated by the passage quoted earlier without the need for further comment. *Ur* means fire, heat, burn, shine, glow, light. It is probably in the light of these fires, these furnaces, these ovens, that Zeus (him again) he of the "wise counsels", notices what Prometheus has perpetrated. Ur, however, was at work some years before Hesiod's report on Prometheus. And who can say whether the pontifical benediction *Urbi et Orbi* might not serve to indicate that the Orb[4] of land and sea is that collection of lights, the collection of many *Urs* which, Zeus, or Ba'al, An, or Marduk could see when they looked down on earth with the same view as an astronaut in a spacecraft today. It is still the case that, in Hebrew, light is *Or* and is written with the same letters as *Ur*, simply positioning a little dot slightly differently on the letter *Vav*, computer permitting (city, in Sum. is IRI, and in Hebr. עיר, ir). Something similar happens in Italian with the close similarity between *Or*ina and the Latin *Ur*ina, and *Or*tica and the Latin *Ur*tica. And the term *Or*igin has undergone the same fate (see Pianigiani). And De Gubernatis (1899, p. 114) notes that gold is called a-*ur*-um because it shines brightly, echoing the term a-*ur*-ora (dawn); adding that in Vedic the Sun is *sûr*, shining; in akk. *ḫurāṣum*.

At the risk of pedantically stating the obvious, I will again stress the fact that, to keep a kiln or foundry working, there is a need for professionals with highly differentiated technical skills to manage the provision of raw material and fuel, the actual burning, and the distribution/ warehousing/ marketing of the finished products. In other words, there is already a division of labour in the era of the Flood (and if we take dear old Tubal-

---

4   Isidore of Seville would not endorse this etymology, because according to him "the orb was called thus in reference to the roundness of its circumference, as being like a wheel. For this reason, a wheel of small dimensions is called *orbiculus*" (Isidore, II, XIV, ii, I); for the *Dizionario Etimologico della Lingua Italiana* the etymology of orb, while connected to Round, is unknown. De Gubernatis (1899) seems to agree with Isidore.

Cain seriously as the Biblical inventor of metalwork, *Gen.* 4:22, "And Zillah, she also bare Tubal-Cain, an instructor of every artificer in brass and iron," I would have to say before the Flood too), and at this point it becomes hard to see how anyone can seriously maintain that the division of labour was a feature introduced by and characteristic of capitalism, unless they're a close relative of Max Weber, or live in the vicinity of Trier (and I'm obviously not referring to St Ambrose).

The City, *Ur*, in akk. *ālum*, but before that, *Ur*uk, is a highly efficient and very extensive organization centred on the Temple of Inanna, the Eanna, and her Priests. It is the priests who, through their officials and scribes, manage the accounts, recording and archiving them; oversee the planning and digging of the canals which comprised such an immense cultural and hydro-geological resource. They take responsibility for the complex logistical aspects of warlike and commercial activities and the whole organization dedicated to the maintenance and deployment of the countless men who do *corvées* and have to be fed, billeted, and supported during their work in the properties of the Sumerian Temple and Empire (and often there were also salaried workers needing to be paid for their work: see Bulgarelli, 2001, pp. 172-75, 189-94; and compare the biographical references in Bulgarelli, 2009). Priests who lend with interest, who make use of fully established exchange mechanisms for evaluations and for goods; who are capable of carrying out cross-border operations, as we would say today; who have properties, flocks, and estates at their disposal and manage them according to absolutely "modern" commercial criteria. It is the priests who invest the sovereign with his Royalty. As they always do, it must be said. All this is made possible by the power of the Temple. The power of Faith and of the economy. And, to sum up very briefly, is made possible by a "first" enormous Sacrifice, or Gift, or collective Oblation: those "voluntary" *corvées* of the faithful *awīlu*.

12.
UR-NAMMU

A few lines earlier, in my list of the city's characteristics, I wrote that there must be "a widespread and certified quantitative system of measures which also includes the qualities of the materials to be measured and the rates of exchange."

It seems easy. As I will clarify later on, in the first universal Empire created by Sumer a single word which defined the universal measure of weight, for example, the *Siqlum*, corresponded to at least four different units of weight, varying from region to region.

There can't be anything astonishing about this for Italians (and for all Nations of the world at least), and in Italy, at least until Unification and in fact for a long time afterwards, and in some cases still today, extremely diverse units of measurement existed in the various statelets which comprised the boot: the *biolca*, the hectare, the *piò*, the acre, to give just a sample; and the Venetian Ducat was different from the Florentine Ducat, just as it was from those of Germany, Scotland, Sicily, or Austria.

Having founded the first Empire in history, the Sumerians didn't find themselves in a different situation, only that the dimensions were bigger, much bigger. And so, for a long time, they put up with different local measures and aimed at unifying them step by step.

A little like us today, trying to unify the decimal system of lengths with the one based on inches.

But even the decimal system of lengths wasn't created overnight: although, one day in Paris, in 1875 to be precise, it was laid down how long a metre should be, how heavy a kilogram should be, how much liquid a litre should contain, and so on.

Why am I discussing this? Because after at least a thousand years following the Flood, during which, instead of exchanging goats and cabbages, people frenetically weighed and measured everything in minute detail, an extremely precise but absolutely fragmented and compartmentalised activity, the First Universal Standard was finally established.

Thanks to Ur-Nammu. Ur-Nammu is, in fact, the founder of the last Sumerian dynasty. The last dynasty and final gleam of Sumerian greatness before the advent of Babylon; and he too was the subject of poems, prayers, hymns, and legends, the protagonist of epic tales.

And his achievements and innovations were collected and recollected, but also elaborated, by Hammurabi 250 years later. And even for Ur-Nammu who, like other legendary sovereigns, will become the protagonist of epic tales and songs, things didn't happen overnight or pop out of some illusionist's top hat: as I have already pointed out, the goddess Nanshe of Lagash, and Sargon and Manishtushu, had all in turn, some centuries before, laid down the basis for *Fair* and *Just* weights. And Noah too, or someone on his behalf, "before" them all.

I will only quote the prologue to the codex of Ur-Nammu (who in *The Electronic Text Corpus of Sumerian Literature* is called Ur-Namma):

> The prologue, typical of Mesopotamian law codes, invokes the deities for Ur-Namma's kingship and decrees "equity in the land". "After An and Enlil had turned over the Kingship of Ur to Nanna, at that time did Ur-Namma, son born of Ninsun, for his beloved mother who bore him, in accordance with his principles of equity and truth (...) Then did Ur-Namma the mighty warrior, king of Ur, king of Sumer and Akkad, by the might of Nanna, lord of the city, and in accordance with the true word of Utu, establish equity in the land; he banished malediction, violence and strife, and set the monthly Temple expenses at 90 gur of barley, 30 sheep, and 30 sila of butter. *He fashioned the bronze sila-measure, standardized the one-mina weight, and standardized the stone weight of a shekel of silver in relation to one mina* (...) The orphan was not delivered up to the rich man; the widow was not delivered up to the mighty man; the man of one shekel was not delivered up to the man of one mina." [Nanna is the god who will later be called Sin, the Moon-god]) (my italics).

The consequence of establishing the standard to be respected, the same one for all, the *just* weight, is very clear:

> The orphan was/will be no longer prey to the rich man, and the widow was/will be no longer prey to the mighty man and the man who is worth one shekel was/will be no longer prey to the man who is worth a mina [!!!].

Moreover, the codex of Ur-Nammu marks the beginning, or rather the 'endorsement' – given that when something takes shape it is because the process has begun earlier (see the goddess Nanshe and her predecessors) – of a move away from the Law of the Talion. This law obviously did not disappear, but new elements were introduced, specifically monetary compensation, money which may not yet have been in the form of the Coin, perhaps, but was certainly coming close to it: here a "silver shekel" or "silver mina" are still units of weight for a certain material; perhaps. This is not the end of the *Bulla* but it isn't yet a Coin, perhaps. My "perhaps" is a joke, since all economists and all numismatists maintain that it's not a Coin. And they all keep repeating in chorus the story of Lydia as the beginning of coinage. But standardising the weight in stone which a silver shekel must have and comparing it to the mina which must weigh 60 times more and then to the Talent which must be exactly 60 mina: what is this if not a monetary system? With the codex of Ur-Nammu, the fair and just coin becomes the basis for resolving, without "bloodshed", the controversies, the injuries suffered, the conflicts of everyday life. Money/ Coin, which begins to replace Blood, becomes the bearer of a Fair and Just "pacification", insofar as these concepts are meaningful in "human" and "historical" terms. Psychologically, we have now left behind Predation and Counter-Predation as the only way to resolve disputes, although such a procedure obviously continues to exist in other spheres, as in the case of the institution of the Ordalia, which is not a Mediaeval Christian invention and continues, unacknowledged, to cut across modern history. Barter pure and simple is left behind and new, symbolic elements appear which have an accepted and recognized value in the fact that they "change hands" and avoid, or tend to avoid, injustices, vendettas, feuds, deaths, and blood. Like it or not, money is this too: it is neither forgiveness nor revenge, but a solution, peace.

With the codex of Ur-Nammu, money assumes the value of a "Counter-Gift" which passes, as in Barter, through a "weighing up" of the relative desire for and cost of a thing, and is transformed into a material element which does not dissolve all that is connected to the injury received but certifies it and attests to the "wrong" which has occurred and the necessity/obligation to make reparation in a measure proportionate to the injury itself. It re-establishes a sort of equilibrium, of peace indeed.

The other thing which I regard as very important is that fixing a unit of measurement does in fact mean only that: fixing a unit of measurement.

Let me explain: in Paris in 1975 it was decided that a cubic decimetre of water should weigh 1 kg. Nothing else. It was not laid down how much 1 kg of water should cost.

It was fixed that this kilogram should also be *equi*-valent to a litre of water and that this litre should be that, and only that, measure of capacity: exactly 1 litre = 1 dm3 = 1 kg; nothing else (admittedly, there a couple of minor additional details such as the temperature and purity of the water and the atmospheric pressure, but they aren't of great importance in this discussion).

It doesn't seem much, but it is the first in a series of incredible steps.

"Only" the "just" has been fixed; this has fixed the *Pacĕre* and the *Pangĕre*, a stable point.

It is a "scientific" barter, an *equi*-valence. (I shan't bother to comment on the recent observations by Felix Martin who, besides "forgetting" the Austrian school, believes that Paris 1875 was the first attempt at creating a single standard.)

That we can still today be debating the fact that a slave is worth less than a free man or whether a woman is worth as much as or less than a sheep, does not change the general sense of the matter.

13.
BERNHARD LAUM

This chapter was to a very large extent made possible by the passionate and intelligent work done over the years by Nicola Parise who, though I have not had the pleasure to meet him in person, I sincerely thank for his intelligence, skill, and tenacity because I don't think it can have been easy to bring together so many scholars and so much highly interesting material at the Italian Institute of Numismatics; Parise's work has enabled me to approach the theses of Bernhard Laum in a stimulating and complex setting; and as with other writers, I can only apologize if I have sometimes "mistranslated/ betrayed" some of his own theses.

From antediluvian times, before Noah, Enmerkar, and Gilgamesh, alongside the highly refined system of comparative weights by which different goods were "exchanged", we find another, older criterion for establishing the Value of something, for trade or otherwise: this is the "meaning" of a thing.

The axe of Gilgamesh – just that hero's axe and not all axes or hoes in general – which in the Epic weighs seven talents (214.2 kg), is important and has value because it belongs to Gilgamesh himself, because he is exceptional, because of the epic adventures he has undergone, the dangers he has overcome, the legendary aura around him, and not because of the axe's craftsmanship or the type of material used to make it; not because of the *societally necessary labour* [1] that went into its construction.

It is in this context that Bernhard Laum maintains that the "initial" concept of Value is the "signficance" which is attributed to a certain object.

Laum is not the first to make this observation, but I think he may be the first to develop it in an original way. The Value of a certain object is given

---

[1] On the subject of societally necessary Labour – and we will see this even more clearly with respect to gold – I feel in tune with the words of Richard Whately, Anglican Archbishop of Dublin (1730-1797) (in Vaught, 1978), who surely read Swift and who was writing only a little while before Marx: "It is not that pearls fetch a high price because men have dived for them; but on the contrary, men dive for them because they fetch a high price."

by its Value to the one who has borne, used, touched it; that is, the Value is given by the importance of the subject who has possessed or used it, by the importance of the undertaking in which it has been engaged; an epic enterprise or something equally important for the community in which the event occurred.

In other words, for a range of scholars, an object has Value as a Trophy, Memento, Impression, or Relic of something or someone who had Value, and it therefore continues to possess in itself something of the origin from which it emerged. And it often also has value as a talisman.

Through a kind of transitive property in some respects similar to the concept of *Mana* reported by Mauss, although he applies this more to the Gift, the value of the action performed, the valour of the Hero, passes to what belonged to the Hero; it recalls him, represents him, remembers him.

It is a Relic, a Symbol of him; his presence in absence.

In the same way, for Laum the Sacred is transferred, putting Girard aside for a moment, from the victim to the objects used to officiate/ celebrate the Sacredness of the Sacrifice.

In this "transitive property" there is an obvious proximity to and possibly coupling with magical aspects. But this condition, fraught with ambivalence, with non-saturation, also determines the proximity to sacred aspects; sacred aspects from which Laum derives the value which the *Obelos*, ὀβελός, spit/skewer, takes on as an instrument of worship and sacrifice.

Nicola Parise (2000, p. 45) makes very interesting observations about this, albeit in a "Greek" context:

> It was in the nature of things that religious mediation would affirm the ox as a measure of value. In a society which would have known new political forms only with the definition of common centres of authority within its own city walls, around the temples of the gods (Odyssey, VI, 7-10), and as an expression of magnificence would have favoured the gift to the gods over the gift given between men, it could only be the cult which gave form to the first abstract expressions of value.

Compare also the impressive passage in *Exodus* 29: 1-37 in which Moses sets out the procedures to be followed for performing the Sacrifice, emphasizing that:

> Whatsoever toucheth the altar shall be holy.

The *Obeloi*, the spits and skewers, are objects used in the sacrifice of the victim and in cooking its flesh, which was skewered and roasted on the *Obeloi*, then distributed and eaten. Always following payment of an equally symbolic sum, which would of course be the Obolos, ὀβολός. And it seems that Obelisk also derives from Obelos. The flesh of the sacrificial and sacrificed victim and the tools connected with it become sacred and sacralised after the rituals, enjoying a religious and sacral value precisely because of being used in a sacred way and for a sacred purpose.

Today we could give different names to these things: for example, Introjection of the victim, or just eating it (the Roman Canon and *Luke* 22:19; lo and behold, transubstantiation!):

> And when the hour was come, he sat down, and the twelve apostles with him. And he said unto them, With desire I have desired to eat this passover with you before I suffer: For I say unto you, I will not any more eat thereof, until it be fulfilled in the kingdom of God. And he took the cup, and gave thanks, and said, Take this, and divide it among yourselves: For I say unto you, I will not drink of the fruit of the vine, until the kingdom of God shall come. And he took bread, and gave thanks, and brake it, and gave unto them, saying, This is my body which is given for you: this do in remembrance of me. Likewise also the cup after supper, saying, This cup is the new testament in my blood, which is shed for you. (Luke 22: 14-19)

Or we could call it Identification with the victim (taking on its characteristics), or Incorporation; all this would apply to the flesh of the sacrificed "animal" or pieces of it.

Laum maintains that, as in *Exodus*, the sacredness of the sacrificial victim extends to everything that has contributed to the ritual: i.e., the instruments with which the sacrifice has been carried out.

In fact, many of the first so-called "pre-coins" were made in the form of a knife, or of axes, both single and double-headed, the latter being the *Pelekys*, πέλεκυς (again, see Parise about this). And Hoes. Hoes which are also useful for tilling the soil with the meanings of fertility and work which are connected to it, but hoes which are, for their part, much like axes, though smaller, with the blade horizontal instead of vertical. From Pettinato (2001, p. 417):

> The golden hoe, with the head of lapis lazuli,
> held in place by clasps of gold and silver
> whose blade seems like a ploughshare of lapis lazuli
> and its point a solitary unicorn on a vast plane.

And as for the Sacrificial Victim, we cannot rule out the possibility that the Sacrifice of the Firstborn, as I noted earlier, also corresponds to the eating of their flesh, something which will later constitute an abomination in the eyes of the Lord but, as in other cases, seems to persist, albeit in exceptional circumstances, as in Ezekiel 39: 18-20:

> Ye shall eat the flesh of the mighty… and drink blood till ye be drunken, of my sacrifice which I have sacrificed for you. Thus shall ye be filled at my table with horses and chariots, with mighty men, and with all men of war, saith the Lord God.

This passage could be interpreted as purely metaphorical, but given the sacrificial context I am writing about, it is not unlikely that the rage and bestial joy of the conquerors could be manifested as a thoroughgoing cannibalistic ritual, with all the significations which such religious-magical-apotropaic-identificatory behaviour can assume.

On the subject of identification: a bull's horns are given pride of place on many warrior's heads, and the fact that they are present in cultures which vary widely in place and time has nothing to do with the family problems of the warriors themselves, often having to fight so far from home, but relate to the fact that the Bull, El, an animal which represents vigour and strength, becomes the sacrificial victim *par excellence* all over the world, even in different periods of history. By "putting its horns on one's head" one takes on its characteristics and "becomes a bull". In this context, it is no surprise that horns are also considered a symbol of Divinity (Seminara, 2006, p. 83). I think the "madness" of Nebuchadnezzar in *Daniel* 4: 32-33 can be interpreted along these lines:

> "they shall make thee to eat grass as oxen…. And he was driven from men, and did eat grass as oxen and his body was wet with the dew of heaven, till his hairs were grown like eagles' feathers, and his nails like birds' claws.

The Italian term *Bue* – *Bos, bovis* in Latin, in English Ox, Bullock, in Hebrew *Shor* – is bit difficult to grasp for various reasons: the first is that the bullock is a castrated bull and, according to the law of Noah, pieces of flesh cannot be taken from a living animal. However, this makes it hard to grasp how a bull could be used for ploughing or transportation in a predominantly agricultural milieu like the Bronze Age. There are various references in the Old Testament to *buoi selvatici*, Wild oxen, *Shor ha bar*, but we have the same problem here too, in that it is not easy to imagine a reproductive activity that would meet society's needs given the lack of the necessary attributes. Nor can we understand why Nebuchadnezzar becomes an Ox, a castrated Bull, unless this perhaps refers to a "suspension" of his power to command. Or else he is dedicated only to the exercise of command, suspending all the rest.

More simply, I believe that the term *Bue, Bue Selvatico*, Ox, Bull, Cattle or Calf, in this and similar cases, still just means Bull.

Both in Sumerian and in Akkadian the same term is used for Ox and Bull, though they are obviously different in the two languages: Bull is GUD in Sumerian, but also Ox, and GUDAM is Wild bull, while in Akkadian Bull/Ox is rendered by *Alpu, Li'um*, or *Miru*, and the wild bull is *Rimu*. Here too the distinction seems to be between the domesticated and undomesticated animal, but this doesn't solve the problem I posed earlier.

In Hebrew, Bull is *Par*, (jokingly, I asked Giovanna, *hamorah shelì*, how on earth the word for Butterfly comes to be *Par-Par* – i.e. Bull-Bull, but we can't work out the reason). As a sign of the Zodiac, however, Taurus is *Mazal Shor*. In fact, all the signs of the Zodiac have *Mazal* before the name of the constellation in Hebrew, but it is still odd that the sign Taurus is not called *Mazal Par*, but *Mazal Shor*. Given that *Mazal* (also) means Fortune, would it be possible to translate *Mazal Shor* "creatively" as a Bull understood as a *Fortunate Ox*?

Who knows whether the animals in question were the Aurochs, the *Bos taurus primigenius* of Linnaeus? Bottéro used this term in his translations.

The terminology used by Linnaeus may however help me to understand that *Bos* and *Taurus* are not necessarily alternative terms.

And in Hebraic mythology there is the figure of an enormous bull, indeed a pair of them, so large that they cannot get into the Ark: this is the *Reem*, very close to the Akkadian *Rimu*.

Julius Caesar too, in *De Bello Gallico* (6, 28), after writing about elks, describes these animals, the *Uri* (Aurochs), now extinct:

> A third species consists of the *uri* so-called. In size these are somewhat smaller than elephants; in appearance, colour, and shape they are as bulls. Great is their strength and great their speed, and they spare neither man nor beast once sighted. These the Germans slay zealously, by taking them in pits; by such work the young men harden themselves and by this kind of hunting train themselves, and those who have slain most of them bring the horns with them to a public place for a testimony thereof, and win great renown. But even if they are caught very young, the animals cannot be tamed or accustomed to human beings. In bulk, shape, and appearance their horns are very different from the horns of our own oxen. The natives collect them zealously and encase the edges with silver, and then at their grandest banquets use them as drinking-cups.

And the *Bull* is Sargon the Great who conquers the world as far as Lebanon and Cyprus, and it is no accident that some authors liken him to the founder of Minoan civilisation, Minos, who, it should be noted had a *Bull* for a "son", the Mino(s)taur. This was because he did not want to sacrifice the most beautiful Bull that had ever been seen, a gift from Poseidon, and Poseidon took his revenge by making Minos's wife, Pasiphaë[2], fall in love with that very Bull. Hidden in a wooden heifer constructed by Daedalus, Pasiphaë became pregnant by Poseidon's Bull and give birth to the Minotaur. (And where is the monstrousness here? In Pasiphaë's "unnatural" desire, even though it was induced in her by Poseidon? Or in the non-sacrifice to Poseidon? Or is it the features of the Minotaur itself, maybe an ancient expression of a quasi-Oedipus who, at the dawn of Greek civilisation, still cannot defeat the "Father" but nevertheless becomes a problem? Soon afterwards, it will likewise be Icarus, and not Daedalus, who succumbs. Or does it lie in the reiteration of the fate which befell Minos's mother, Europa, seduced by Jupiter who turned himself into a Bull in order to get close to her and seduce her?) And the Bull is Hammurabi and all the Kings of Babylon before and after him. Bulls, formidable *Aurochs*. The goddess Inanna frequently calls Gilgamesh "my Bull", desiring union with him and being ignominiously rebuffed; and the Heavenly Bull will be the sacred animal sent by the gods to punish mankind, and specifically the arrogance of Gilgamesh and his scornful rejection of a union with Inanna, a union which would have entailed a self-sacrifice, being himself the sacrificial object, as Gilgamesh furiously points out. The Heavenly Bull will be defeated in an epic encounter by that very Bull-Gilgamesh:

> Enkidu confronted the heavenly Bull, and took it by its thick tail
> Enkidu held it with his firm hands
> Gilgamesh, like a heroic butcher,
> struck the heavenly Bull with a firm secure hand:
> he plunged his sword between its horns and the tendons
> of its neck.

---

2   Pasiphaë is cited by Dante, *Inferno*, XII, 11-13: "*e 'n su la punta de la rotta lacca/ l'infamïa di Creti era distesa/che fu concetta ne la falsa vacca*" ["and on the edge of the chasm was stretched out the infamy of Crete, conceived in the false cow"]; *Purgatorio* XXVI, 41-42, "*e l'altra: 'Ne la vacca entra Pasife/perché il torello a sua lussuria corra'*" ["and the others called out, 'Into the cow goes Pasiphaë, so that the bull may run to her lust'"]; *Purgatorio* XXVI, 86-87, *"il nome di colei/ che s'imbestiò ne le 'mbestiate schegge"* ["the name of her who bestialised herself in the bestial wood"].

The Bull is *Apis*, an Egyptian divinity representing sexual potency and creativity. The Bull is *Jupiter* who seduces *Europa*. The Bull is the animal of *Mithras*. And *Shiva* goes forth, riding a Bull, Nandi. And in far-off times *El* is the name of the "great bull-god *El*," as I noted earlier. And El remains the prefix of a great many names of God (*El*-ohim, *El*-yion, *El*-oah...).

Just as Hector's and Achilles' armour continue to maintain their value as having belonged to Men of Valour, so the skewers and spits for the sacrifices and sacred rituals maintain the Value of the sacredness of the sacrificial rite for which they were instruments.

Therefore, it is not only the flesh of the victim consumed, eaten, offered, burned (and paid for) which is sacred but, according to Laum, also whatever is connected to the rite itself, and above all, the Obelos.

Laum is definitely in good company, since, as I noted a little while ago, in Exodus 29: 37 we find, "Whatsoever toucheth the altar shall be holy."

And Laum ups the ante even further:

> I therefore affirm that in ancient Greece only sacred instruments could become and did become money (in Montepaone, 1997, p. 79).

Leaving aside the fact that Laum "is mistaken" – though he's in happy and numerous company – in making this process start with the sacred rites of Greece in the 7th-6th centuries BCE, a period in which, according to the "classical" version, the first coins appeared, I nevertheless think that the line his reasoning follows is exactly right and very interesting.

Numismatists, anthropologists, and economists have long debated these topics, taking as read the cultural horizon in which they have made their various reflections, one limited to ancient and classical Greece; and Laum, a child of his time, is no exception. Scholars have often gone as far back as Lydia, but considering it a kind of cousin-province to Greece, without ever reflecting that Lydia could have been influenced by the Mesopotamian world long before adopting Athenian, or Spartan, or Persian garb. The caterpillar and the butterfly all over again. And there have long been divided judgements and evaluations about the scope of the first coins, always considering the first to have been Grecian or Lydian – which portrayed objects or animals, often sacrificed and hence a "totem" of the *Polis*, often representatives of the city's tutelary deities – but without bearing in mind (or so the official studies have led me to think) that the "Pelasgians" were not yet born and Greece was only a remote, primitive, and desolate moorland when Gilgamesh went to *Kur*, Lebanon, the land of the Cedars, or headed off towards *Dilmun*, a happy isle, maybe in the Persian Gulf, or plunged into the *Abzu*, the Eritrean sea and the abyss in search of the plant of youth. The *Abzu* is not Hell or the Land of No Return, as suggested by Jeremias and as I have already noted, but is the Abyss which Gilgamesh must cross to reach the Innkeeper *Siduri*, heading in the direction of Venus.

I have already mentioned the introduction of Weights which, even before Enmerkar's time, before the Flood, become the rule of exchange, although initially with all the complicated apparatus this required, and they introduce, or arise from, the concept of Justice and Equity, a requirement evidently regarded as an important function in the management of conflicts.

All of which was two thousand years before the Greek *Obelos*.

Two thousand years: not much from a geological viewpoint but it is the same distance as separates us and our nuclear reactors, internal combustion engines, computers, cell phones, and my lovely aged scooter Vespa Primavera ET3, from Octavius Augustus Caesar!

Or, if this example seems too celebratory of modernity, we could remind ourselves that over two thousand years we have lost much of the moral rigour and sense of responsibility with which the first Roman tribes founded Rome itself (see Mommsen, 1856). Or else, and here I am miming the words of Bruno Leoni, in two thousand years we have lost the profound good sense by which the Roman Senate refrained – and thus maintained for millennia the structure of Roman justice which continued with Byzantium – from kicking the shins of the Common Law which regulated the lives of individuals, leaving the *Traditio* to govern the *Mos* rather than a plethora of intrusive laws which would increasingly circumscribe personal liberty, as we have constantly seen in our own time, an intrusiveness perpetrated with the most fantastical justifications and highly obscure objectives. The observations which von Mises makes about Bureaucracy (1944) are of great interest in this regard.

A lot has happened during this arc of time: caterpillars have been born and a great many butterflies have taken flight. Millions, billions, billions of billions of times.

From a shamelessly Eurocentric viewpoint, there is much of interest not only in Semerano's reflections, but also those of Martin Bernal in *Atena nera* (1987).

But there is another question. Can we say, with relative certainty, or at least some kind of hint, that something changed two thousand years before the *Obelos*?

Was there a moment of transition which might corroborate this hypothesis?

I take my cue not from a historical source but a literary one. For historical and metrological observations, I recommend instead the very interesting article by Bobokhyan (2006); I have also been stimulated by many of Bulgarelli's observations (2001, 2009).

My discussion starts from the comparison which must be made between two poems both of which are about Gilgamesh, arming himself in preparation for two epic battles which await him: first, the preparation/arming of Gilgamesh in a Sumerian version and then in the Palaeo-babylonian version. Comparing the two texts, we notice that something important has changed.

There may be a thousand years, more or less, between the two poems. Maybe more, maybe less. What seems important to me is that we can observe certain things BEFORE and that the same things are differently presented AFTER.

If there were only one year between the two compositions, the difference they present would still unequivocally show that, whatever the lapse of time between BEFORE and AFTER, some change has occurred.

I like to think that the ancient Sumerian version could still have been "in fashion" at the time of Ur-Nammu, the last Sumerian dynasty, and the Babylonian version may be contemporary with Hammurabi, the founder of the first Babylonian dynasty. So I imagine this change, like other changes I will look at later, as taking place in the 250 years which separate the two sovereigns.

The structure of the poems, the dialogues, and the heroes' endeavours against powerful enemies are very similar both in terms of their testing of human limitations and of the acts of which the heroes are capable. Naturally, there are also a great many differences, as Pettinato observes.

Between the Sumerian period (3000-1890/1830 BCE) and the Palaeo-Babylonian (about 1890/1830-1500 BCE), the periods when the two poems were written, together with many differences of content, style, composition, and metre which I am not in a position to explore in depth, there are some which have struck me. One in particular.

This is how Pettinato (2004) presents a Sumerian text, *Gilgamesh and the Heavenly Bull* (my italics):

> To defeat the bull...
> Gilgamesh, to defeat the bull...
> put the armour of fifty mina [on his flanks]
> girt around him *his sword of seven talents and thirty mina*
> [grasped] his [battle] axe [of seven talents]...

This is Pettinato, 2004, transcribing a Palaeo-Babylonian text, *Gilgamesh and Humbaba*, Yale tablet, my italics):

> They went straight to the armorers.
> The craftsmen sat and pondered.
> [Then] they forged great cleavers
> they forged axes of three talents each,
> they forged great swords
> whose sheaths were of three talents each,
> javelins of thirty mina for their thighs;
> *the blade of the swords was thirty gold mina.*
> Gilgamesh and Enkidu each carried
> ten talents of arms.

In the Sumerian poem there is a description of Gilgamesh's weapons which considers their weight, in talents: 15 talents and 20 mina; of bronze, a suitable material for battle.

The breastplate weighs 25 kg; the sword 229.5 kg; the axe 214.2 kg.

In the Palaeo-Babylonian poem the sword is thirty mina *of gold*.

We are still in the Bronze Age, and it seems clear that a sword MADE of gold is not very suitable for fighting the monster Humbaba, or Lilith, or Agga, King of Kish, or the Heavenly Bull. So we need to think that the differences between the two descriptions have some other meaning. Something which goes beyond the forging of the material used by the craftsmen to prepare the weapons.

In the Sumerian poem, Gilgamesh's sword WEIGHS seven talents and thirty mina (229.5 kg) and in total he carries an overall WEIGHT of 15 talents and 20 mina of weapons: that is 469 kg, presumably in bronze.

In this way the poem underlines the exceptional strength with which Gilgamesh is able to brandish the sword and all the rest. It suggests the Strength, Power, and Vigour of Gilgamesh; his heroic Valour. And his weapons acquire the Value, with which he, the Valiant One, endows them, as I pointed out earlier.

By contrast, in the Palaeo-Babylonian poem has a VALUE of 30 gold mina, or at least 15 if we consider this the total for both Heroes. And each of the two heroes, Gilgamesh and Enkidu, is wearing the equivalent of 10 gold talents: i.e. armour with a VALUE of 10 gold talents, or 306 kg of gold. This still needs a lot of strength but it is clearly something different.

In the time which passes between the Sumerian and Palaeo-Babylonian poems, the terms of reference have changed.

Gold, Sum. KUG.SIG, Akk. *ḫuraṣum*, and silver, Sum. KUG.BABBAR,

Akk. *kaspum*, have become the paradigm to which the poem refers, the unit of measurement to which all weights and objects – i.e. *merx* – must refer.

It is no longer necessary to express a standard for each item that is exchanged and to compile endless lists of equivalences between the myriad different kinds of goods. All of this can be mentally transformed, straightaway, into something else, something very special: it can be expressed in gold (and silver).

Transformed, changed in form, and hence exchanged; not transubstantiated.

And in that lapse of time, in these fantastic 250 years, in the interval between the late Sumerian and Palaeo-Babylonian versions of the two poems, the oldest known codices of Laws come into being (and they clearly don't just spring up like mushrooms). There are four codices which characterize this transitional epoch: Ur-Nammu's, Lipit-Ishtar's, Hammurabi's, and that of the city of Eshnunna. (Other collections of laws exist, but I am citing these for convenience of explanation.) These are codices which "follow", interpenetrate, and sometimes simply repeat each other; and in these very Codices in which the first solemn declarations of Equity and Justice appear (again Bottéro, 1987 p. 106, in a very simple and modern piece of reasoning, claims that in the Babylonian empire there were essentially two fundamental concepts on which interpersonal relations were based, obviously without idealizing them as unfailingly respected: *Kittu* and *Mêšaru*, which he translates as Equity and Justice) so also appear the Sheqel and the Mina.

In other words, the Coin makes its entrance.

All in this very brief arc of time.

There is also a fifth "codex", though Bottéro does not really accept its definition as a "Codex", that of Manishtushu, which is less a Codex than a deed of acquisition of land, cited by Einzig (1949) but which I was unable to obtain by any means until Dr Bulgarelli kindly sent me a copy (ed. the Chicago Oriental Institute: however, a translated fragment of this deed can be found in Bulgarelli, 2001, p. 118-120). I cite this deed, drawn up by the son of Sargon the Great of Akkad, because, according to P. Einzig, who referred to it for completely different reasons from those which prompt me to introduce this information, in this deed which he believed precedes the codex of Ur-Nammu by 600 years, whereas it is now considered 200 years earlier, prices, transactions, accounts, taxes, and commercial exchanges are already being calculated in silver Sheqels. Thus:

"Some 1.500 years before the supposed invention of coinage by Lydia, silver ingots in Babylonia were stamped in many instances with the image or superscription of the god whose temple is supposed to have guaranteed their fineness" (P. Einzig,p. 240)

It is also interesting that Einzig cites later Hittite codices, from around the fifteenth century, where the base unit in which prices are calculated is the Half-Sheqel (pp. 218-219) and this seems very interesting in relation to what will emerge later, given that after Hammurabi's death the Hittites were already so powerful that they could sack Babylon. This fact, connected with the Hittites' adoption of cuneiform, and the fact that Cappadocia remained an Assyrian colony, and not only in commercial terms, and the fact that Harran is not very far from Cappadocia, gives a highly complex and detailed setting in which to understand the importance of the Half-Sheqel as a monetary unit until the arrival of Sennacherib (see below for N. Parise's observations on the Annals of Sennacherib).

But in this way we have already reached a point between the middle of the third and middle of the second millennia, and so we need to turn back again. To the *Corvées* in the age of Enmerkar.

14.
ADDED VALUE OR ORIGINAL CAPITAL

On the meaning of Added Value, and the definition and critique given to this concept, I refer again to Benedetto Croce (1896) and, as an optional extra, to some observations by Vaughn (1978) about the theory of Value in John Locke.

Whether one wishes to see the strength, the power of the Temple, or of the King-High Priest who governs the City, not the Polis, as the result of hundreds, thousands of Potlatches which I would call "vertical" (I wonder if Mauss would like this term) or as the result of voluntary *corvées* over centuries, in both cases (which would not necessarily be mutually contradictory), the fact remains that in *U*ruk, but also in Lagash and also in *U*r, Ebla, Nipp*ur*, Kish, Eridu, and eventually Ass*ur* and Niniveh, such an accumulation, such a critical mass of riches was created that scholars unanimously believe the Temple became splendid.

Temple construction receives investment while the houses of the worshippers, for example, do not change much (see Liverani, 1998, p. 32: in the 3,000 years before the reign of Enmerkar, the mean extent of a citizen's family home for 8-10 people remained "constant" at around 100m2, and also for a long time after Enmerkar, while "just" in the years of Enmerkar's reign, or rather straight after the Flood, the Temple, having remained fairly stable in size, like private dwellings, grew from 20-30m^2, to 1500-2000m^2; the Royal Palace grew in the same way but, it seems, a "few" years later; my summary); an enormous quantity of riches goes in the direction of the Temple; it is not directly redistributed among families or to the immediate benefit of individuals, those who provide the *corvées*, the "food producers," but it has secondary repercussions which are equally, if not more, important.

Not only does the Temple become splendid, but it has ever greater riches available to it and also functions as a bank, offering loans with interest, and it even has its own clearing houses (the places where modern banks exchange the cheques issued by other institutions); clearing houses not only for cheques written on clay, but also for Goods in all commercial categories; and clearing houses equipped with stables and warehouses, edifices in which a wide range of goods are changed and exchanged, as the British archaeologist Sir Leonard Woolley astutely intuited as far back as the early twentieth century.

As an institution, the temple generates employment, culture, rituals, ceremonies, academies, and astronomical studies which regulate agriculture. All of this reverberates across the community, creating an impact, a principally cultural and cultic return for the society (rituals, studies of geometry and mathematics, writing schools, academies, scribes, technical personnel, surveying, registries, hydraulic operations, factories, immense archives, priestly hierarchies) and only later, and to a lesser extent in terms of investment, will there be an impact-return in terms of economic "redistribution", at least for the "food producers"; while for the "tertiary services," that is the priests, officials, scribes, guards, ritual officers of various kinds, staff for the ordinary and extraordinary management of the Temple, we have a "return", a direct and highly substantial economic "redistribution" (which it would be better simply to call a salary/wage), given that the increase in the Temple's functions and extent requires ever more staff for the Cult and its organization.

Indeed, I believe that this is the beginning, or the first seed, of the phenomenon of inflation, not in terms of money or speculation, or financial wizardry or 'creative accounting', but in the number of staff required for the production of goods relative to the number of staff required for the production of "bureaucracy" and "services", a "simple" problem of proportions and costs which the taxpayers must bear, as Adolph Wagner and Frédéric Bastiat intuited (as the previously quoted Pseudo-Xenophon may also have done, this time teaming up with Aristotle) in order to maintain, as happens today, the frightful plethora of *misthoi* for public duties as in Periclean Athens; and it would also be necessary to consider that Aristotle's antipathy towards money is not only snobbery or reaction, but is justified by the decadence he observes around him and which the moderns, perhaps because they are too like those Ancients, do not take into consideration. See the Athenian Constitution. A great many authors have expressed their views on bureaucracy, from Marx and Weber to Klitsche de la Grange; but not today).

In the light of what I have said so far, in practice the dominion of Enki-Inanna-Enmerkar over the World, is generated by the "riches" which the "voluntary" Sacrifice-Gift-Oblation of the faithful enables the Temple to sequester; and in this Sequestering-Organizing-Redistribution (Mauss's Giving-Receiving-Exchanging, but seen here in a religious light) the

Temple becomes the centre of the community's cultural, commercial, economic and, obviously, religious life.

The faithful taking time out of their employment and Sacrificing-Giving-Offering it to the gods becomes, from an economist's viewpoint, a way of maximising the Temple's profits by discharging, redistributing, spreading the costs onto the nuclear families of those very faithful who are providing the *corvées*. (A bit like what happens in the Amish communities when building granaries, but continued over centuries by thousands of people; or as among charitable volunteers.)

I quote Liverani (1998, p. 34) again, who says, of Enmerkar's city:

> And the Temple was the only institution able to convince the producers to give up substantial amounts of their labour [corvées, not produce) to the benefit of the community and its leaders, in their guise as a divine hypostasis.

The *corvée*. From the Latin *corrogatio* (invitation) and *corrogatum* (called together, gathered together, put together) present in eighth-century French as *corvada*, and by the twelfth century as *corovée*.

Many scholars who have written about *corvées* more or less agree on one fact: if we allow ten people in a family to work on, let's say, a palm grove, or a field, or herds, only one of them is recruited for the period of the time devoted to sowing, reaping or harvesting, or draining canals, as necessary for managing the Temple property or maintaining public works.

The family of those called to the *corvée* would feel the strain, but it was doable for nine people to take on the work normally done by ten. Hence, a partial and tolerable loss of labour for the productivity and needs of the family themselves. True taxes, payable in cash, only appeared much later, with Lipit-Ishtar, a little before Hammurabi.

That there was one less mouth to feed during the precise time needed for the work was at the charge of the Temple; so, no slavery, no forced labour, and no Ben Hur chained to a trireme for years.

This obviously concerns ordinary citizens and, maybe, some *awīlu*.

It is not unlikely that the institution of the *Decima*, tithe (tenth) in English, as an equal "fiscal levy" which remained in force until not so very long ago, derives from this proportion.

In Hebrew, the term is *maaser*, "tenth", i.e. the Decima/tithe.

Jokingly, but only in part, even today the *ḫāwiru* are obliged to pay up, despite not having the protection of the Court; and they pay much, much, much more than a tenth!

Whatever the case, the picture which Biblical tradition and iconography have handed down of *corvées* in Antiquity, typically of Babylonian and Egyptian slavery, seems to have been very different from what those people actually lived through.

Obviously there were slaves and paid workers acquired as a result of military conquests and slaves who were ex-citizens most often reduced to slavery as a result of unpaid debts, though not necessary for the whole of their lives; such bondage often lasted for the length of time needed to work off the debts accrued (the Sabbatical Year, every seventh year, or like the Jubilee, seven times for seven years, i.e. every 49 years, which for convenience became 50 in Christianity, also had the meaning of establishing a limit to "slavery" for debts, and there are numerous references to this in the Bible).

I don't want to go into the merits of Liverani's fascinating 2007 thesis, previously proposed by Winckler, about the historicality or otherwise of the captivity in Egypt, with the system of slavery from which Moses frees his people, which has been handed down to us for centuries (obviously Yerushalmi and many others do not share this hypothesis and their response might be, "If you say so!" as Yerushalmi says, humorously, to a student at the start of his book), a thesis which is in some respects also maintained by von Rad (1962, p. 26) for whom.

Israel was the name of the holy federation of tribes established in Palestine after immigration [that is, after the captivity in Egypt],

but as for the Sumerians – the free citizens of Sumer, that is – things worked a bit differently from what has been passed down to us about the Egypt of the Pharaohs: no whips, no beatings, enough food and clothing, and money for those on wages. Naturally, as I have noted, the institution of slavery was present, but it is not directly concerned with the concept of *corvées*.

Whether it is understood as a calamity or as the fulfilment of divine tasks, labour, specifically that dedicated to *Corvées* for the Temple and/ or Palace, it was an obligation which came directly from the gods. Citing Pettinato (2005, p. 322) again, he translates the Sumero-Akkadian texts directly, referring to at least three literary traditions which came, over the centuries, to form the Sumerian theogony. Here I will only quote a fragment from the school of Eridu, which returns in other versions and on other occasions, albeit slightly modified, to give the general sense:

> Enki, the wise, the intelligent, the aware...
> ...
> when the gods had to provide their food and drink
> with the hard work they were obliged to do...

...
the creator, he who forms all things, sent forth his Sigensigshar [vital spirit]
Enki models his arms and forms his chest;
Enki, the creator, makes his wisdom enter into his creature;
then he says to his mother Nammu:
"My mother, to the creatures which you will bring into existence, assign the *corvée* of the gods as their task."

In different ways and in other traditions, *corvées* were connected to the gods, religion, and faith; and to "sacrificing" one's own time and energy for the gods.

At this point, taking the discussion to the next stage and expressing myself in an economist's terms, it would seem that after centuries and centuries of "unpaid" work, the blessed Surplus, the so-called "original capital", has been established. But this is really not the case.

In reality, at the beginning/in *one* beginning-*Bereshit*, we have Faith which, when channelled, literally moves mountains; and these really are mountains if we think about the dimensions of the ZiqqUrat, the towers of Babel in baked brick. And I once again stress the potency of that "myth" and its social, cultural, economic, and technological implications, as well as its religious ones.

But how we are to define Faith (and Religion)[1] today is a subject I'll leave to the Giants.

---

1   I don't want to close the discussion of Faith/Religion without mentioning some reflections which Bottéro (1987) offers on p. 220 in the chapter entitled 'What is a religion?' which I consider very fine, especially in the light of his personal history: "I prefer, and I don't hide it in the least, to use the evidence in order to move beyond the social and, in the first place, I try to identify the religion in the individuals who practise it: each of them, as is understandable, comes up with their own definition, whatever it may be, according to their original conditions, without it ever losing its fundamental features. But any ontological analysis must stay close to things, to their essence, to what we call the natural order, abstracting from the material and existential achievements and from their unfolding in space and time. We will perhaps have a better understanding of what religion is if we compare it to love. Love is founded above all on the feeling which forcibly takes us towards another individual of our species, in whom we detect, more or less obscurely, a completion and enrichment of our personality and our life. Having moved towards this other person in this way, love prompts us first of all (always, as I said, in terms of the order of nature and not the order of time) to know them better, at whatever level this acquaintance has occurred; then, depending on what links us to them and to the idea we have formed of them, to adopt a behaviour towards them which expresses this idea and the emotions which animate it."

15.
WEALTH, MONEY, COIN, AND GIFT ARE THEY ALL THE SAME THING?

At this point, before I go on, having introduced the theme of wealth – i.e. the enormous flow of Work-Wealth which goes to the Temple – I think it would be useful to at least try and see if it is really necessary to maintain the terminological confusion-superimposition of Wealth, Money, Coin and, in certain respects, Gift.

I have already made some observations about this in what I said on Jeremias and Freud.

Now I will try to make some others, of a slightly different kind.

I must admit, I'm not all that keen on getting involved with bronze axes used as an investment-prestige-symbol of royalty in the Bronze Age (even though axes cut, like knives); nor do I feel much like going into the concept of the "Pre-Monetary", so beloved of numismatists, referring to loaves of copper and Cretan bronze; or making a further investigation of Bernhard Laum's *Obeloi*, from which derive the *Obolos* and the *Obelisk*; or the *lebetes*; or Achilles' braziers or the *Æs Rude*, *Grave* or *Signatum* of the Latins; or the forms, moulded and not coined, anthropomorphic and zoomorphic and all sorts of other shapes, such as shells or hoes, which the "first units of currency" took on and the question of whether or not they possessed a "soul"; i.e. the *Mana* (the Giants go to town on this too).

I would rather simply suggest that the term Money could be connected to the ancient and venerable Sanskrit *Dhana*, wealth/ money. Although the transliteration seems similar, this is different from the concept of gift, *Dāna* in Sanskrit, with the same root as *Dare* (Pianigiani; in English: *to give*), and which in Latin is *Munus* (munificence, remuneration connected to *Munero,-as, -avi,-atum,-are*, but also to *Donum*, and *Dono,-as, Donare* and *Do,-as, Dare* also have the same root.

This will be a coincidence, but *Munus* is also one of the Sumerian terms for Woman.

Fortunately, since it had already been invented, we have a Sanskrit term for "coin", *Dinara*, which is obviously related to the Roman *Denarius* and the Arabic *Dinar*, and refers not so much to Coins in general as to a specific "cut" of a coin with this name; so this Sanskrit word would appear to be not so very ancient, to the satisfaction of Semerano.

Many Giants derive the term *"moneta"*, "coin" in Latin and Italian, from Juno Moneta, Juno the "Admonisher" (?) in whose temple, and in whose honour – 100 years after the Gaulish attack on the Capitoline Hill and the Brennus episode (around 390 BCE), that of the *vae victis!*, Marcus Furius Camillus (*non auro sed ferro recuperanda est Patria*), again on the Capitoline Hill – the *Officina argentaria* was set up in 296 BCE for the minting of Roman coins; which again relates to elements of an ethical-moral character – the *Monitus*, the admonition/ warning given by Juno.

Maybe I'm a distant descendant of Brennus, but this explanation has never convinced me. While the sense of Number survives in the Greek term *Nomisma*, the story of Juno's earthquake and the Capitoline geese (390 BCE) definitely overshadows the Twelve Tablets, *pecunia, pecunia moneta, pecunia numerata, pecunia præsens*, the *Aes, -ris* (hence 'aerarian') *Signatum* of Servius Tullius (574 BCE), the *Nummus*, and lastly the *Nomos* itself.

I think the apparent similarity between the two Sanskrit terms cited above, but maybe not only that, may, long before Freud, have prompted a certain tendency to consider the Gift as related *ab ovo* to Wealth, and in turn Wealth would have been rather hastily associated with Money. But if, for example, I write the adjective GREAT, and a verb like GREET is only one syllable different from it, this does not make them mean the same thing.

I also think that this "similarity" to the Sanskrit is not by any means irrelevant to the observations made by Mauss, by many anthropologists, and by psychoanalysis itself; if wealth-money were directly connected to the gift, there would inevitably be questions like: Gift in restitution? Gift to be distributed? Gift offered? Gift received? Received by Lot? From the gods? From the Community? For one's own merits? By Destiny? Gift which brings one closer to or further from truth and virtue?

It is possible that, to the 'man cub', the enormous "potency/ wealth" which adults have at their disposal might in fact seem to fly in the face of the immersive wretchedness, the wretchedness and penury to which he finds himself condemned (my interpretation of *Envy and Gratitude*) and that this sets in motion the expectation (demand) to receive, gratis, a part of such opulence (you are so rich that you MUST give me things since I'm so poor!).

What then? We only give what we have in abundance? Do we occasionally give something by *"depriving"* ourselves of something? At other times do we give something because we've made a big effort to "do more"? At other times do we give with the "secret" hope of receiving, like Puss in Boots?! We should ask Epicurus, or Seneca, or the Cathars, or maybe Max Weber, but also Melanie Klein. We are faced with too many questions. Too many questions and too many Giants crowded onto a bridge that is too narrow, too rickety, too slithery. Too much of everything. Not today.

All I am stressing today is that the Coin appears much, much, much later than Money, and Money in turn appears much, much, much later than Wealth.

And I'll confine myself to throwing out one suggestion: I might here be thinking of Wealth-(Riches), cousin not child of the Sanskrit *Dhana*, as something which defines the size of a person's estate; where Money, far from being *Monita* – and in this respect I disagree with many Giants and with the traditionally accepted etymology – I think it is instead to be understood, at least at first glance, as linked to *Munus*, gift, or rather the act of Giving, and perhaps to the Sanskrit *Dana*. *Munus* is also office, duty, tribute, in this way being connected to the meaning of the *corvée*; and *Munus* as *Donum* connected to *Do,-as, Dare*, hence *Munerata* (from *Munero,-as,-avi,-atum,-are*: i.e. to give as a present, to donate, to recompense) and linked to the *Manus*: from an economic viewpoint, a wealth, a spendable, liquid value (Keynes would say, a high-liquidity assets), cash, *pecunia praesens*, 'dosh'; and in Akk. *nadānum*, to give, (root *ndn*) is a verb and hence tends to lose the initial *nun*.

From G. Semerano (who transliterates differently from me):

> *Moneta-*, æ (…) Corresponds originally to Akk. *manû*, Hebr. *Mānāh*, (count, weigh; Akk. *manū*; Hebr. *māne*: minah; also with the sense portion, fraction; Gr. μνά: hundred drachma coin; ten silver mina are worth one gold mina; see *manus*, hand; the natural tool for digitation; Akk. Manûm (count) (Semerano).

I'll add that Counter in Hebrew is, *moneh*; I think we'll meet him again.

All in all, in this context, there would turn out to be no *Juno Munerata* or *Juno Munerans*, and definitely no *Juno Moneta*; again in Semerano, under the entry 'Moneta[Coin]-attribute of Juno', there is a reference to Livius Andronicus, who translates *Juno* as Μνημοσύνη, Mnemosyne, a Titan and daughter of Uranus Pianigiani, in his *Dizionario etimologico*, notes that the final *-eta*, of Moneta may recall a possible Greek root and adds that Moneta would be related to the Greek *menuo*, μηνύω, I indicate, I declare, and *menutes*, μηνυτής, indicator. And he continues along these lines, noting the Doric *ma-yuo* and the Sanskrit *man-ya-mi*, connected to the Latin and Greek, for show-indicate.

And he also borrows the Hebrew term for Counter, Greekifying it as needed: *moné + eta*.

I think that the connection given by the Indicator (μηνύτης) who Remembers (i.e. *Mnemosyne*) and Indicates Quantity/Weight (μνα) is not so far removed from what we will see later on, as we will with the Libra, being identified as Mina, as a coin (*moneta*): i.e. μνα, mna, i.e. a *m(o)na-eta*. And in Latin, *Monē-re* does not entirely lose its sense of Indicate, even though the modern "sensibility" modulates its meaning into Admonish. *Mina, Mna, μνα,* is derived from the Sumerian MA.NA 60 shekels formerly present in the codex of Ur-Nammu.

Hence there could be a Juno who is indeed Moneta, but in the sense of Mina, the coin stamped in Greece, and earlier still in *Ur*, which has nothing to do with *Monitus*, at least with the commonly understood meaning; hence, a *Juno* MA.NA, a *Juno μνα,* i.e. a *Juno Mina*.

Mon-eta. Mon-naie. Mon-ey. Mon-itus. Mon-e-re. Mon-è. MAN-A. Min-a. Μν-α. Mon-ili (Necklaces). Other references to Mon-eta could be found in the structure of the Cypraea; but not now.

The gesture of the gift/ passage of property not necessarily connected *ab ovo* with wealth-money, but rather a *Munus* connected to the act of the gift, to the means by which one gives: which is the *Manus*. *Manus* in turn connected to the Computation/ Quantity/ Value of the Gift itself.

Thus, a Moneta connected to the act of separation which is enacted when the exchange takes place between the tobacconist and me: I separate myself from the green banknote (the cigarettes cost me €100) and with my hands I offer him the money; a green banknote which is also *the indicator which remembers* the value that is at stake. It seems to be a gift. It is a "gift" but one entailing a corresponding, reciprocal act, not unidirectional and self-denying. And the tobacconist does the same with me, separates himself from the cigarettes and hands them to me. It is a contextualised gift and precise in its value, its modalities, and its effects. A gift effected in that moment and in those conditions.

It still has in it all the characteristics of the Gift, but there is a judgement, a still more equal and just evaluation which maintains it: thus it is still a Barter, a more complex and precise one.

It all happens between hands which exchange things. Hands which Count/ Weigh/ Compare.

The concept of *Moneta*/Money being derived from *Monitus* and *Monēre* could, in my opinion, still be viable to some extent insofar as the *Monitus*, the Indication, coming from the divinity, in this case Juno, relates back to a sense of transcendence and in some ways reveals, reconnects to, and confirms a religious origin. Like Shamash, Juno Indicates and Sanctions Value, Indicates the danger connected to not respecting Value, and does not admonish/reprimand the Capitoline Centurions and Geese about the enemy who is approaching, telling them they shouldn't be asleep: oh, sleepyheads! Or he warns about the approach of an unavoidable earthquake: come on, move, you lazy lumps!

Returning to Maffei's observations: if a hasty translator, or one with no interest or awareness of it, wanted to give the Sanskrit term *Dhana* a modern meaning, it wouldn't be too much effort to translate *Dhana* with some modern equivalent which gave the idea of Wealth, and such a translator could easily use the term Money, which is ready and available.

If he was a Neolithic translator, Money not having yet been invented, he could translate the term Wealth with something like "so many arrowheads" or "so many bearskins"; or, as we will see in the case of Abraham in the Bronze Age, he could translate it as "so many head of livestock"; the term *Pecunia* comes from this, from *Pecus*.

And so on. (Servet is not much in agreement with this derivation.)

And Benveniste holds that *peku* is an archaic term of unknown origin which would have indicated moveable wealth and then comes to define smaller livestock; hence, according to Benveniste, it is the small livestock, the *pecus*, which derives its name from *peku*, moveable wealth, and not the other way around, as is commonly accepted. Perhaps cognate with Gilgamesh's *pukku* and *mekku*?

Another small element on which I would like to dwell for a moment, given that it has very present-day implications, is the concept of Gift in the sense that is so dear to Mauss: the *Potlatch*. The *Potlatch* that Mauss is so fond of does not, and cannot, refer to economic matters, and Mauss never says how the wealth that is offered-consumed-destroyed in the *Potlatch* was generated. He goes no further than the observation that the economic relations between different groups would have been established by means of festivals and exchanged goods, even over a period of years, the intention being "not to lose face" for the head of the clan, and for the clan itself or we might say, in more modern and precise terms, for the chiefdom. His 1914 work, *Les origines de la notion de la monnaie* argues along the same lines, though here a "concrete" element is introduced, in the form of the market; and the value of gold is considered for the "faith" in its concrete and material benefit which may derive from it, and no longer just in terms of "face".

I think instead that the *Potlatch* can assume important symbolic valences within the concept of Predation, seen as a quick way to increase one's own Wealth: simply stealing them from someone else. (No, I won't talk about Fromm this time.)

But the *Potlatch* cannot assume a particular valence in the creation of Wealth.

A further small polemical point: if I am "constrained-bound-obliged" to exchange a Gift, can it still be called a Gift?

The first form of "transfer of property" of something (objects, food, women-*munus*, slaves, animals, etc.) – i.e. Predation, which increases some people's Wealth at the expense of others – often has the unpleasant secondary effect of unleashing some small form of "resistance" on the part of the robbed, the prey, or the preyed upon; "resistances" which, sooner or later, inevitably generate a role-reversal between Predator and Prey.

This ancient form of "enrichment" is still present in ancient Rome, amply testified by the Rape of the Sabine women mentioned earlier. Now I come to think of it, it's still in force on many buses today. And on second thoughts, not only on the buses. And not only in Rome.

Anyway, in this context of Predation, and only in this context, I think the *Potlatch* assumes an "economic" importance in the sense that it is more "economical" if we play at preying on each other, taking it in turns, without any actual killing: that is, by avoiding Blood and by ritualising the procedure (I'm not thinking in constructivist, but decidedly counterphobic terms, of reactive formations by which, after centuries of reciprocal refinement – but centuries in which, for example, the women abducted by groups that were necessarily not very far away, start preparing food in the way they had learned before being abducted, in one field or another – this slow "move to uniformity" in dietary customs; techniques for tanning skins; sewing methods; aesthetic taste in the design of clothing with different colours, fabrics and techniques; not forgetting their ways of caring for infants, children of the new unions; together with the mother tongue; the history and stories from the mother's side; the songs from their group of origin; all become something which is grafted into the new group and determines an ever-greater "cultural" proximity between the belligerents (a bit like the plot of the Fifties film *Seven Brides for Seven Brothers*); and the procedure for exchange slowly changes from being ritualised to being Sacralised. In any case, even in this form of exchange – whether in Predation/ Counter-Predation, or in the *Potlatch* – there must be the root of what will subsequently (according to Mauss) and in parallel (according to me) become barter, a form of exchange and a ritual form that is "shared and accepted" or else "imposed and undergone", a set of rules-times-values-behaviours on both sides, both opposed clans, both the Prey and the Predator.

Together with the ability and willingness of both sides to respect such rules, even when they may seem, or are, harsh, violent, and cruel.

However, I think it would be laughable to regard all of this as the economic foundation of a process of "Wealth"-creation.

What do the groups, Clans, Tribes live on between one *Potlatch* and the next?

The attribution to the *Potlatch* of an intrinsic capacity for generating wealth is too much like children selling off their old toys for money (toys which had been bought by their parents in the first place for a much higher price than the one the children will charge), too much like *Farmville* et al, interesting games in which I only need to have virtual friends who will give me a virtual tractor so I become virtually richer and cultivate my virtual factory, giving virtual food to virtual animals and people, and in my own turn eating virtual bread and drinking virtual milk.

But without allaying my real hunger. There is little *Virtus* in this virtual play. The adjective introduces a misunderstanding by drawing on one of the late meanings of *Virtus* understood as a "potential" element, an element yet to be expressed, having a force which may achieve concrete expression. But *Farmville* is not and never will be a concrete Farm; and so my Hunger will always be a concrete hunger.

Overall, the meaning of *Gift/Potlatch* in Mauss is too much like *Farmville* or, if you like, a chain letter where, like a paradigm of all economic crises, Derivatives always get grafted on, and thanks to which someone always wins: i.e. always the Bank. And many people lose: i.e. everybody else.

And if Food of the concrete variety is not a game which refers back infinitely to another game, we have to ask the question: Who builds the first real tractor? And how much does the real tractor – the one that ploughs the real earth, to produce real bread – really cost?

Of course, I can give a real tractor to my cousin for his onion field, but I think it's important to remember that the theory of the spontaneous generation of life left the stage a while ago, thanks to Lazzaro Spallanzani (1722-1799) or, if you prefer Lucretius: *ex nihilo nihil*. And so, all those *ummannu* (as they think of themselves) who talk with supreme confidence about redistribution of wealth, of the Gift, or Social Justice (as in the Incomes Policy, implemented in various forms since Antiquity to mark the crowning of a new sovereign with Debt-Forgiveness across the realm, or on the return of the Sabbatical Year, and still present in the Lord's Prayer: *Et dimitte nobis debita nostra*) and make "ethical" calls for equity of distribution, whether these are applied to the most recent financial legislation or the Eighteenth Dynasty of Egypt or to Joseph's warehousing of grain, should know that this isn't a virtual game we're playing or an exercise in creative translation; that we aren't living in some State from the Golden Age where rivers of milk and honey flow spontaneously, a Land of Cockaigne where sausages grow on trees to be freely picked. The grain which will be distributed in Rome during the Annona must first be cultivated and paid for by somebody. Only afterwards, when it's needed and when all the previous stages have been carried out, can it be redistributed (in Egypt and in Rome, by paying for it; by donations, a redistribution paid for individually or by Maecenas, or Pharaoh, or the Senate always comes about after payment, maybe at "political" price, when it suits the distributor, more often at an increased price given the scarcity of the commodity, but always paid for in money, land, personal property, work, etc., since someone has spent money to store up "grain" and prevent "famine").

And the simplest way to "forgive" someone's debts, in all ages and under every sky, is by Royal Decree, decreeing the spoliation/predation (a more elaborate regression to ancient Predation?) of someone else's wealth: for example, putting other individuals' liberty and goods under the jurisdiction – i.e. property – of the barons when they have to be "fleeced", individuals who, in the circumstances, have no choice on, pain of death, but to bow their heads humbly. This is, in other words, sequestration/ confiscation/ expropriation of other people's goods (in Fromm's sense this time: robbing-depriving somebody of their own property, but by Royal Decree) just as, confining myself to the historical sphere, Louis X of France, for example, did in 1311 according to Attali's account (2002, p. 267), confiscating *private* goods from (*depriving*) the various "*Lombards, Juifs et autres usuriers*", who could in turn, being goods and not people, part of the Lord's *property*, also be "granted" by one nobleman to another on payment of substantial sums; or debts were "forgiven" by simply being left unpaid, as they were by Edward III of England (1327-1377) who, having got into debt with the Florentine banking family the Bardi, refused to honour it; or like Charles VIII of France (1470-1498) who imprisoned Jacques Coeur, to whom he owed a great deal of money, and expropriated his wealth; and as Philip IV, the Fair, Louis X's predecessor, did with the Templars, using the excuse of their presumed heresy. Naturally all done in compliance with the Law. And obviously with Morality, as is suggested by Luther's (1553, p. 200) bizarre observations about the Jews:

"And yet [the Jews] live on the goods they have stolen from us."

A law designed for the specific purpose of a brutal and inhuman predation. The example of what happened in Nazi Germany, but not only in Germany, in terms of systematic spoliation of the property and dignity of entire sections of the population, most notably the Jewish population, but not only them, remains the most concrete evidence of the concept. Jeremy Bentham (1787, p. 59), in a classic illustration, writes:

> Indeed, the easier method, and a method pretty much in vogue, was, to let the Jews get the money any how they could, and then squeeze it out of them as it was wanted.

Under the aegis of Law and Legality, obviously (and obviously both Upper Case). In other words, it is all too often the *awīlu* depriving the *ḫāwiru* and not the other way round. And this is how we know who the real parasites are (Charles Adams, 2005, addresses the long history of this very fully and interestingly).

But if we think about it carefully, *Farmville* does have an economic value, not virtual but concrete: an enormous economic return for... its inventor, and of course the Bank. What a coincidence! Everything just as it should be.

And so I think we can definitely say that the Gift can be part of the economy and in certain circumstances is an important element of it: take Charity, for example, an ancient form of redistribution of income, in all its varied expressions; or, from a transgenerational viewpoint, the legacy which someone may inherit from previous generations, a legacy which may consist of credit, but also of debt.

However, this Gift of Inheritance only means that others have worked, or not worked, to be able to leave it. And it is the result, the effect of a wealth-creation, or dissipation, and not its source. For an individual, the legacy received may also represent something to be invested, if it is credit, and may then generate further wealth, but he remains essentially in the condition of a child who lives on pocket money, however lavish, given by his parents if he doesn't *main*-tain and increase, *motu proprio*, what he has received.

In the strictly economic terms of wealth-creation, the Gift does not and cannot found anything since it only offers, only donates what has already been created. Broadly speaking, this holds true in psychoanalysis as well: we can only "donate" what we have worked through/ acquired/ done inside ourselves before offering to our patients.

And as I stressed a little while ago, the generation and germination of life or wealth, is NOT spontaneous. Detailed studies of "different" Economies – creative, mutual, Islamic, cooperative, fair trade, responsible, eco-sustainable, "freak", hippy, virtual, narrative, or what have you – functioning on the basis of Cowrie shells or Wampum or pigs, pepper, salt, etc., would take these pages on a long detour, and I would rather leave that task to professional historians who will, I hope, be much better at it than I am. I'll say nothing about so-called Sustainable Degrowth, which I believe to be a poorly veiled, concrete and tangible expression of a regressive individual suicidal tendency disguised as a "sweetened" collective euthanasia by the promise of a "paradise" even more fantastical and ideological than the *Potlatch*; an ever-present death-wish, but especially "live" in time of transition/ decadence (*peccatores, improbi homines, vos pœniteat*!!!).

Though maybe a glance at the Cypraeidae.... No, I've already said no to that.

But having said this, I cannot allow myself to ignore the *Bullae* containing the previously mentioned Abnu, used since the time of Enmerkar's temples and, long afterwards, by his descendants. With the *Bullae* and by means of the *Cretulae*, seals in use since 4000 before the Flood, the *Abnu* constitute a formidable repertoire. And they too have evolved. They are no longer mere "stones" or "tokens", but lend themselves, symbolically speaking, to describing, or rather representing actual materials, commodities, goods; and this will foster the process I will address shortly, the birth of Money. They have become precise and dependable tools, standards weights, and not simple tokens.

But first, another spatio-temporal digression, leaping forward 2000 years. To Rome.

16.
ROME

There's a thick fog coming down. You can't see a thing. I trip over something.
A distant flash of lightning, then a muffled thunderclap which fades into a very strange silence.
Given that my name is Benini and not Benigni, I feel like Massimo Troisi in *Non ci resta che piangere*.[1] After a moment that seems to last for an eternity, the fog lifts, slowly.
The sun peeps out and... I find I'm in Rome. Long, long before the wars with Alba.
I brush off my Latin and look at the people around me, who regard me with suspicion, struggling to understand what I'm babbling in my strange language, and unable to understand why I am dressed this way, but I manage to talk, more in gestures than with *consecutio temporum*, with some people I meet in a square. There is a market, buying and selling. Goats and broccoli everywhere, obviously. Naturally I see no *Aurei*, *Sestertii*, or *Asses*.
Nobody knows Servius Tullius, or his address, or his mobile phone number.
Librare, Comparare, Placare, Venum Dare.
Venum Dare. Besides the "venality" which clearly characterizes the traders in various goods, I dwell on a first gesture that I observe constantly, as if hypnotized by its simplicity: all the transactions go through the hands.
Vēnum, accusative of *vēnus*, (different from *venus, -eris*, physical/carnal/fleshly love) derived from *manus*, as Semeraro shows.
Not only Goods and copper pans, goats, cabbages, scales and weights, but also the Hand.
Giving by means of the hand. Receiving by means of the hand. Giving the hand to confirm a transaction that has just happened. The *Manus*: it is by means of the Manus that I make Gifts, conduct business, offer, receive,

---

[1] The 1984 comedy film known in English as *Nothing Left to Do but Cry*, starring and directed by Roberto Benigni and Massimo Troisi.

count, hold, let go.

And then my eye is caught by a little group of people who seem to be putting on a show, performing something, though they aren't actors. They are enacting a ritual which seems to have something ancient about it. It centres on the *Manus*, but not only on that.

I discover that ancient Roman Law, reflecting the ancient Roman idea of Right, effects a very important distinction between *Res Mancipi* (Taxable Patrimonial Estate, which requires a *Mancipatio* before it can be accessed by its title-holder and concerns properties that are *pretiosiores*: i.e. houses, land, livestock, etc.) and *Res nec Mancipi* (Assets available by *Traditio* for the use of the title-holder): this is my summary of Aranjo-Ruiz and Filippo Gallo.

It almost looks like the distinction between Money and Coin, that I shall be making shortly.

After another leap of a thousand years, it is interesting to discover that, through a long evolution, the statute of *in Jure Cessio*, merging with the statute of the *Res Mancipi*, in turn absorbed the *Res nec Mancipi*, generating over a millennium the primacy of the written contract to be placed in the hands of the Civil Authority and dealing with the various forms of Weighing in public, and with witness testimony which, in ancient times, was only obtained by means of the Libra, *per aes et libram*, and performed by the *Libripens*, the weigher of the libra, who gives the *raudusculum*, a thin strip of copper, as a symbolic pledge of the transaction which has taken place (the receipt of the time, perhaps?).

And it is precisely this "little show" that those people are putting on in that square.

The transaction through the statute of the *Res Mancipi* was achieved by the purchaser pronouncing the ritual formula which was spoken when acquiring a slave, for example:

Hunc ego hominem ex jure quiritium meum esse aio isque mihi emptus esto hoc aere aeneaque libra.

> (I declare that this man belongs to me according to the law of the *Quirites* and he was purchased by me with this bronze coin and this balance [i.e., according to all the legal formalities]).

This formula reprises modes set up in antiquity for transferring property of one type of *Res* between one person and another (the acquisition of the tomb for Sarah, for example, as we will see later) and derived in their turn from the ancient *Mos* which had generated the *Jus*, and will in turn determine the *Lex* which will be sanctioned by the famous Twelve Tables of bronze displayed outside the Senate in Rome. Over time, the consolidation of the statute of *in Iure Cessio* – a written contract deposited with the Authorities – will follow the same course, but this time on parchment instead of a clay tablet, and in the presence of a civil authority, not a religious one like that created among the Priests of ancient Babylon: 3,000 years before!

I will not dwell on the strange fact that in this declaration, which is public and follows its own ritual, it is the purchaser who claims to be already the legitimate possessor of something he is still in the act of buying, by sacred rite actually derived from the god Quirinus, although all this refers back to the fact that exchange occurred in a religious context, as it had done for Enmerkar: but it is astonishing to rediscover – centuries, millennia after Enmerkar – the religious context, the invocation of the god (Quirinus), as the basis of an exchange of *res*, made with weights, by means of a scribe – here, the *Libripens* – who declares, sanctions, and certifies an agreement established in a solemn manner and before witnesses; and it is astonishing to come across – here too, thousands of kilometres away and thousands of years afterwards – the weights and scales, as for Ur-Nammu.

All of which is connected to an act which by its nature recalls Handling, Taking in Hand, *Man*aging directly, being grown-up enough to conduct transactions with one's own Hands by means of *eman*cipation, and to *man*age one's own belongings, one's own property. (From *Manus agĕre*, to do things with the hands, and/or *Manus gerĕre*, to operate/direct.)

What I am interested in emphasizing is the concepts of *Manus* and *Capere*, which seem well represented in these statutes and, while lending themselves to an infinite number of interpretations, are indissolubly linked to *Manus* and *Capere*, to Taking in Hand.

I also find the expressions *Mancipio Accipere* and *Mancipio Dare* very interesting. (And psychoanalytic implications of and reflections on the following topics would also be of great interest: Availability of a good to an individual, Emancipation, Property, Sale, Purchase, Capture, Appeasement, Touching Money with one's hand, and the Coin.)

But why did I want to make this jump to Rome? Because I wanted to see, in the most complete way possible, Barter, Merx, Weights, Scales, Coin, and Hand at work. But also *Comparare*, *Venum Dare*, and *Placare*. All at the same moment, in the same square.

All as 2,000, 3,000, 4,000 years before, in Sumer, a world now lost and forgotten, but without it Quirinus would not have had his *Libripens* and his *Bi-lanx* available to him.

Largely identical behaviours, for all the chasm of time that separates them; identical in expression and sacred manifestation in the presence of witnesses, sanctioned by an *Arbiter*, here the *Libripens*, while 3,000 years earlier it was the *scribe/ weigher of silver* who, in great pomp, announced to the four winds that an important sale, corresponding to the *Res Mancipi*, was about to go ahead, that he would call witnesses to the public square; he announced the act of sale three times, summoned three senior citizens to give testimony, demanded three copies of the contract written on tablets, and invited anyone who might be opposed to the sale to come forward (compare also Bulgarelli, 2009, p. 12, note 4). Astonishing.

All closely connected to a ritual, public, intoned, performed, official act; all practised by *e-Man-cipated* people who were in a position to take possession of their own property *brevi manu*, without a *tutor*, who held their destiny in their own *Manus*. Free Men, *ḫāwiru*.

And *Manus* is directly connected both to the *Munus*, which is the *Donum* made with the *Manus*, and to the *Venum*; as I said a little earlier, the *Merx* is a *Donum* which becomes *Venum Datum* and the *Donum* is *Merx* (or *Mers*) which becomes *Donum Data* (donated).

These are still actions in which the Hand is directly involved. Actions which, in turn, are directly concerned with and carried out and sanctioned by scales, weights, certificates, public declarations, writings, and scribes. All operations which are necessarily to do with Counting, with the Akkadian *manū*.

Before "taking off" on another leap through time, a couple more words on the formula quoted earlier, which ends with "AERE AENEAQUE LIBRA".

I don't know whether Virgil gave any thought to the matter. He certainly knew Latin better than I do. He couldn't let such a macroscopic detail escape him. Certainly, the idea of making the Romans descend from a noble line, as they taught me at school, is an excellent reason for resorting to the expedient of connecting them to the nobility and valour of Troy.

But could it possibly have escaped Virgil – and me, for sure, till now – that, as in the ritual and sacred formula first reported as being used in the most remote periods of Roman proto-history, before Romulus and Remus, bronze was also called *Ænea*?

And Bronze, a word we also use today to indicate a certain historical epoch represented by its discovery and use, isn't it perhaps also the principal means for creating pre-coins and axes and breastplates and swords? And doesn't it, with the necessary division and specialization of labour before Capitalism, melt at 1,000 degrees Celsius, which is also the temperature needed for firing bricks?

I don't know if Virgil wanted "preconsciously" to make a connection with a Bronze Age, *aenea proles* in Ovid (*Metamorphoses* I, 125) which he couldn't have known and in which he might not have any interest – not like me, at least – (though he was extremely familiar with the Ages of Gold and Iron; and Hesiod also "knew" the decline from the Age of Gold), but it remains a singular circumstance that the founding Hero of Rome turns out to be so close to Bronze, so close as to share its very name.

And the first currency units are bronze, not only in ancient Rome but across the whole Bronze Age. The scales are made of bronze. The swords are bronze, stronger and more resistant than those made of soft iron. The Twelve Tables are bronze. Bronze is the steel of Antiquity, primitive titanium, before enough fragments of stars, *Sidera*, fell from the sky to be gathered and melted together, and before men were capable of compelling fire to unleash a 1,500 degree heat: Sider-*Ur*-gy. Like a decayed nobleman, bronze still preserves some of its ancient prestige today; discoloured, aged, long past its best. But still Bronze; the same Bronze which Lycurgus dulled with vinegar to diminish its desirability (Xenophon, *The Constitution of the Spartans*). Perhaps Virgil wanted to pay Rome a yet greater homage than the attribution of noble births: I like to think that Virgil wanted to emphasize that the Law, that which we understand as Right, that of the *ḥāwiru*, solid as Bronze, solid as *Ænea*, was forged in Rome; and Rome, born from *Æneas*, is founded on *Ænea*, Bronze; Bronze as money, as Weights; and on the Libra, also of Bronze. So there we are: Libra, Bronze. Right. The Goddess with the Scales. Equity. Justice.

Noah again. As I recalled earlier, half seriously, half facetiously, one of his laws demands the institution of tribunals as a sacred act in one of his commandments with impartial third parties as judges, just as the Weight must be impartial. The Scales as a symbol of Equity and Justice against fraud and theft. Just Weights against oppression and abuse.

The Romans too were followers of Noah. That is, AENEAQUE LIBRA. Bronze and Libra. The very things cited above, in the formula of the *Mancipatio*.

The term that indicates the instrument for weighing, Libra as Balance-Scales, gradually becomes a word that indicates what is being weighed, the Weight itself, and no longer the weighing instrument; and later still, it slowly becomes the term designating the standard to which all weighed objects must conform. And then *it takes off*, becoming an abstract, freestanding unit; it will become *Lira/Livre*, acquiring a life of its own. Nations have had recourse to this unit of currency in the most diverse periods, customs, and circumstances, from Charlemagne to Turkey, from Italy to

Great Britain.

But where does the *Lira*[2] come from? I don't mean the Lyre, music instrument, but the Lira-Libra-Scales.

It comes, of course, from Libbra, Libra, derived from *Librare*, the action of weighing that was performed in the beginning by the *Libripens*, the weigher of/with the *libbre*, the descendant of the *Euphratesians'* weigher of silver. In Italian today, as a distant memory of past glories, all we have is *Allibratori*, bookmakers. *O tempora, o mores!*

But *Librarsi* also means to take off lightly into the air! The great Theodor Mommsen (1956, vol. 1, chap. 14, p. 258) writes a delightful comment on Italian weights and measures:

> To this end, nature offers us as the most immediate reference points, for time the passage of the sun and moon: that is, the day and the month; for space the length of man's foot, which is easier to measure with than the arm; for weight, the load which a man with outstretched arm can lift (*librare*) on his hand, which is why the unit of weight is called libra, *pensum* weight [from *pendo, penděre, pensum*, to cause to hang down].

In Latin, weight is *Pondus,-ěris*, hence the English Pound, the *Lira Sterlina*, the £.

A friend of mine who writes poetry (which personally I don't much like, but don't tell him) has played with words to invent a possible etymology for Librare: *Liber per aerem*. Not bad, for once.

But now it's late, I must go back to where I started from.

---

2   There is little apparent connection with *Lira* meaning 'ploughed Furrow' – the line which defines the two sides raised by the plough or, in other traditions, the ploughed-up earth itself, also known as *Porca*. The two *Lirae* do not seem to have a common origin, just the loss of the b in Latin: i.e. *Li(b)ra* becomes *Lira*. This Lira, understood as Furrow, is the root of *Delirium*, Straying from the Furrow. For Lyre understood as a musical instrument, Pianigiani derives it from the Greek, λύρα; in Romance languages the Greek υ often assumes the sound i, hence the confusion. Lyre is in fact connected to the root -RU, resonate; Sanskrit RU-TA, sounded, played; Latin RU-MOR, noise.

17.
THE BULLA: I.E. MONEY

Passing through another mist, I abandon ancient Rome and once again pick up the thread of the Sumerians and, finally, the Abnu. I'll pass over the whole part relating to the Cylinders, which are a kind of half-way stage between a *Cretula* and the *Bulla*, used one after the other by both the public administration and by private individuals, rather like frankings saying MADE IN ITALY or FRAGILE followed by a picture of a glass that you can still see today on packing cases being sent abroad, and I'll move straight on – really, this time – to the *Abnum*, an Akkadian term for stone, rock.

I quote Liverani (1998) again:

> The Susa (2500 BCE) sequence has provided the most significant indications about the use of the seal (specifically on "cretulae", blocks of clay which seal doors and mobile containers), and about the use of small tokens (i.e. the Abnu) of clay or stone which symbolize physical entities (people, animals, objects) and the use of numerical signs. In the Susa sequence we pass from the enclosing of tokens inside clay *bullae* covered with marks made with seals to the impression on the outside of the *bullae* of imprints made by the tokens inside, and lastly to the abandonment of tokens since the external imprints plus the numerical signs are sufficient to transmit the desired information. In this final stage the *bulla* will become flattened and give rise to the "tablet".

With the *Bulla*, we come to the fusion of Number, Sign, and Carved Word[1]; a scribe with a "simple" clay tablet is able to certify, from thousands

---

1   A note on the union of Number and Letter and then also the Word: given that in the Hebrew alphabet each letter possesses its own numerical value, it would be interesting to look in depth into the Qabalah but, besides my unpreparedness for the task, the exegetical, occult, speculative, and esoteric aspects would take us a long way from what I'm writing, even if I were "only" to look at the Sefer Yezirah and the Zohar, and not the endless literature about them. Anyway, it should be remembered that these aspects, which suggest great attention to Number and Letter, are a highly significant presence in the tradition of Hebrew studies. I'll add

of kilometres away, that a given individual is the legitimate possessor of a certain quantity of goods (fields, house, number of slaves, animals, objects, or other merchandise), giving further details (specific commercial features: male/female slaves, adults, children, names, provenance, age; or rams, lambs, ewes, or goats; quality of textiles and yarn, provenance, destination, etc.). All without the need for such goods to be carried around physically.

"Just" a tablet. And this is the case both for trade and for landed property, both in tenant farming and in loans for use, etc.

The Temple(s), the entire complex and interconnected economic-religious system of land management across Sumer, as far as Aratta, guarantees compensations, valuations, and conversions by means of these very Tablets which are real accountancy, real certifications and guarantees, establishing extremely refined standard criteria which deal not only with the features of the goods themselves, but also the value, their seasonal availability, possible losses, profits, percentages of credit and debit, etc.; and if all this is certified by the Temple, and therefore guaranteed by a god, it has to be accepted by the other Temples, and all the more so by human beings. The god, the gods, are its guarantors.

The Tablet generated by the Bulla + Cretula + Abnu, brings Money[2] into

---

that "our Sumerian philosophers developed a doctrine which became dogma throughout the Near East, the doctrine of the creative power of the divine word. All that the creating deity had to do, according to this doctrine, was to lay his plans, utter the word, and pronounce the name" (Kramer, *The Sumerians: their History, Culture, and Character*. Chicago, CUP. 1963, p. 115). And all of this was being done long before the Hebrews. Seminara's observation (p. 103) is also interesting: "Sargon II claims that the perimeter of the city he founded (called Dar-Sharrukin, 'Fortress of Sargon'), 16.280 cubits, equated to his own name." I believe this is the first example of that "science" of correspondence between numbers and words which, in Rabbinical exegesis, will be called (using a Greek word) *gematria*, sending its root back even before the Creation: "When God was about to create the world by His word, the twenty-two letters of the alphabet descended from the terrible and august crown of God whereon they were engraved with a pen of flaming fire. They stood round about God, and one after the other spake and entreated, 'Create the world through me!'" (Ginzberg (1909, 1937), *The Legends of the Jews*. Baltimore, Johns Hopkins UP. 1998, I, p. 5-6). This is followed by the account of what God said to each of them, beginning with the last, Taw, and ending up, obviously, with Aleph.

2 "In this sense, money has been defined as 'abstract value'… in a similar fashion to the sound of words, which is an acoustic-physiological occurrence but has significance for us only through the representation that it bears or symbolizes" (Simmel, 1900. *The Philosophy of Money*. London, Routledge. 2004, p. 118). I have found this definition interesting because it also establishes a direct connection, albeit narrative and literary, between Money and the Sound of a word,

being in the form of a clay Tablet, a clay Cheque (or if we prefer, a clay Banker's Draft, a clay Credit Note, a clay Banknote; the first to propose this hypothesis, which later became no hypothesis, was Sir Leonard Wolley at the start of the twentieth century); Money is issued, certified, and guaranteed by the gods or at least by the organization which legally represents them. Money certifies the Wealth that the individual has certified at the Temple (I believe that the superimposition of the two terms, Wealth and Money, occurs at this moment). Here too, it is not Italian merchants who invent the Banker's Draft or the Credit Note; they "simply" reinvent it.

Caterpillars and butterflies again. The same reasoning holds for Attali's argument that the Children of Israel would have invented the Credit Note during the Babylonian captivity in the VI-V centuries BCE: they also reinvent it. Or, more simply, go on using it.

So, in this case too, we cannot claim that the mental leap necessary for the creation of a "fiduciary coin" or the increase in symbolic capacity that is made necessary by its use, to which Dr Schimmel refers, is the prerogative of modernity. The Sumerians, Akkadians, Gutians, Amorites, Egyptians, Hittites, Assyrians, Babylonians, Chaldeans, Martu, Elamites, Phoenicians, Philistines, Jews, Canaanites, and all the other protagonists of Mesopotamian history, had this symbolic capability.

And searching for it, attributing it, thinking of it, only as a characteristic that only we moderns have acquired (after losing the virginity of Barter? after the Reformation had made us more intelligent, more capable of symbolization and abstraction?) just seems to me like the usual narcissistic error, fundamentally a piece of arrogance with infantile characteristics which, like the Ptolemaic theory – another profoundly human construct – continues to make its presence felt on various levels. Just as for Ptolemy – though the whole world thought as he did – the Earth was the centre of the Universe, so each generation, one after the other, thinks it's the centre of history and, deep down, better (or worse, which narcissistically is the same thing) than all the generations that went before. Why Ptolemy? Because every day, in every newspaper, in every weather forecast, in everyday language we continue to say, "The Sun rises, the Sun sets…" An imperishable immediate perceptual fact, but in reality the narcissistic fact of believing that the Sun revolves around me, taking no account of the Copernican revolution.

---

as at the start of my exposition, where the Sounds of the word become incised and visible Signs: that is, written Word. Thus, in Simmel's metaphor, Money changes from Sound to Sign, visible in the form of a clay tablet. But also: from the Sound (-Money) we pass to Sign, Incised (-Coin).

Anyway, through long, complex transitions and by making the initial choice of a Sacrifice (that's to say a religious matrix), a Gift, an Oblation (continued over time), and Faith have gradually generated a totality of situations through which the deities are able to sanction (sanctify), certify (render certain), guarantee (protect) the issuing, saving, covering, collecting, and exchange of terracotta "Cheques", of Money. (It seems like a description of the functions of a modern Central Bank. And all of this – as we'll see very shortly – long before the Biblical Judges: but let's not tell Attali that.)

It is an act of faith, dedication, and love, and first of all it is an act of Faith in the god(s) themselves. And it is, in turn, a Symbol, a pact which commits and binds both contractors, both the human individual and the god(s). And one of the meanings of Symbol is Pact, Alliance.

The Money-Tablet is a Pact (and a modern cheque is also a Pact): a pact between men and a pact with the god(s). Anticipating what I will describe later on, let me stress that in Hebrew, Pact is *Brit*. And a Pact with a god is a Pact with God! And this is its sense in *Brit Milah*, Covenant of Circumcision, *Milah*, from *Limol*, circumcision. Dr Icin, who is my Hebrew teacher today, doesn't agree, but I like relating *Milah*, Circumcision, to *Milah*, Word. Today the two utterances are written exactly the same way, but a Yud used to mark the difference, not appearing in "Word"; they have always been pronounced identically, however. I like to see *Milah* as the sacred commitment one makes in giving one's flesh being superimposed onto the sacredness of giving one's word.

In any case, the Money-Tablet, pact – and symbol of the pact – with the god, still needs the god in order to be fully reconverted into goods, and has little freedom to fly by itself. It is exactly like a modern cheque which can pass from hand to hand, be transferred and exchanged with other things, always guaranteed by the institution which issued it, but it must return to the same institution in order to complete its cycle.

Only the Coin, as a reserve of autonomous value, say the Giants, can aspire to freedom, since, once it has been issued, it contains its own value within itself without having to return to the Temple. Obviously I'm referring to the "early coin" of precious material, which is obviously different according to time and place (copper, bronze, iron, silver, amber, gold; materials which in turn reflect the technological and specifically metall*Ur*gic level of the society in which they are adopted) and not to the present-day fiduciary coin under forced circulation. (Besides, the Gold Standard is not so very far from us in time, and it's by no means certain that isn't still much more present than might seem to be the case. There is something appealing about

Sismondi's excoriating observations on the brand-new paper banknotes, called *Assegnati*, which replace gold and "send nations to ruin." In *Faust* Part Two, Act One, *Dark Gallery*, Goethe seems of the same opinion when he makes Mephistopheles, who has just invented banknotes, call them "paper ghosts of florins.") "Is the present-day fiduciary coin the child of forced circulation?" Here we are, back again with Ptolemy, the Sun, and Narcissism. Hendin (2010, p. 46), referring to Judea in the 1st century BCE and quoting Mørkholm and Grierson, writes:

> While all manufactured coins were fiduciary, the bronze coinage "differed from that of coinages in the noble metals by the fact that the profit to the issuing authority was much greater, the bronze being used as a token coinage of very little intrinsic value..." It is clear that the bronze coins were widely used in daily transactions, silver coins were far less common.

So, in the Judea of the first century BCE, the coin has already taken on the characteristics of the fiduciary coin and is issued by authorities who profit from its low "intrinsic" value and from Feudalism; and this too is not an invention of modern Capitalism.

(As for the English term 'coin', I find Murdin's (2012) observations frankly bizarre: between pages 88 and 92, making a free and easy use of Latin, etymology and geology, he manages to string together a particularly juicy string of "pearls of wisdom".)

Returning to the "primeval" coin, to the genesis of the very idea of the Coin, it might seem that, once the need for a new instrument had been revealed, man – Enmerkar, the High Priest of Marduk, Marduk alias Bel, or a stand-in for him – set out to study a new type of credit card, a new investment fund or particular derivative suitable for the emerging Countries. I don't believe that's how it was. In fact, I think the opposite, that the emerging need to identify new forms of faith, the profound necessity for a change of register tending in the direction of monotheism through the search for Equity and Justice or, if you prefer, *Maat* – whatever the reasons for it, sets in motion a range of complex moves which generate new symbols and a new mythopoeic capability, as yet unexpressed, which over time comes to take concrete form in truly new actions and by means of original instruments unlike those they derive from. Just as the *Bulla-Tablet-Money* appears at the moment when the Sumerian Pantheon begins to assert itself as Universal and is the economic, social, political, and religious expression of the New Faith – sanctioned by the Sumerian triumph over Mesopotamia (the beginning of the end of a fragmented polytheism varying from city to city), a triumph well represented by Enmerkar's poem which extends as far as Aratta – an innovation replacing the old conception in which each city did its own thing, and in the absence of a global plan, whenever there was a change of dynasty, the Monarchy passed to the service of the new order and the new god; just as Enmerkar, with Writing and "the clay cheque" is a good representative for the beginning of the end of enolatry (in which one deity becomes predominant over theother, an "intermediate" stage between polytheism and monotheism) so, a thousand years after Enmerkar, around the twentieth century BCE, the Coin, the agile new instrument added to Barter, Weights, Scales, and Money, makes its entry into history and says that something new is being born, or has just been born. And this something says that one deity is starting to assert himself overwhelmingly over the others in the emerging Pantheon. He may be called Shamash, Ba'al or Be'el, or Marduk, and will later have many different names, but around the twentieth century BCE the need begins to arise for a divinity who, for various reasons, is acknowledged as superior to the others. And this is a further step in a clearly monotheistic direction (Tomislav Vuk wouldn't be quite so sure). It will take many centuries, but the tendency is traced not only through Israel but, before that, through Babylon. As proof of this, I quote a passage from Liverani (2007, p. 226) about Marduk in the time of Josiah (second half of the seventh century BCE: it is interesting that, as Schama (2014) records, during the reign of Josiah the scroll of the Mosaic law was "found", sanctifying the uniqueness of Yahweh!) but not forgetting that Hammurabi had previously honoured Marduk:

> Urash is Marduk of planting, Lugalidda is Marduk of the Abyss
> Ninurta is Marduk of hoeing, Nergal is Marduk of battle
> Zababa is Marduk of War, Enlil is Marduk of accounting
> Sin is Marduk who illuminates the night, Shamash is Marduk of Justice
> Adad is Marduk of rain...
> *Marduk is everything.*

It is said that Mohammed, at Mecca in 630 CE., seized by a frenetic afflatus and religious impulse, yelling the famous *Allāhu Akbar*, "God is great(est)", accelerated his monotheistic turn, destroying in the blink of an eye all the 360 (!) deities present till then in the Ka'bah, but evidently considered "minor" in comparison to the "greatest."

Given that Abraham also came up with something of this kind (as we will see shortly), it is not the case that the march to monotheism moved in a straight line, the way flowers grow, and certainly not in a rush. And above all, it is not the case that the only Monotheism is Mosaic.

18.
THE EPIC OF THE RING

The Coin, the Coin-object, the Coin-tool – this is the thesis I want to maintain – is a consequence-index of the birth of a new conception, a new faith.

However, this assertion is too bald, almost dogmatic.

I will try to explain myself better: in the pages that follow, I will focus on what I consider the "final" part of a long process which precedes the birth of monotheism; a process which gets under way long before Hammurabi-Abramo. This process develops in a parallel way between Royalty and Pantheon. Put extremely briefly: before we have the appearance of a King who, setting Girard aside for a moment, overcomes a thousand terrible difficulties and trials by means of his Strength, Courage, Skills and Tenacity, and is therefore a demigod. Subsequently, the idea of royalty evolves and the King is "required" to combat injustice, to right wrongs, to bring peace and wellbeing to the kingdom: i.e. an Arbiter who is *Super* (*partes*). And we end up with the idea of a King who, being firm but fair, perfectly embodies the role of *Paterfamilias*, who devotes himself exclusively to those he loves: thus, a father but also a man.

In parallel, in the beginning, we have a god who moulds, combats, defeats darkness and created light and life out of nothing; then a god who makes equity and justice triumph, and is in the end a Good Shepherd for his flock.

If we were to read this sequence, which is already present in the epic of Gilgamesh, as if they were in a comic book, we would have (Kings and Divinities) Nembo Kid/Superman, Superhumans with super powers; then we'd find The Phantom and/or Batman, very strong Humans but without super power; and finally, with an apology for the irony which I wouldn't to be mistaken for blasphemy, we would Phileas Fogg, human, neurotic, fragile, in love with a princess and quarrelling with his servant. And now I will address the Paterfamilias and the Good Shepherd. And Phileas Fogg, naturally.

And every faith has its own rituals and sacrifices, and by its very nature founds, seeks, and finds new Symbols and Covenants. It appears, if we bear in mind the literary parallel I made a while ago and which I have been expounding up till now, at the start of the second millennium and develops within what must necessarily be a wider flow, putting down roots deep into the past in order to branch out vigorously into the future.

The end of the third millennium BCE is a period in which, as I have suggested, we can look for the birth, at least metaphorical if not strictly historical, of a new religious phenomenon: the emergence of a striving towards the idea of monotheism. The gods cede the leadership of the Babylonian Pantheon to Marduk, and Marduk/Bel becomes the undisputed chief who emerges from all the representations of the divine progeny.

Not that a creator god hadn't existed before. Sometimes there was one, sometimes not. There was An, but he was a kind of immobile entity who minded his own business. Sometimes the gods' decisions about the world were collective, with all the negotiations of the residents' association that we're familiar with, and especially throughout Greco-Roman mythology. Scorn, spite, little dramas and great tragedies were the order of the day even among the deities of the Fertile Crescent. Sometimes the fate of Man and the world were already decided in titanic battles in which the bodies of the vanquished became "building" material for the World, like Tiamat.

At other times the World had already been built, ready-made: houses, rivers, and mountains existed, but no men. At other times, as we have seen, Man was the result – to be honest, not always a welcome one – of the deities' disputes. It is between the middle of the third millennium and the start of the second that Utu-Shamash becomes, in a sense, the Zeus of the Fertile Crescent; the genitive and ablative of Iuppiter – *Iovis, Iove* – resemble Iavè (I've written it like that on purpose) to an impressive degree. And Utu is the divinity who presides over the Standard weights and measures. In Greek Zeus, Ζεύς, in the genitive becomes Διός, Dios.

Arriving slowly but inexorably at the Marduk of the Babylonians, through whom, while maintaining all the religious, literary, poetic, ritual production both in the great celebrations and in the small everyday invocations, the whole theogony comes to be re-manipulated, revised, reorganized, and harmonized so as to present Marduk, i.e. Babylon, as the Saviour of the gods themselves and, therefore, Babylon with her god who saves the Mesopotamian world, while Gilgamesh starts to wear Babylonian dress instead of Sumerian. Although many Giants won't agree, it is in this picture, in this flow in a religious direction towards monotheism that what, in my opinion, is the Coin appears. It will be another coincidence, but from

The Epic of the Ring 223

being the son of the male god NANNA (the Moon), UTU (the Sun) "takes power" in this same arc of time, and NANNA is transformed into Sin: still a very important deity but now beginning to be a bit subordinate.

And so, while in order to come into being, the coin "needs", or rather is the expression of, a new flow and a new faith, the one that is tending towards a definite monotheism (just as the Tablet which becomes Money is clearly connected to the new politico-religious set-up given by the construction of the Sumerian Realm-Empire which enables the unification of standards for weights and commerce, also unifying the cult) it is inevitable that we should ask ourselves what this new faith is, why it comes into being and generates another covenant which, being a "reserve of autonomous value", is necessarily a "smaller" covenant, easier to use than the Tablet-Bulla-Money. The new covenant does not require Archives, long-term accounts of income and expenditure as a guarantee for merchants who are abroad for years at a time; or cycles of harvest and sowing which last at least a year; calculations of interest at the end of the season; balances of palm tree plantations which yield nothing for three or four years and on which you pay the rent only in the fifth year; percentages and comparative totals of the births and deaths of livestock. It establishes new rules for the restitution of a dowry in cases of marital infidelity or fictitious adoptions which help to expand the value of landed property, monetising it. This is no longer a matter of enormous tablets certifying stocks of copper, or diamonds, or thousands of balls of wool, or thousands of skins of oil or head of livestock or food to be harvested, stored, cooked, distributed among thousands of the faithful who work for the Temple.

The Coin contains in itself all the same divine-institutional certification as Money does: indeed, it makes this still more necessary. However, it can perform all the functions listed above, though only among *ḫāwiru*. It is a "little covenant" for every day, at least to begin with. Later, it will take multiple forms. Such a covenant finds concrete expression in a unit of weight, the GIĜ (Sumer) and *Siqlum* (from Weight, exact Akkadian equivalent of the Sumerian GIĜ, related also to GIN) and divisible into ŠE (grain/kernel, referring specifically to barley, but also indicating roundness) which is the *Gaggaru/Gugguru* (in Akkadian) from which derives the Hebrew *Kikkar* (which will become the Talent) always consisting of 60 mina (1 *mina* = 60 *gin*, 1 *gin* = 180 *še*, in Sumer; 1 *manum* = 60 *siqlum*, 1 *siqlum* = 180 *uṭṭatum*, in Akkad).

Kikkar from the Hebrew root KKR (round, rotating).

And Gaggaru-Kikkar is the name of the first "coin" to appear in the history of humanity, at just this time, around the twentieth century BCE.

In Egypt during this period or perhaps a little later, and coinciding with the rise of Hittite power which in various ways controlled Egypt between the thirteenth and eighteenth dynasties (1781-1550 BCE.), appears the *Shat* (sometimes transliterated as *Shaty*, sometimes *Kat*, band), a fraction of the *Deben* (1 Shat, 7.58 gr = 1/12 Deben; while the *Uten* will be a multiple of the Shat) which has the same characteristics.

And Gaggaru-Kikkar is translated as Ring. Because it is a Ring.[1]

---

1 On the subject of the Ring as a means of payment, Williamson (1894) writes: "Gold earrings carry a further allusion to the use of pieces of metal with a known and acknowledged weight for the purchasing of goods or, if that is not necessary, simply as ornaments" (p. 14). And on p. 15: "The ancient Egyptians are represented in the portrayals of the period weighing rings of metal, gold and white gold [i.e. silver], while keeping to hand vases which contain piles of those rings already weighed, each of them having in all probability its own value;" and he adds (p. 16) the Biblical reference in which "the servant of Abram gave Rebecca 'a golden earring of half a shekel weight'" (Genesis 24:22) observing, "The coins [money] used by the children of Israel when they went into Egypt to buy grain could indeed have been in the form of a ring.... This use of the ring as coinage [money] and its frequent use at the same time as an ornament for the person, which represents material and disposable wealth, is widespread among many eastern nations and remains the case today" (I am quoting from memory). We might jokingly think of this as being like what happens when, more in jest than seriously, we say to the waiter in a restaurant, "I've left my wallet at home, I'll give you my watch instead," and pretend to slip it off. Bresset (2007) cites several Biblical passages in which rings appear, including the one just quoted, but also Genesis 41:42 (the Pharaoh who gives a ring to Joseph and puts a gold chain around his neck); Exodus 32:2 (Aaron who organizes a delivery of earrings); and especially Job 42.11: "every man also gave him a piece of money, and every one an earring of gold", i.e. Flakes and a Ring. However, I will add that, in reference to Job, the term used and translated as "piece of silver" or "piece of money" without excluding its general meaning, is *Kshitah*, which Williamson transliterates as *Kesitah*, though it should be *qshitah*. Besides Ribbon, Flake, or Piece of money, this term also means Rill and the Jewish Encyclopedia defines it as "pieces of money, moulded in the form of a ring," which would connect back to *Pecus* and *Pecunia*, to the Lamb of the Sacrifice, to the Abnu which take the shape of what they represent, and to the *kaniktum*, the witness to Fair and Just weights, which is also often created in animal shapes and of a fixed weight. Obviously, this further means that *Qshitah/Kshitah*, Flakes/Coins were used at that time, with an acknowledged definite value and a use in the exchange/offering of Goods. *Kshitah* is probably derived from the Akkadian *Kiššatu*, Indemnity for a lost object, Release from slavery, Compensation; a situation which necessitates a gesture of restitution of something in order to re-establish a state of equity; hence, once again, *Placare/Pacare/Pagare*/Pay. A reminder that Piece of money could also indicate a Portion/Fraction of a coin – i.e. Mina, as I pointed out earlier.

The Epic of the Ring

It is round, "empty", a bent "string" of metal, a "ribbon" of silver. Sometimes of gold.

In Sumerian we also find the term *Gur* which, besides referring to a measure of volume, also often refers to a measure of barley, which I would call Great Gur, GUR MAH, a unit of greater volume, equivalent to about 250 litres of barley (but some authors make it 500 litres; and it is possible that such variations depend on historical, economic, politic variables…) and we also find a GUR LUGAL, a Royal Gur. *Gur* (implicitly a small Gur) is the basis of the term Gaggaru/Gugguru (and a few Giants will probably take exception to this, but I am a "gardener", not a Numismatist/Linguist; I think Freud may have made a similar witticism about himself); and *Gur* also means rim, edge, border, margin, circle, circular object; and the verb to rim, border, hem, circle (from the Pennsylvania Sumerian Dictionary). And, among its many meanings, *Gur* also has "Sphere, Circle, Ring, Loop, Hoop" (see also Halloran).

The Ribbons, Rings, *Shat* and *Gur*, *Gaggaru/Gugguru, Kikkar* are related to *Kaniktum*, a weight with a specific mark of value (from the Akkadian *Kanākum*, to seal), certified, guaranteed by the god (often the god Shamash, the Sun, who represents a first step towards monolatry) and then, by the Sovereign; the *Kaniktum* literally serves to weigh the *Kikkar*. (A great many Giants have got their teeth into the subject of metrology too; personally, I have found some observations by Madden, Saulcy, and Williamson very interesting.)

Jewish cuisine today still has the *Kikkar Lechem*, the round bread with a hole in the middle, a kind of doughnut; and *Kikkar* also means 'town square' in Hebrew.

And *Kikkar* is also the root of *Cicer, -eris*, chickpea, the pulse, also meaning *roundness* (see Semerano). Perhaps the story of the lentils cited by Jung, cotechino permitting, is more to do with this roundness than with difficulty in digestion and "defecating coins".

Sédillot (1989) devotes a brief chapter to this aspect, a rather sketchy and very terse one entitled *Du lingot à l'anneau* (pp. 51 ff.) which it might to some extent be possible to connect to a later chapter in this book.

Hendin (2010, p. 66) presents very similar observations to those of Williamson, Madden and many others:

> Thus, gold bracelets and earrings became storehouses of wealth, worn by men and women alike. It was not uncommon for this wealth to be called upon when it was needed for commercial transactions or even for early fundraising.

And Javons (1875, p. 24):

> A passion for personal adornment is one of the most primitive and powerful instincts of the human race, and as articles used for such purposes would be durable, universally esteemed, and easily transferable, it is natural that they should be circulated as money.

Bressett seems to have the same opinion (2007, p. 24):

> Gold ear-rings and bracelets that are prevalent throughout the ancient world were decorative items that represented the wearer's wealth and status, but could also be used as money whenever necessary.

Mander and Notizia (2009, p. 86), while being critical of the concept of the Coin, add:

> "As previously described, the most substantial part of the silver reserves in the treasure-house of Puzriš-Dagn [a site founded by Šulgi] was devoted to the creation of luxury goods which were dedicated to the various divinities of the Sumerian pantheon or donated to members of the politico-military élite and to ambassadors from foreign lands. The most interesting group of objects of this type is the so-called silver rings. According to the texts in the archive, these silver artefacts weighed between 1 and 10 shekels.... Today, objects like those described in the archive of the treasure-house of Puzriš-Dagn and also attested in texts from the capital Ur, have been found in northern Mesopotamia, in the Diyala region. They are in the form of a spiral and show no inscription of any kind.... Such objects cannot legitimately be described as 'coins' considering that, in the first place, there is no evidence that they were accepted as a measure of value on the basis of their shape and, in the second place, that silver, in whatever form it was exchanged, will have been weighed until the latest phases of Mesopotamian history. It is possible to claim, however, that through the instrument of ceremonial donations by the king to various dignitaries or foreign personages, goods were assigned which not only guaranteed the recipient some prestige by virtue of being gifts from the sovereign, but also a capital to be spent as they wished, given the standardised natures of the silver rings."

As I have suggested, weights, measures, reparations, compensations, and general guidelines come to be standardised under Ur-Nammu and then Shulgi, as is recounted in the Prologue of Ur-Nammu quoted earlier, to be sanctioned by Hammurabi "a few" years later.

From rings we pass to the golden Mina and the silver Shekel, which are not in any different from the Gold Sovereign, or the Pound Sterling of the last century with its own fractions and fractions of fractions, including the halfpenny and the farthing. And what does the Bank of England do when it creates the gold Pound? Doesn't it establish the relationship it must have with the silver Pound and its fractions? And doesn't it perhaps establish the Weight, the quantity of gold that must be present in it? And isn't the Weight of the gold it contains directly connected to "third" standard weights – whether expressed in grams or pounds makes little difference – which are as valid for gold as they are for onions? And – as I've said before – what makes all this any different from what Ur-Nammu created? From the codex of Ur-Nammu:

> He established the measure of the sila of bronze, standardised the weight of a mina, and standardised the weight in stone of a silver shekel in relation to a mina.

From the MI.NA, a unit of weight, from the Gur, a unit of weight and a measure of volume, from the Siqlum/Skekel, a unit of weight, we move, using the same terms, to identifying monetary units.

Not forgetting that the god Shamash is almost always represented with a Ring/Circlet in his hand and a Staff, as I said before, strangely "similar" to the Mekku and Pukku of Gilgamesh. It is also an interesting fact that "a few short years" later, the god Ahura Mazdā, perhaps remembering Shamash, is represented with one arm pointing upwards, indicating the one and only true god, while with the other hand he holds a circlet.

And so begins the epic of the Ring (though, here too, many Giants will disagree; but maybe I'll have Frodo and the Lord of the Rings on my side;[2] and maybe Shamash as well; and certainly the Greek/Lydian Gyges).

And another epic begins: that of the particular and evolved circlet: the Moneta, Mνα-ετα.

And not only in Mesopotamia but in the other great civilisation of the period, the other superpower of the time, Egypt.[3] And likewise in the

---

2 "Three Rings for the Elven-kings under the sky,/ Seven for the Dwarf-lords, in their halls of stone,/ Nine for mortal men doomed to die,/ One for the Dark Lord on his dark throne,/ In the Land of Mordor where the shadows lie./ One ring to rule them all, one ring to find them/ One ring to bring them all and in the darkness bind them/ In the Land of Mordor where the shadows lie." (Tolkien).
3 Madden (1881, p. 2, note 15) observes: "The earliest payments, purchases, and even salaries in ancient Egypt appear to have been made or paid in *Utens* of copper." Uten is the name given to multiples of Shat. Madden adds, "The '5 kat' [standard unit of weight like the *kaniktum*]... formerly in the possession of Mr. Harris of Alexandria (...) is made of grey-green stone, known as Serpentine of the Desert. Upon the top of it is engraved a vertical band of hieroglyphs, which Mr.

nascent Hittite Empire.

As a result, this apparition concerns the whole of the then known world.

Or at least, it is what interests us here, since the Coin will only appear in the rest of the world after another thousand years!

But why did it begin with a Ring? And where in concrete terms, does it come from, this Ring that will evolve so rapidly?

At the moment, I have no idea. Maybe.

Edwin Smith has read 'belonging to the sun'. Madden refers to the reign of Thutmose III, Eighteenth Dynasty, and the grandfather or great-grandfather of Akhenaton. The text which Madden reports, carved on the 5 Shat weight, which he transliterates as 5 Kat, "property of the Sun [god]," can definitely be superimposed onto the definition/ concept which Ascalone and Peyronel (2006) report, referring to the Abnu of the god Shamash/the Sun.

19.
A VERY QUICK AND (ALMOST) CANONICAL HISTORY OF THE COIN

The best-known coin in Mesopotamia and its area of influence in the time this book refers to, let's say for convenience the time of Ur III (a thousand years after Enmerkar, in fact) is the *Siqlum-Shekel-Sheqel* (8,50 g), a word that comes from the Hebrew *Lishqol*, to weigh, and the Akkadian *Šiqlum*. And according to many authors the most widespread is the half *Siqlum* (4,25 grams).

It should also be specified – I mentioned this earlier – that, depending on the geopolitical area, there are at least four standard *Siqlum* (Syrian, 7,8 g.; Syrio-Palestinian or Egyptian, 9,4 g.; Anatolian, 11,7 g.; Assyro-Babylonian 8,5 g.; Ascalone and Peyronel, 2006).

For ease of explanation I'll make the Assyro-Babylonian my approximate standard.

The approximation is naturally my own, given that in the period, as is now clear, materials were weighed in "approximations" of thousands of a gram (1 *Siqlum* = 180 *Še*, i.e. 8.5g divided by 180 = g 0.047222); like many others, Hendin has a different view and holds that the oscillations in the weights may have been as much as 3 grams, although I can't see how that could be possible on this kind of scale, which allows for 1/180 of a Sheqel.

Then, we are told, the coin will appear in the Western Classical world in Lydia, with King Croesus no less, only a thousand or fifteen hundred years later: i.e. around the 7th-6th centuries BCE. Some even guess at the ninth century BCE.

Geographically, Lydia is half way between Assyria and Greece.

In the legends about the birth of the coin in the "Greek" world, besides Croesus, we find King Midas, the river Pactolus, and the Golden Fleece (I recall that flakes of gold were initially collected in a sheepskin immersed in a flowing river and used as a sieve, a filter to catch the flakes of gold naturally present in the current: that is, gold was collected, not excavated).

Among the best known and most important Greek coins, we have the *Statere*, from which comes the Latin *Statēra*, scales, which takes us straight back to the concept of Weight and its certification: in other words, the

Greek equivalent of the *Kaniktum*. (And maybe this sense of the English term Standard is closer to *Statere, Stadera* than to *étendard*, as the Oxford English Dictionary maintains; but that's not a question for today.)

In the Attic standard, the *Statere*, Στατήρ,-ήρος, also called *didrachma*, weighs 8.5g like the Assyro-Babylonian *Siqlum*. And the drachma, Δραχμή, -ής, being half a *didrachma*, weighs 4.25 g, like the *half-Siqlum*. (It is not clear to me how Saulcy, 1854, p. 18, the great nineteenth-century numismatist, deduces that the half-shekel is equivalent to the *didrachma*, but perhaps that's just a typo.)

In China, it seems that money appears around the 12th century BCE in the form of axes and hoes as in the Bronze Age Mediterranean, but two thousand years after Babylon, but only in the fifth century (a bit like the West) do round coins appear, and here too there is a hole in the middle (a bit like in Mesopotamia, as I'll illustrate better shortly, but in any case long after the Sumerians). Chinese coins are also round with a "square" hole in the middle, a combination which, the numismatists and ethnologists say, refers to the union of heaven (round) and earth (square). In other words, the contain a concept of Transcendence. There's no sign of a god anywhere, but there is the union of Earth and Heaven, a heaven that is the domain of the gods. The Chinese word or ideogram "China" is written as an empty quadrilateral: i.e. "Earth", since the Chinese believed that the Earth was square (perhaps giving us "the four corners of the world"), adding a vertical dash which bisects it, representing "The Middle Land". Some believe that the square "hole" is to make it possible for coins to be threaded on a string for portability, which makes sense from a practical point of view but doesn't justify the invention of the coin: it's more a justification for inventing the belt. In the same period, Greek coins show a mysterious square on the 'tails' side of the coin, making the two units of currency very similar from a certain point of view.

A few centuries earlier, still in China (8th century BCE), "coins" are moulded in the shape of a knife and, as an in-between stage (7th-6th centuries) coins appear which are round, with the usual square hole in the middle, and small blades on the outer edge.

Circular coins with a hole in the middle will appear at different moments later on in the reproduction of the various Caterpillars, and in various parts of the world, including the Celtic regions, just as coins with the most diverse shapes will appear in other parts of the globe and in eras far distant from each other.

In addition to specialist numismatist sites, you can just click Ring Money on Google to see what production of Rings and Ring Money, and how much, is present in all historical periods.

Rings, knives, blades, heaven, earth, divinity, transcendence, sacrifices, symbols, covenants.

One last small note taken from Parise (2000, p. 114) about the common conviction that Lydian coins were among the first:

> Thanks to the testimony of the Annals of Sennacherib (E 1, VIII, 9-19), according to which *half-shekel* coins were frequently minted in Assyria, it seems possible to conclude that in the upper Tigris valley drops of metal were in circulation with a determined weight, issued under the control of the king.

And in a fuller argument to which I am referring:

> And while the coin was not yet explicitly named, the reason was perhaps to be found in a *stubborn error of perspective made by modern scholars*. But if this was how things really were, the earliest use of coins may very well be ascribed to the Assyrian Empire (my italics).[1]

In a 2009 work written by Nicola Parise's students and dedicated to him with great affection and gratitude, a work with the significant title *Obeloi* (eds. Camia and Privitera), Maria Emanuela Alberti, writing about Rhodes and the Aegean, concludes her fine contribution as follows:

> The standards referred to were all eastern.... However, it seems clear that in this historical phase [9th-8th centuries] the Minoan-Mycenaean traditions had declined and the most widely diffused standards were those of the more dynamic Levantine societies, probably in continuity with the tendencies of previous periods. And these eastern units with a long tradition, once assimilated, will constitute the basis of many systems of weights and money in historical Greece. In this connection, it is essential to recall the rediscovery of a weight from a set of scales at Pithekoussai (early 8th century), a ring of bronze filled with lead and weighing 8.7g: a mass equal to what will be known as the "Euboean statere" but is in fact the heir of the Mesopotamian shekel.

We are no longer among the Sumerians, but Sennacherib is the son of Sargon II whom he succeeds in around 705 BCE. According to some scholars, Sargon II takes this name specifically to suggest continuity with his "ancestor" Sargon the Great, Sargon of Akkad (2270-2215 BCE) who would actually be the founder, setting out from the coasts of Lebanon and

---

1   It is likely that Professor Parise was not aware of the revolutionary discoveries by Aglietta and Orléan.

Cyprus, of what will later become the Minoan civilisation, precursor of the Greek. Let's not forget that the previously mentioned Manishtushu is the son of Sargon the Great; and in connection with the possible conquest of Crete by Sargon, it is interesting to observe how two, admittedly contentious authors, Albright and Waddell, during the nineteen-twenties, point out the various names which Manishtushu assumes in the Eastern Mediterranean: in Sumerian he remains Manishtushu/Manishtusu, in Egyptian he becomes Menes or Manium and in Minoan, *Minos*; curious. On which subject, Semerano also derives the name Minos from the same root as Moses, i.e. *child*, a term of Egyptian/ Middle Easter origin (and others, including Freud, make this association; but Semerano does it in a slightly different context); and given that Sargon the Great had conquered everything, ahead of Alexander, but from east to west unlike the Macedonian, he too "going to Hell" according to Jeremias, this hypothesis would reinforce the influence of Mesopotamia in the acculturation of the "Pelasgians" and in the foundation of Minoan civilisation, from which the Mycenaean and the Greek cultures will originate. If we add to this the fact that the "upper Tigris valley" is situated between Colchis and Lydia, the possibility that Lydian coins are not the first is significantly increased.

I have mentioned the *Obeloi* and the *Drachma* in connection with Bernhard Laum. Still from Parise (2000, p. 13):

> The skewer, *obelos*, gave its name to the best-known fraction of Greek monetary systems (*obolos*); and the bundle of skewers, *drachmē*, to the corresponding fundamental unit. Orion (Sturz, pp. 118, 19-22) and the *Etymologicum Magnum* (613, 12-15) took from Heracles Ponticus (cf. 152 Wehrli) the story of how Phido of Argos, once the skewers used as an intermediary had been withdrawn from circulation, had dedicated a certain number of them to Hera, replacing them with a silver coin minted in Aegina. Plutarch (*Lysander*, 17, 5) and Pollux (IX, 77) testify to the continuing use of the skewer as a measure of value and a means of exchange before the coin [Plutarch makes Lysander say, 17, 5: "In ancient times little iron rods, sometimes bronze, were used as coins"].... Principally based on the inventory of the *hiera chremata*, Ἱερὰ Χρήματα, of the Thespians (SEG, XXIV, 361), it is believed that the tripod, cauldron, and skewer formed a kind of triad of pre-coinage currency and that, when necessary, they were dedicated together in the sanctuaries: the cauldron on the tripod and the bundle of skewers in the cauldron.

Parise suggests that we suspend this hypothesis; whatever the case, I think *hiera chremata* is hard to translate, except completely tendentiously as Sacred Usury, Divine Usury, or Great Usury. Later, on p. 16, after noting

Gernet, another "heterodox" author little cared for by Dumézil, Parise writes in a manner close to Desmonde's reading:

> In fact, as B. Laum demonstrated, "skewer" does not mean only the sharp iron rod onto which portions of meat are slipped for roasting, but also by metonymy the portion or portions of meat slipped onto it. And the contrast between concrete and abstract which the actual circumstances of the utensil do not let us glimpse, becomes absolutely evident if what is to be taken into consideration is instead the meat and the ways it is distributed.

Indeed, in Brescia when someone is invited to *"mangiare lo spiedo"* [literally "eat the skewer"], a typical dish in these latitudes, in general this refers to birds, all the better if their hunting is prohibited, *"mombolini"* [strips of pork loin], potatoes, and meat skewered and cooked, and certainly not to the iron rod holding them; and yet the skewer, *Obelos*, is still the rod of metal that will generate the *Obolos*, and this is the meaning I want to stick to.

> Round the blazing fire
> and the crackling skewer
> stands the hunter whistling
> on the threshold to admire
> among the reddish clouds
> as flocks of black birds
> like exiled thoughts
> migrate in the twilight.

(Giosué Carducci, St. Martin (in *Rime Nuove*, 1861-1887)

By contrast, my cousins in Bergamo could have used a frying pan as *hiera chremata*. But, setting aside culinary matters, (on the etymology of this term, from the Akkadian *qullû*, to glow, burn, see Semerano; roasted, burnt, in the Concise Dictionary of Akkadian) I wanted to pause over these words of Professor Parise's which, picking up from Laum and developing the discussion more broadly, clearly emphasizes in great detail the sacred matrix from which the Coin arose. Even though his whole discussion is positioned between the 9th and 7th centuries BCE: that is, 1,000 or 1,500 years after Ur-Nammu. And again, in Greece.

What is true for the skewer, *obelos*, is also the case for the *drachma*, which takes its name from *drassomai*, δράσσομαι, "hold in the hand, grasp," and whose "value" is fixed at 6 *obelói*, and likewise for the *statere* which becomes a *didrachma*, the equivalent of 12 *obeloi*, as I've said;

from the tool which helps to do something, the accent shifts onto what is done with it; the name that indicated the utensil passes by metonymy to the finished product realized by the utensil.[2]

---

2   On this subject, in *A History of Measures*, a book from some time ago (around 1925), Livio C. Stecchini, a "slightly" heterodox scholar also quoted by Desmonde, writes, "The Hebrew term is *kikkar* from the root KKR meaning 'to oscillate, to move in a circle'. The Hebrew *kokkar* as well as the Greek *talanton*, refers also to the pans of a balance. The Hebrew and the Greek term are used to describe a ritualistic object consisting of a small disk representing a scale pan; I have determined that coins developed from this object. The meaning of the Akkadian *gaggaru* is related to this last meaning of the Hebrew *kikkar*." It is clear that Stecchini is not referring to the Skewer, but nevertheless to the tool (the pan of the scales) and not to the material being weighed.

20.
THE SACRIFICE AND THE NEW FAITH

The Sacrifice is certainly "violent" (from *vis*, force, strength) and it has a heap of connotations and features that are often lurid, savage, bloody; but I'm not going to address Girard's Mimesis or Mimetic Desire, nor bother to attempt a fusion of Marxism and mimetic theory as the advocates of the Penny Black have tried to do by playing with words and attributing "sophisticated" meanings to abstract notions which, as I have already hinted, smell of the lamp.

I think you can see where I'm aiming for without the need to trouble Weber, Girard or Money-Kyrle any further. It doesn't seem important here, for now at least, to distinguish between Taxes, Extortions, Stamps, Duties, Exactions, Excise, Annona, Tributes, obligatory "Gifts", Tzedeka, ritual Almsgiving, institutionalised Robbery and the rest; by Sacrifice I mean the voluntary act by which one makes an offering with a sacred function: that is, the function of rendering something sacred: any act aimed at entering into contact with the deity by offering something proper to oneself. "Proper" in the sense of "personal", "private", "mine", no one else's.

Whether we mean sacrifices aimed at eliminating and hence sanctifying the scapegoat (Girard), or expiating one's own guilt, or corrupting, buying, placating, ingratiating oneself with the gods through a preventive or reparative *captatio benevolentiæ* (my summary of Money-Kyrle), or whether we mean "simply" apotropaic or magical features, the elements necessary for the organization of the sacrifice are always the same: the faithful, victim/sacrificial object, deity, celebrant, blood, and knife. Not all invariably present, but all invariably present at least in symbolic form. Bernhard Laum, discussing the sacredness of the Sacrifice and the sacredness of Money, says much of interest, although as I mentioned, it should be backdated by at least 1,500 years:

> In order to gain historical knowledge about a form of civilisation, it is always necessary to go back to the point where a specific word has been coined to designate a specific form, since only at that point is there a high probability that the form of the thing and the meaning of the word are in harmony. *The essence*

of an object is enclosed in the form in which it is outwardly represented and in the word which expresses its intimate meaning (Catarzi, 1997, p. 25; my italics).

If the sacrifice of a "special" animal or plant or object is capable of generating the Totem and hence exogamy (even though it seems to me that exogamy was the normal condition of Sons as far back as the time of the Father of the Horde – that is, before the Assassination – since the Father was the lord of all the Horde's women and he compelled his sons to leave the group) with all that it entails, including the *Jus primæ noctis* hypothesized as existing as far back as Gilgamesh; if the sacrifice of a "special" person is capable of generating Royalty (Girard); if the main instrument of the sacrifice itself, the knife-axe-hoe-stylus applied to clay, is capable of generating the first "pre-coins", the first form of Writing and thus giving History its beginning; if the sacrifice of a bull, which has in the meantime become the sacrificial animal par excellence across the world, a shared universal symbol of strength and potency, going back to the Heavenly Bull which Gilgamesh will defeat – a Bull we will often meet in the epithets given to Gilgamesh, which recur constantly in his epic, and is also applied to Hammurabi in the middle of an enormous list of epithets in the prologue to his Code (*"I am the wild bull who annihilates his enemy"*, as we have just seen abundantly) – generates in turn the first letter of a new alphabet (even today the letter A upside down represents a bull's head and is also the origin of the Hebrew *aleph*; and the sacrificial Bull is also present far away in ancient China); if the Sacrifice/Offering of part of one's own Wealth (Youth, physical Vigour, Pre-eminence, muscular Strength, Time, Devotion, Love, Faith) through the *corvées* generates the Tablet-Money, what comes to be sacrificed in order to generate the *Gur-Gaggaru-Kikkar*-Ring? These seem to be almost "equi-valent".

In accord with the hypothesis which emerges from these "equivalences", and taking into account Bernhard Laum's indication just quoted, what comes to be sacrificed should be something new. And in order to be *sacer*, it has to be something personal, private, mine, must be dedicated to the deity (compare this with Mircea Eliade, 1937, p. 42, who, on the basis of a bad translation of the *Enûma eliš*, introduces the concept of Self-sacrifice which Marduk would make in order to create man; and men, in a general sense, imitate the sacrifice which the deity himself makes, a sacrifice connected to the Creation). It must be something small, to avoid having to return to the Temple as the Tablet does. And having the form of a ring, it must refer to a Ring. Form and Word coinciding, says Laum. In the chapter dedicated to him I quoted a very decisive assertion of his:

Therefore, I assert that in ancient Greece only *instrumenta sacra* could become and did become money.

Now I think I can answer the previous question: what is the Sacrifice? In the Bible the word Money appears for the first time in *Genesis* 17:12. No, not exactly. The word "money" is the translation of *Kesef*, which literally means Silver. And, since Carolingian monetising is yet to come, the French and Spanish echoes must be rejected, at least historically speaking, though it is a fact that *argent* in French is used to indicate Money/Coin (while in Spanish Silver/Money is also *plata*, and a plate is round, as are Rings; I am only partly joking, given the engraving reported by Stecchini a little while ago). The point is that Silver-*Kesef*, "buying something with silver", Enmerkar's "giving for/in exchange of silver", is different from buying something with the "tablet" of clay-money, or receiving a kid in part-payment for work done for the Temple, and it is different from bartering. Here, in *Genesis* 17:12, *Kesef* means money, i.e. coins; not clay tablets, not payment in kind, and not barter. It means Coin, since, as we will see later, Avimèlech will pay Sarah 1000 *Kesef*, 1000 coins (*Bibbia*, ed. Rav Disegni). And so, in *Genesis* 17:12 the Coin appears for the first time in the Bible. "Simultaneously" in the code of Ur-Nammu, reprised by Hammurabi, and before that by Manishtushu, the *Siqlum* and the *Mina* appear, understood as Coin and not "simple" unit of weight. Hence I do not understand why Attali (2002, p. 47) should write:

> *Pour la première fois, la monnaie apparaît au détour d'un verset du Livre des Juges (15:9), quand il s'agit de payer à Dalila le prix du piège qu'elle tend à Samson* (The coin appears for the first time in a verse of the Book of Judges (15 :9), which tells of Dalila being paid to set the trap for Samson),

adding that the *Sheqel*, Shekels, appear a little later in the same book of the Bible, *Judges* (17:12), which should in fact be 17:10. A few lines earlier, still referring to *Judges*, claims:

> *La monnaie n'existe pas encore*
> (The coin does not yet exist)

And a little later, addressing the Classical horizon, he adds:

> *Mais, par-delà la métaphore, il ne s'agit ici que d'un anachronisme : on le verra, la première monnaie au monde n'apparaîtra que cinq siècles plus tard, en Asie Mineure.*

(But, except as metaphor, this is a mere anachronism: as we will see, the first coin in the world will not appear until five centuries later, in Asia Minor).

Apart from the fact that the first reference is an error, because Dalila's payment appears in *Judges* 16:18, not 15:9 (but I think that I too, as I type, will have got a lot of references wrong), I find it surprising that Attali can believe that the *Sheqel* first appears in *Judges*! After *Genesis*, after *Exodus*, after *Leviticus*, after *Numbers*, after *Deuteronomy*, after *Joshua*? As Money-Coin-Shekels-Talents, the Coin appears 135 times in the Bible before the verse indicated by Attali. It's inexplicable. And yet more inexplicable is the fact that in *Judges* 16:18 (and not 15:9; yes, I know, I'm a bit of a pedant) what Attali indicates as Coin is *Kesef*, exactly the same term as we find in *Genesis* 17:12 with Abraham. (There would also be a very important mention of gold, the river Pishon and the land of Havilah as early as *Genesis* 2:11). Moreover, silver in Akk. *kaspum*, has the root *ksp*, which is not so remote form *ksf*, all the more so since the Hebrew *Pei* is a dual letter used to render both P and F: in the case of *kesef*, it is a terminal F *sofit*.

All of which takes into account only the Bible, written, let's say, very generally speaking, from the 8th century BCE, and without taking account of the Akkadian *Siqlum*, which is more ancient by at least 1,500 years.

Whatever the case, I don't wish to spin out the references to Hebrew and Sumerian: instead, I'll extend "Sumerian" to include the whole of ancient Babylon (i.e. exactly what Professor Jeremias did; Babylon which continues to use a modified cuneiform script despite the changes of dynasty and population shifts, and their reciprocal consequences. Oh God, I'm getting perilously close to the concept of *Euphratesians*!). However, in Hebrew, the Sun for example is *Shemesh* (I apologize for the transliteration, an apology which needs to be extended to many of the transliterations I've used) and in Akkadian the sun god is called *Shamash*, as the sun is in Arabic. And for millennia the dynasties of the Semitic people who took over from the Sumerians considered it fundamental to preserve the appellation "King of Sumer and Akkad" for their own sovereigns. And they continued, again for millennia, to use clay, knife, and cuneiform script, although it was a cuneiform in constant development until it became a kind of alphabetical cuneiform in the Persian Empire.

And as late as the sixth century BCE, a Babylonian wonders "What kind of scribe is it who doesn't know Sumerian?"

And what's more, Abram the Chaldean comes from *Ur* (like Mathematics, the sexagesimal system, Astronomy, Astrology, Magic, and Sarai who reads the future and is a wonderful kind of witch princess), from *Ur*, which besides being the name of the City, also means Blaze, Fire, Light, Furnace; and it is in one of these furnaces, in *Ur*, that Abram's brother, Haran, dies, as I will explain more fully in a little while. And reading the comments of Rashi de Troyes, the great Mediaeval Biblical exegete I introduced earlier, we could also see in this death the conflict between the old Faith and the new Faith (now it is Inanna who is "old" and the transformation of *Utu/ Shamash/ Marduk/ Ba'al/ Be'el* into *Yahweh* that is "new") well represented by the familial dimension which this conflict takes on: Terah, father-old-world-ancient-idolater on one side; Abram, son-youth-new-monolater-monotheist on the other; and in the middle, Haran, who tries to be clever and dies for it. It's as if there were no room left for middle ways: there's either Idolatry/ Polytheism or Monolatry/ Monotheism.

And while *Or* in Hebrew means Light, including electric light today, and in the past also the flame of a lamp that "burned" oil, in the Second Book of Esdras it is also the root, for example, of the name of the archangel Uriel, Light of God, who returns, note, from Babylon. *Or*, אוֹר, written in exactly the same way as Ur, changing only the point which marks the vowel ו on the Vav – thus אוּר. And somewhere between divination and magic, on Aaron's chest we find the *Urim* and the *Thummim*, Lights and Perfections, stones on his breastplate which shine when questioned by the High Priest seeking divine responses; all of which is connected to the Babylonian Tables of Destiny (see the *Jewish Encyclopedia*). And *Ur* – and I've written this before, but I'll say it again – still understood as Sumer, is the place where almost all the myths and stories present in the Bible were created, starting with *Adam* made from *Dam*, Blood, and *Adamah*, Earth. Adam could be the masculine form of Earth, earth given life with Blood; earth, clay which is often Red, i.e. *Adom*: so perhaps Man is connected to Clay/ Bricks/ Civilisation? Baked earth? Tablets perhaps? And maybe they are all derived from the more ancient Adapa, "first" legendary man of Sumerian tradition who, from fear and deceit, "refuses" immortality. Adapa is also indicated as one of the *Apkallu* (Pettinato, *Mitologia sumerica*, p. 432). In Akk. Blood is *damum*, and *ada(m)mu* is the colour red, and *adamatum* is red earth; but Battle/ Conflict is also *adammû* and, still in Akk., we find *adāmum* meaning "activity, business matter". There is a delightfully prudish translation of the Sumerian ADAMA in the ePSD as "a dark-colored bodily discharge".

Unlike Jewish tradition, in which Adam is born already "perfect", already circumcised like Noah and other especially important personages, Sumerian tradition makes the first man actually the eighth: that is, the outcome of the eighth attempt in a challenge between Enki and Ninmaḫ. After a series of "Men/Creatures" all the work of Ninmaḫ (*ibid.*, pp. 407ff), and all characterized by serious deformities (weak hands, over-sensitive eyes, misshapen feet, idiocy, inability to *u*rinate, no reproductive capacity, lack of genitals both male and female), Enki finally creates a hairy, feeble entity incapable of standing unassisted, called UMUL to emphasize the fragility of the human being. In the Sumerian language U.MU.UL does indeed mean "sickly creature" (from the CAD).

In another Sumerian myth, Man is born from the sacrifice/killing of the gods called the Alla by the other gods; the blood of the Alla will give life to Ullegarra and Annegarra, who will have the task of standing in for the minor gods in everyday work and in the *corvées*. And from *Ur* come the creation myths which imagine Man, Hebrew *Ish*, from whom Woman, Hebrew *Ishà*, is generated, as created with the same earth that is used for manufacturing bricks, baked in the furnace and no longer 'raw'.

It may seem a tangential fact, but cooking bricks is no trivial matter; it means "sacrificing" wood and coal, differentiating their use, "taking up time, heat, technique" for something "after", for an idea of Eternity which simple, everyday, malleable, raw clay cannot manage. And the same applies to tablets: the tablet takes on the sense of transmissibility into the future, the sense of an enlarged view of time, not that of everyday survival but of a legacy/ memory to be left to someone we do not know; and here too we have a different Time from simple everyday life.

And the Bull-*El* that is sacrificed has to transcend the immediate and seek the eternal: what could be more "stupid" than an agricultural society "throwing away" a tool that is essential for working in the fields, a precious and very costly tool, the source of the energy which drives the plough and transportation, and generates still other forces (calves), generates milk, utensils, clothing, and meat, and multiplies, likewise multiplying one's wealth? What could be more stupid, considered solely from an economic viewpoint? But *Ur* always has its eyes fixed on the Temple; this is Enmerkar, gazing upwards to the Infinite, the Eternal, from the height of his Ziggu*r*at. And it is *Ur* that originates the narratives of the Universal Flood, the Earthly Paradise, the genealogies of the Patriarchs from Noah to Moses (the various King Lists), not forgetting the Jeremiads (see the *Lamentations for the Fall of Ur* and for that of *Akkad* in Pettinato, 2001).

Now, it seems clear to me that there is some affinity, continuity, shared striving towards the Infinite between *Ur* and the Jewish people, and I believe it may be helpful to take that into account in examining these subjects. The name of Jerusalem itself seems attributable to *Ur*, as Liverani

(1999, p. 18) also points out, but he is not alone, since in the el-Amarna Letters, written in cuneiform, Jerusalem would be *Ur-u-Salim*, merely the City of Salim (in Sumerian, city is IRI; in Hebrew, *ir*. But THE City, in Sumerian, is UR). In my opinion, Josephus Flavius, takes a different view, drawing on his Hellenistic learning, and since Peace is *Shalom* – but in the Greek version of the Bible, the Septuagint, it is rendered as Σόλυμα – derives Jerusalem from *Iero-Solima*, referring it seems to a place two stadia from Jerusalem where Abraham would have been received by Melchizedek, then King of Sodom (not yet corrupted, but allied with the Patriarch). From this etymology, Josephus Flavius turns the simple, modest town of a certain, unknown Mr Salim – Jerusalem/ *Ur-u-Salim* – into a lofty-sounding Sacred City of Peace/ City of Sacred Peace; an *Iero-Solima* rather different from what emerges in the el-Amarna Letters.

See also *Genesis* 11:2 which speaks of *Shine'ar*, which is Sumer and it is in Shine'ar that the universal language is spoken, the sacred language according to Rashi. And the Hebrew names of the months directly reflect those of ancient Akkadian: the Akkadian *Nisanu* corresponds to the Hebrew *Nisan*, the Akkadian *Ayaru* to the Hebrew *Iyar*, *Simanu* turns into *Sivan*, *Du-Uzu* into *Tamus*, *Abu* into *Av*, *Elulu* into *Elul* and so on. I have already mentioned some of these things in relation to Jeremias, but it seems important to brush the dust off them now.

After this long and seemingly disorderly digression, I come back to the subject. I was talking about *Genesis* 17:12. Abraham, *Ur* of the Chaldees, and the Covenant with God.

God has made many covenants, *Brit*, with man; this will be done out of trust; or because He is more stubborn than a wild ass, more stubborn than a ḫāwiru; or because he can't resign Himself to the idea that He's created such a bungler so unlike His image and likeness. I don't know why, but He's made a lot of covenants.

With Adam, with Noah, with Abraham, with Moses (in the *Concordances* on the Vatican website, Alliances in the Old Testament are mentioned fully 200-220 times). Christ too renews the Alliance by means of the Eucharist (in the New Testament, 25-30 times). But with Abraham he makes one in particular: Isaac, *Ytzchaq*, which means "he will laugh", must be circumcised[1],

---

1 The *Brit milah*, "Covenant of Circumcision" [but as we have seen, this is very close phonetically to Covenant of the Word] was required of Abraham by God as a sign of the eternal bond between Him and the House of Israel (*Genesis* 17:7). This is a *mitzvah*, commandment/duty, a unique one in that it is imprinted on the flesh of every Jewish male: we are told that King David, being observed bathing unclothed, was unhappy not to be able to fulfil any commandment at that moment, when he

circum caedere, on the eighth day after birth, and Abraham too, in his nineties, had been circumcised the year before (and I don't think *Avraham ytzchaq*, that "Abraham will laugh" about this). Ytzchaq won't have been very happy about it either, but there's always primary repression (which doesn't come without a cost; maybe it would be interesting at this point to return to the concept of partial object and the presumption of disconnectedness from infancy, but I'll do that another time), and I do think that, for all the official declarations made at the time and the son's well-known good disposition towards such a father, even *Ytzchaq* (Isaac) *Ytz'aq* (will cry out). This is, of course, the *Brit Milah*, Circumcision. In fact, God had already given Abraham another covenant which is often disregarded. It is a "strange" covenant (*Genesis* 15:8-11), in which Abraham wants to know Sarah's ovulatory-generative situation:

"Lord God, whereby shall I know that I shall inherit it?"

A question to which God "replies" by requiring the sacrifice of a heifer, a she-goat, and a ram, all three years old, to be divided in half, followed by a turtle dove and a pigeon. And Abraham :

laid each piece one against another: but the birds divided he not.

---

remembered the *milah*, and was heartened by this. The aim of this story is to teach us that even a man with nothing, not even clothes to wear, still has the opportunity to acquire the merit of circumcision. The *milah* must be performed on the eighth day after birth during daylight hours: if by mistake the circumcision is carried out before the eighth day or during the night, it is not valid and a drop of blood must be spilled after the wound is healed. In accounting terms, the day of birth is considered as day one, and so the *milah* happens on the same day the following week. The *milah* consists in the performance of three normally distinct actions: 1) the *milah* proper, the cutting of the foreskin, the skin which covers the glans; 2) *peri'ah*, removal of the underlying mucous membrane; 3) *metzitzah*, sucking the blood from the wound. Tradition explains the choice of the genital organ as the site of circumcision with the fact that, when he gave the commandment of the *milah*, God had instructed Abraham to be "perfect" (*Genesis* 17:1) (in the King James Version; blameless or whole or complete in most modern versions) and the male foreskin is the only part of the body that can be removed without causing mutilation. Therefore, implicitly, it is only after the *milah* that man can call himself perfect. Another explanation attributes the *milah* to the wish to moderate the sexual appetites. The *mitzvah*, the duty to carry out the procedure, falls onto the father; if he is not able to perform it in person, he usually employs an expert to carry it out in his place, the *mohel*, who acts as the father's delegate/deputy. See also a more complete and detailed description in the *Jewish Encyclopedia*. One observation: the earlobe, for example, could be removed without mutilation. Who knows why this was not chosen? Or the hair. Or the fingernails (for the cutting of hair and nails, see Caetani, 1914).

The Sacrifice and the New Faith

There follows a deep sleep for Abraham, with terror and darkness. Then God foretells the Egyptian slavery for 400 years and other things. At the end:

> behold a smoking furnace, and a burning lamp that passed between those pieces. In the same day the Lord made a covenant with Abram (*Genesis* 15:17-18).

But God does not seem to answer Abraham's question, or at least we don't know the answer. Therefore, it is not given us to know how Abraham managed Sarah's ovulatory situation; that he did succeed in dealing with it is indisputable, given the birth of Isaac, but we don't know how. I wouldn't like to seem disrespectful, still less blasphemous, but this passage from the Bible is very "strangely" constructed: the usual exegesis understands the words "whereby shall I know that I shall inherit it/ that I must have possession of it?/ that I shall possess it?" as referring to the Land which God promises to Abraham. But if that is the case, it is Abraham who, by asking for explanations, risks being disrespectful and blasphemous, since God's promise alone should be enough (Moses will be punished for far less); moreover, what happens next does not answer Abraham's question "whereby shall I know". And so, prefer the hypothesis – a bit of a bizarre one, I know – that I have just formulated, and I certainly don't want to go into the symbolic significance of the animals being sacrificed (turtle dove and pigeon); but the meaning of "inherit/ know" is intriguing nonetheless. The whole ritual recalls something markedly pagan, suspended between magic, polytheism, monotheism, and augury, which seems more applicable to the promise of a son than to the moment, which Abraham will not see, when he takes "possession" of the Land. If I wanted to play the "mini-Bion" I could jokingly say, "One day a dictionary came into my consulting room and after a long silence said in Hebrew, '*Iarash*'. I did not understand, and asked, "What?", and it said, 'Inherit.' Thinking about this for a bit, disregarding the fact that it might or might not *inherit* something, I said, '*In-her-it*', and it went away satisfied." This has nothing to do with "ice cream" and "I/scream" as in the celebrated passage from Bion, or his playing around with the term *at-one-ment*, in which we forget that a little-used English verb, to atone, expiate, hence atonement, expiation (not "thinking with one mind"), but the verb *Iarash* does indeed appear in the text from the Bible (*Genesis* 15:8-11) and is indeed rendered as inherit. But when I playfully split this verb à la Bion, I get my bizarre reconstruction. However, a brief observation needs to be made about pigeons.

From Caetani (1914, p. 105):

> In the Ka'bah, pigeons were and still are sacred and untouchable in the Middle East, being the sacred bird par excellence dedicated to female deity of love and fecundity.

And in *Leviticus* 12:6-8:

> [The woman who has given birth] if she be not able to bring a lamb, then she shall bring two turtles, or two young pigeons; the one for the burnt offering, and the other for a sin offering: and the priest shall make an atonement for her, and she shall be clean.

Eliade (1937, p. 61) refers to Hittite rituals of bird sacrifice (pigeons, turtle doves) specifically connected to good auguries linked to childbirth; and so I think my observation might turn out to be a little less off-beat than it may seem when viewed in a purely humorous light.

In fact, what seems an innovation, Circumcision – and in some respects is, at least for the Hebrews given that until then it does not appear to have been among that people's customs (Noah isn't circumcised and the covenant begins with Abraham. But von Rad, 1972, p. 36, has a different opinion: "The nomadic progenitors of Israel already practised circumcision") – could also be seen as an enormous attempt at integration which von Rad himself locates in the transition from a nomadic or seminomadic population to a sedentary agricultural population.

At the start of the second millennium BCE, many populations around Abraham were adopting circumcision. The Egyptians certainly did, as the highly celebrated images at Saqqara demonstrate (and if Moses was Egyptian, how on earth, for the very reasons Freud illustrates, did he forget the circumcision of his son? And his own?). And was Moses noble or not? Did he stammer or not? Did he have other speech defects (I beg your pardon, I should have written, 'was he otherwise phonetically challenged'?). Was he killed by his own people or not? Is it possible that another Moses turned up, someone with the same name, taking over his inheritance, role, charisma, and objectives after the Murder? And where did he come from? Madian? And as for Madian, which of the two do we mean? Yerushalmi's work on these questions is very interesting, especially from p. 123 onwards. It is said that Pythagoras, "a thousand years later", had to undergo circumcision merely in order to enter Egypt, so it is very strange that Abraham was not circumcised before!

The Chaldeans are believed to have circumcised themselves, but not all scholars agree on this.

And in the mysterious Harran, in Armenia[2], it seems was indeed performed. (On the mysterious Harran, Saporetti (1985, p. 14) regards Harran not as Damascus, as the majority of exegetes do, but a still more northerly city, Carrhae in Latin, where Crassus met his death in 53 BCE at the hands of the Parthians who, we are told, poured molten gold into his mouth to highlight his lust for gold.) For Freud (*Moses*, p. 335), in the time of Moses, long before Abraham:

> it may safely be presumed that the Semites, Babylonians and Sumerians were uncircumcised,

although Herodotus (II, 104. 2-3) writes:

> The Colchians certainly appear to be of Egyptian origin.... The inhabitants of Colchos, Egypt, and Ethiopia are the only people who from time immemorial have used circumcision. The Phoenicians and the Syrians of Palestine [from the Greek viewpoint, Israel means the Syrians of Palestine] acknowledge that they borrowed this custom from Egypt. Those Syrians who live near the rivers Thermodon and Parthenius, and their neighbours the Macrones, confess that they learned it, and that too recently, from the Colchians."

As for the Sumerians, we do not know; some say yes, some say no. (Buccellati is of the latter opinion.) Personally, since the ancient Sumerian pictogram representing Man, reported in Kramer (1958), indicates a clearly circumcised penis, I would incline towards yes.[3]

Practically the whole area that once was Babylonia, plus Egypt, Ethiopia,

---

2 "Circumcision was well known also among the Chaldeans on the mountains of Armenia and Kurdistan, as is well documented by the clay tablets rediscovered by Sir [Austen Henry] Layard in 1849 in the ruins of the palace of Nineveh (from 1600 BCE, it is believed) and collected by Ashurbanipal" (Professor Paolo Santoni Rugiu, Past President of European Plastic Surgery, 2004). This information is also reported in the book by Calcagno (2009) where we can find an extended description both of the almost global diffusion of the practice of Circumcision, and of the sometimes bloodcurdling procedures by which circumcision is/was carried out.

3 From left to right, the evolution of the term Man, from the initial pictogram to the ways it was written in Sumerian cuneiform (summarised from Kramer, 1958).

the whole of Turkey, and Syria adopted circumcision, a rite that will also be adopted by the other great Semitic population, those who will become Muslims.

Strangely, Avram, going to Canaan and then to Egypt, makes a ten-year stop in Haran (it is not his brother; here the H is a *het* not a *hei*. Furthermore, the double r, i.e. Harran, appears in some texts, while at other times it is just Harran; the same is true of Avram's brother), possibly the homeland of Terah, Avram's father, and certainly a Chaldean and Amorite city (which reappears with "the oaks of Mamre" in Amorite territory) where God meets Abraham, *Genesis* 18:1), hence Semitic and in the area of influence of old Sumer, but lengthening his journey by at least a thousand kilometres, and Harran stands very near the Hittite kingdom, which will soon afterwards conquer both Babylonia and Egypt. And Babylonia had only just been through this at the hands of the Amorites. Remember that there are two captivities described by the Jewish people. Maybe more, if we want to take other features into account, but still in two places, Egypt and Babylon.

But Abraham's inspired innovation is important here, just like the inspiration of Enmerkar: as an indication that something has happened. Overall, it isn't important to know the name of the seminomadic shepherd who used a sickle to cut the first notch in history on his staff (Zipporah, wife of Moses, will use a sickle to circumcise her son; some say, to circumcise Moses himself), nor to know if whoever carved the first words really was called Enmerkar. It is important to know that something happened around that time. Just as linear time isn't of much importance in the unconscious and in dreaming, here too "historical" time and space aren't very important. What is important is what happened, how it happened, and why it happened, given that it continues to happen. And the more it happens, the more it changes what is happening. Like the syncretism of the Jews who, borrowing all their juridical, mythological, and religious precedents, did not stop at copying-and-pasting but, by dedicating to one god what the "idolators" spread over an infinite number of divinities (the method that Mohammed will copy), this syncretism changes religion itself and the way it is perceived, introduces new meanings, rituals, customs, values, creates traditions, nexuses, developments, all entirely unknown to previous religions. This syncretism becomes foundational of something absolutely original. And it necessarily introduces and generates new symbols, new covenants. A new *Brit*.

Among other things, Bruno Bettelheim observes that, in the populations which adopt it and in the most varied parts of the globe, circumcision does not appear to be linked to religious elements, but to those connected with rites of passage, whether lay or societal, or concerning changes in role, and to elements involving "bisexuality" and/or the envy of one sex for the reproductive-sexual attributes of the other; he provides some starting points for reflection which, while detaching himself from the "simplistic" Freudian metaphor of the Horde and the bloody, castrating Father, enrich the debate about the genesis, the possible meanings, and aims of the practice of circumcision.

Taking all this into account, we might think that Abraham (-Hammurabi-Ur Nammu) performs a dual operation: it seems that first he adopts a usage which does not belong to him but to the whole world around him, making it his in order to integrate himself in this way within the cultural climate that surrounds him, and then conjoins the rite of passage, now also his, to the religious element; a religious element which immediately acquires markedly monotheistic characteristics while still remaining anchored to "pagan" features; or perhaps, starting out from markedly monotheistic requirements, it immediately acquires religious value. In my opinion, both assertions may be considered sound. From the little boy who can only stay in the house of woman to the Man who chooses a Woman. And then another leap, taking this "practice of maturity" from the profane to the religious. From the little child to the man, and from the man towards "God". From the social to the Sacred. And in a similar way, albeit in the deeper past, we move from the offering of "gifts" to the community (*Potlatch*) to offering gifts to the protector god of a single group (Parise, 2000; see above) and then to the gods shared between several groups, and then to the offering of the gift to a single God, god of everyone and everything; from the horizontal to the vertical, at first a fragmented horizontal, then the absolute vertical, the sense of the Sacrifice-Gift-Oblation changes (from Idolatry, through Enolatry, or the Enotheism of Müller, then to Monolatry, and onwards to Monotheism). And there is a change in the means by which the Sacrifice is carried out: that is, a change in the offering of what is sacrificed. Of course, bulls, goats, and other animals keep being sacrificed, but something much more personal and tangible also comes to be sacrificed. There is always blood, but a different blood, Attali would say.[4]

---

4   As I observed earlier, Jacques Attali (*Les Juifs, le Monde et l'Argent*, p. 31), writing about the plagues of Egypt, calls them "dix sanctions économiques" and goes on, "Les mots mêmes qui les désignent recèlent des messages à caractère économique. Ainsi l'une des plaies, le sang, est-elle nommée par le même mot, dam, qui désignera plus tard l'argent (damin); le sang et l'argent inséparables depuis le sacrifice d'Isaac." In a note in Dr Schimmel's book there is an identical observation about the terms Blood-*Dam-Damin-Dmei* (contraction of *Damin*)-

There is still a knife, but used in a different way.

As I mentioned earlier, perhaps the initial reasons for the Offering-Sacrifice-Gift-Oblation to the deity are the same as those that Mauss glimpsed in the *Potlatch* (that species of redistributive paradise which subverts all the rules of creation and the accumulation of wealth and reserves); and perhaps the *Potlatch* itself is just a gigantic enactment of disavowed stranger-anxiety; or an extreme identification with the aggressor; or a fantastical expression of reaction formation; or an equally gigantic dramatization, a gigantic acting out, of the mimetic identification Girard speaks about; or a representation of a primal "protection racket"; or else the gift is "only" the continual re-offering of what all living creatures find themselves receiving "free of charge", biologically and psychologically, from their parents (I don't think it's by chance that the god/Shamash/God/Yahweh is called Father and Innana/Ishtar, or whoever stands in for her, Mother).

Perhaps all of these together. They do not generate wealth but permit the

---

Money. I don't know if one writer has influenced the other, but in any case *Damin* or *Dmei* does not mean Money *per se*, since the term for *Argent* is *Kesef*: *Dam-Damin-Dmei* indicates Money as "small change", that we part from "unwillingly" (?) as pocket money for the kids, *dmei kis* (which then takes on the specific, personal meaning of *argent de poche*); advance, *dmei qadima*; a tip for the waiter, *dmei shtiya*; protection money, stake, corruption, *dmei hasut*; alimony, *dmei mezonot*. In contemporary Hebrew, Money is indicated by *kesef*, and the coin by *matbea*; banknotes are *shtar*; financial activity is *shuk abursa*; the minting of coins is *lehatbìa*; and wealth is *osher*; I love the idea that happiness is also *osher*, spelled differently but pronounced the same way. What's more, in almost all languages, referring to various aspects of Money, we use expression such as "they've bled me dry," "I've run dry," "they're leeching off me," "the government/banks are vampires," etc., taking us back to ideas of money as vital fluid, not necessarily understood as blood, though blood is the most important representative among the vital fluids, and so in think that Attali's observation, without being unquestioningly endorsed, can to a large extent be shared because it indicates a vein of sense that is interesting in any case, but I also think that Attali has greatly enjoyed himself playing "kabbalistically" with the letters of the Hebrew alphabet and that his pyrotechnic wordplay on pages 39 and 40 may lead more to confusion than to study of the problem if taken too far. Being an educated and intelligent person, he has identified the meaning which reconnects Money, *Kesef* to Desire, *Kasaf*, burning desire; as well as to the very important meaning of Money as a pacifying element, replacing the Talion and the shedding of Blood in retaliation for Blood shed. Returning to the sentence I quoted above, I don't understand why Attali writes "qui désignera plus tard l'argent" when the plagues of Egypt happen during the Egyptian captivity and under Moses – long after Abraham and Isaac – while going on to speak about the sacrifice of Isaac which happened long before the plagues of Egypt. Strange.

possibility of escape from predation; like exogamy, perhaps. I suggested this a while back, but I say "exogamy, perhaps", since in the Horde, if the Father is castrating and brutal, and castrates/eats his children, then his children – if all goes well – must escape if they can, and in escaping they are by definition compelled to look for exogamous relationships. Therefore, as I have suggested, exogamy seems to me to be an antecedent condition to the original homicide, not a consequence of it.

Whatever the case, and wherever it comes from, with Avraham-Hammurabi we are present at a potent and formidable "single, vertical, internal *potlatch*"; no longer horizontal and no longer dispersed among hundreds of divinities, but directed in a single, powerful direction.

I say "internal potlatch" because in the new faith, every thought and every action must be related to God, the one and only true God; and so a great quantity of energy must be expended; internal, not necessarily material, energies of course; and indeed, the sites of Avraham's sacrifice, and for a long time after him, will consist of no more than some humble stones and a bit of earth, and no longer enormous and majestic buildings like the Ziggurats. Of course, as I was saying, the group of the faithful does not surrender its subsidiarity, quite the reverse; and many of the ancient traditions and old habits continue to persist (one need only compare the Code of Hammurabi and the Bible; and the famous Law of an Eye for an Eye, art. 196; or the story of Hagar and Ishmael described exactly in articles 145-146 of the Code of Hammurabi, or art. 25 of Ur-Nammu as an element of "Private Right"; *private*, not *robbed*); there is still collaboration, indeed it is strengthened, as is the group identity, but the dedication-offering-sacrifice-gift-oblation is now addressed solely to G-d; to the one and only true God. I also think it is no accident that, with different nuances and different sensibilities, Avraham is considered the founding father of all three principal monotheistic religions; there may be many reasons for this, but the fact remains that they all converge on Avraham. It doesn't matter if the story says it happened here or there, to Punch or to Judy, to Tom or Dick or Harry from Bologna. The important element is that something happened, there or thereabouts. Something important. That Abraham in person or someone on his behalf, in that era or centuries later, in a story that is necessarily as apologetic as it is foundational, can have appropriate something previously written-spoken-done-started by others before him, around him, near him, doesn't matter a jot. All religions have always done it. What seems important to me is that through Abraham, the Chaldean from *Ur* (or Abraham from Ur of the Chaldees) and his "double" Hammurabi, is present at the founding of the initial phase in the birth of a people, a religion, a faith, a new way to conceive of History, the World, relations between individuals. And a new covenant is founded. It will take at least another three thousand years for this covenant to be completely affirmed, but the new covenant has been founded. And as is obvious, and as I have already said, though they preserve they attachment to the tradition which precedes them, new covenants need new symbols. And new covenants do not only concern the relationship with God but also govern relations between men, given that covenants, oaths, are made by touching the "thigh" of him to whom the promise is being made: "Put, I pray thee, thy hand under my thigh" (*Genesis* 24:2).

Now, how does the thigh come into this? Well, it really does: before leaving for his expedition in search of a wife for Isaac, Eliezer makes a covenant with Abraham, and to sanctify it he touches Abraham's thigh; he doesn't shake his hand, or spit on the ground, or swear on the heads of his children or on his honour; and the same goes for Joseph when at the request of Jacob, who wishes to be buried in the Cave of Machpelah near Hebron, where Abraham and his other forefathers are buried, he touches his father's "thigh", albeit hesitantly, while swearing solemnly to fulfil the request: and he doesn't say "Scout's honour" or "Cross my heart and hope to die." But the same delicate expression "touch the thigh" becomes more explicit in the words of Joseph reported by Graves and Patai (1980, p. 3: and clarified on p. 229, n. 7) drawing on various Hebrew sources; words which say, referring to *Genesis* 47:29-31:

> It is not fitting for a son to touch the circumcision of his father. Nevertheless I swear to you, by the living God, that you shall be buried in Hebron.

While on the subject of translator-traitors and the way they render concepts: in the Bible of the Conferenza Episcopale Italiana we find Jacob asking, as Abraham did before him, to sanctify a promise, saying:

> "Deh! se ho trovato grazia presso di te, metti la tua mano sotto la mia coscia", e Giuseppe risponde: "Farò secondo la tua parola", ma Giacobbe si secca e con tono imperativo esclama: "Giuramelo"; e Giuseppe glielo giurò.
> ["I pray that if I have found grace with you, put your hand under my thigh" and Joseph answered, "I will do as thou has said," but Jacob grew angry and in a commanding tone exclaimed, "Swear it to me," and Joseph swore it to him.]

This is the sense of "touch the thigh": touch the circumcision, which is a symbol of the covenant with God, to sanctify the solemnity of the covenant between men, between *ḫāwiru*. Josephus Flavius adds specifically that touching the circumcision is a mutual act between the participants in the covenant, not a unidirectional act but one in two directions. Touching the thigh, i.e. touching the circumcision, thus becomes a symbol of the Covenant of Covenants, that made with God; the circumcision is touched, not the thigh and not the penis, as Friedman (2001) maintains. Graves and Patai (1980, p. 188, n. 3) add a further observation which I think it is important to cite:

> in Biblical Hebrew, covenants were not 'made', but 'cut' (*Karath B'rit*, Genesis XV, 18; XX, 27 etc.).

(*Karath* is to cut off, to cut down, in Biblos.com.)

Touching the Thigh, that is, touching the Circumcision as an element to be "cut", *Likrot*, sanctifying a covenant, a *Brith*, *Karath Brith*, a Covenant Cut, "cutting" something that is to be "cut", becomes a gesture which sacramentally binds a man to his promise. Indeed both men are mutually bound to the promise they exchange in the name of another promise, that promise which is founded on another still more solemn promise, that *Brit* made to God and required of Avraham by God, the *Brit Milah*, a promise literally carved onto his own skin, carved into his own flesh. I don't think apotropaic features are the only ones to be seen in these circumstances, in this oath-taking; to me it seems rather the sublimation of a threat (if you don't keep your word I will emasculate you, spoken by the one who solicits the promise in having himself be touched) or it is a pledging of something that is most precious to you (I am putting what is most precious to me in your hands/ under your hands/ you have responsibility for my life/ you have the responsibility to honour a commitment which calls God as a witness…); and none of these hypotheses seem to contradict one another. Another brief observation: when Jacob wrestles with the Angel (or with God?) (*Genesis* 32:25-6), in order to get away at the end of the fight, the Angel (or God?) cheats, plays a trick; he cheats and "touches socket of his (Jacob's) hip"; so that Jacob will limp from that day onwards, having been offended in that part of his body: the sciatic nerve, according to several commentators. And that is why, to this day, the Jews, or at least those who acknowledge the religious value of the dietary practices inculcated by tradition, do not eat that part of the body of animals destined for the table.

Leaving aside the fact that Jacob was not exactly an animal destined for the table, but perhaps as I have speculated this alludes to the human sacrifices of the first-born, and even though Jacob is the second born, he is the one who, using the trick with the red lentils (back come the lentils!), received the birthright (the name Yacov, "he will follow", besides being connected to the heel, *aqev* and to "coming out after/ he has followed," is connected to the fact of being immediately second born: that is, "he who will follow") the point that I find very interesting is that in Hebrew, one does not reinforce the concept by repetition, saying "hip-hip", but *Iarekh*, "palm of the pubis". Dr Baruch Avezov, who has helped me in this and on other occasions in addressing many etymological aspects of the Hebrew and Aramaic languages, has given me this suggested definition of the Biblical passage about Jacob's hip, or the end of his femur, pointing out to me that what is rendered in modern versions of the Bible as "hip" in the Hebrew Bible is "palm of the pubis" or "pubic triangle". Since the Bible, though a precious tool, certainly cannot be considered an anatomy manual, especially when speaking about struggles between humans and Angels (or God?), and given the shamefacedness of tradition – i.e. "hip" for Jacob and "thigh" in other situations, to which have been proposed the alternative terms "pubic triangle" or "palm of the pubis", or rather, "circumcision": in other words, "circumcised foreskin" – I believe, or at least it is not out of the question, that the episode of Jacob's fight with the Angel (or with God?) can be interpreted as meaning that here too we are presented with another gesture that has sacred elements: the benediction/ promise which Jacob/ Israel asks from the Angel (or from God?), with elements of a solemn promise about Jacob's future and hence the future of Israel, absolving him from his past misdeeds and renewing/ re-confirming a promise made by God Himself to Noah, Avraham, and Isaac, as well as to Jacob; and so I think that "touching the pubis/ pubic triangle" can be seen as the equivalent of the gesture made between Eliezer and Joseph, but this time made by the Angel (or by God Himself?) who touches Jacob's circumcision, that cut thing,[5] as a promise and oath.

---

5   On the "cut thing": since gaining some familiarity with the Hebrew language, I have always found that the seventh letter of the alphabet had something "strange" about it, without being able to grasp what made it so strange. This letter is called *Zaîn*, and is written ז. Obviously, it can become quite curved depending on the way it is written, which gives it great variety, just as is the case with Latin letters if they are written, for example, in gothic or italic, printed, or upper case. Each Hebrew letter is a symbol and assumes a great many valences and meanings while preserving its own numerical value: in this case, of course, 7. Thus the letter *Zaîn* can come to represent the days of the Creation, the seventh day, the Sabbatical rest, the lights of the Menorah, and a great many other things. Mandel (2000) reports a variety of meanings that are attributed to the letter ז: Weapon, Dagger, Javelin, Dart, Arrow, Sceptre, God's Staff, and many others. Similarly, *Lamed* is originally noted as the symbol for an instrument, the shepherd's crook which

Theodor Reik (1951, p. 238) also connects hip and penis before going on, in the light of *Totem and Taboo*, to argue that the whole combat would represent the totemic meal in which it is God-Father who is wounded (emasculated) and not Jacob; hence the prohibition against eating that part of the body would represent the prohibition against eating the father's penis.

Given its concave form, the hip, *Iarekh*, recalls the letter *Kaf*; and, since *Kaf* is the term for spoon, which is also concave, it takes on the meaning of "spoon of…", as in "spoon of the hand" or "spoon of the foot", indicating those bones that suggest concavity. Therefore, since spoon is *Kaf*, pronounced the same way as the letter of the alphabet, *Kaf*, the iliac bone comes to be called the spoon of the hip. Rashi speaks of a "ladle" (Commentary on *Genesis* 32:26, p. 273) and not "spoon", but the sense is clear.

This bone, as my doctor and friend Dr Dal Pozzolo has pointed out to me, is connected to the sciatic/iliac nerve, and one evening at dinner, after I had tortured him too with the Sumerians, the Bible and the rest, he asked me, "Do you know its anatomical name? It's called Unnamed," and this was apparently the case throughout Europe until the Second World War. I have to say I found this name simply astonishing in the light of what I am

---

Prods, Guides, Teaches, and thus becomes the first letter of the root for *Lelamed*, to teach, which becomes *Limed*, "he taught", but at the same time has exactly the same composition as the letter *Lamed*. I don't want to venture into the Kabbala: I've already made clear my lack of qualification for that, but a while ago, perhaps riding the wave of what I'm writing now, I realized, by myself, what was "strange" about the letter ז: I remembered a circumcised penis. Then I told myself that what I was writing had started to condition my visual perception. However, when it was explained to me during a lesson that *Zakhar* means Male (in Akkadian, *Zikarum*), and that for the first letter of "male" (and certainly not by chance) the letter *Zaîn* was chosen, a kind of weapon/ phallus which represents the Male symbolically, as we saw a little while ago, my eyes popped out of my head. *Zakhar* also means "he who has remembered" and the noun *Zekher* is Memory. They didn't go in much for subtlety in Antiquity, and a another, not secondary, characteristic of the Male is, not too bizarrely, that of "leaving a memento", a Memento in the biological sense: Sperm/Child. When all is said and done, we find the same lack of poetry in the term for Female (probably, *ab ovo*, human female, but now referring to the Female in general) which comes to be called *Neqeva* (Neqev-ba): that is, "hole inside her". Not very different from the Latin concept of *cunnus*. My curiosity aroused by this "revelation", I asked some more precise questions and discovered to my enormous surprise that *Zaîn*, exactly renders the vulgar Italian – equally vulgar in Hebrew – "*cazzo*","*prick*" or "*dick*" in English, "*Schwanz*" in German, "*bite*" in French, "*polla, pito*" in Spanish) which, now stylized as ז, seems to be an exact echo of the Sumerian pictogram reported by Kramer above.

writing: Unnamed, Unnameable; I think there may be more connections with sacredness than we know what to do with. How does the sacred bone come into this? It's not so far removed? I'll ask him. (Maybe Josephus Flavius could help me with this, since sacred bone in Greek is *hieron osteon*, Ἱηρόν Ὀστέον; I'm joking, because the sacredness of this bone seems simply to be connected to an error in translating from Greek to Latin where *hieron*, which also means *big*, has been translated as *sacred*.) Since I'm getting close to sexual matters, it would be interesting to broaden the discussion to include the concept of Nakedness, present in so many passages of the Bible, but I think that's better deferred to another occasion. It would also be interesting to make a detailed examination of the Sumerian term HAŠ, especially HAŠ$_4$,"thigh", which is behind HAŠ.DUG, to have sex, literally "talking of/between thighs". This term/ expression/ metaphor of the "thigh" was already present in Sumerian, where we find many phrases alluding explicitly to matters of sex, until we come to the dying Enkidu (see also the translation by Bottéro,1992, *L'Epopée de Gilgameš*, tab. VII, col. iv, line 3) who, having cursed the sacred prostitute/Courtesan, Dispenser of Joy, Šamḫat, blesses her, wishing her "*(l)imḫaṣ šaparšú*", that a man "may beat his thigh (to arouse himself before the encounter with you)." For now, I will make do with having sketched out the setting behind the sacrifice and the faith. This is not directly about Circumcision, but it is about Cutting: I've just recalled a memory from when I was a child in the Celtic and highly attractive valley where I grew up, perhaps because it was so full of "fibbers". I remember that when we made a solemn promise to another child, a promise that was naturally of vital importance, with all the solemnity and seriousness that children are capable of – "I promise I'll bring you the marbles tomorrow," or "I swear that when the kittens are born, one will be for you" (and someone would straightaway self-righteously say, "Don't say swear!"). Or "After Mass I promise we'll go and steal cherries, even though Mamma doesn't want me to go out because I've got homework to do," or "I promise you'll be in goal for the next match" – there was always a third boy making a solemn gesture with his hand, "cutting" the handshake of the two parties to the promise, as if dissolving the covenant which had been established and become so strong as to bind the two parties for all eternity; and so, in order for it to be honoured, a third party was necessary to dissolve commitment so indissoluble that it would obviously have turned the two protagonists to stone, making it impossible for them to honour their solemn promise. Who knows? Handshakes without the Cutter's gesture were not so binding, whereas those with the Cut were absolutely binding and mandatory, on pain of heavy ostracism and blame from the other children.

21.
ON WHY, IRONICALLY OR MAYBE NOT, IT IS DIFFICULT FOR THE COIN TO HAVE HAD A GREEK (OR ROMAN) ORIGIN

A little earlier I quoted a sentence by Bernhard Laum for the second time:

> Therefore, I assert that in ancient Greece only *instrumenta sacra* could become and did become money,

in other words, the *hiera chremata*, Ἱερὰ Χρήματα; and I claimed he had made an error in location and date. Now, following a different path, and joking a little, I would like to explain the "deep" reasons why Laum's thesis about the birth of the coin in ancient Greece is hard to support (though, in my opinion, his idea about the *instrumenta sacra* should be preserved and endorsed). I'll start with Mommsen (1856, p. 4):

> *Language thus proves to us that, at some unknown period, from the same cradle there issued a stock which included the ancestors of the Greeks and Italians; that subsequently the Italians branched off; that the Italians divided into Latin and Umbro-Samnite stocks.*

Mommsen continues:

> *The commonality of all the most ancient terms relating to agriculture proves the intimate connection between the two civilisations: ager, ἀγρός; aro, aratrum, ἀρόω, ἄροτρον; hortus, χόρτος; milium, μελίνη; rapa, ῥαφανίς; vinum, οἶνος.*

And in another, very poetic passage he adds (I quote from memory):

> The Italians and Greeks are like twins who turned their back on each other long ago, one looking East, the other looking West.

These observations – but Mommsen makes many others – seem sufficient to introduce what I am about to say. The Latins too, and then the Romans, like the Greeks, sacrificed the Bull in the *Feriae latinæ*, on Mount Alba and sacrificed it to *Jupiter Latiaris* (Jupiter Father of the People of Latium). The parallel between Hellenes and Italic peoples which I wanted to emphasize by quoting Mommsen and the sacrifice of the Bull is a narrative expedient with a precise purpose: I am playing, though not overdoing it, with the concept of the coin as it takes shape in my work. The thesis I am trying to bring to light is, essentially, this: it is only with the sacrifice of something extremely personal, extremely delicate and precious, that the Coin comes to life as the new symbol of a new alliance. Its own value can only be derived from *instrumenta sacra*.

A Coin based on a very personal sacrifice.

But the Greeks and Romans performed sacrifices, didn't they? So why couldn't they be the ones to invent the Coin? Why couldn't Lydian coins be the first in history? The proof I want to present is, let's say, indirect; it is a negative and ironic proof. Ovid, who is not a professional historian and is a shifty character besides, does nevertheless write some relevant things about this. Not so much because of the things in themselves, but because in writing those things he knows he can count on the attention of the readers in his time who will find what he writes entertaining, or attractive, or comical, but not absurd, because then they wouldn't have been understood, or blasphemous because they would have been rejected: all of which denotes a non-opposition, a non-extraneousness to the cultural environment of his time.

In the *Fasti*, a poem Ovid never finished (fortunately), he makes a panorama of the Months and Days of the Roman calendar, making poetry out of their sequence, their festivals, and the divinities assigned to them, as well as reconstructing the mythologies, etymologies, and stories of the events, names, words, and situations encountered on the way. In Book III, line 277ff, speaking about Numa, first King of Rome, Ovid sings the great legislator's praises:

> At first the Quirites were too prone to fly to arms;
> Numa resolved to soften their fierce temper by force of law and fear of gods.
> Hence laws were made, that the stronger might not in all things have his way,
> and rites, handed down from the fathers, began to be piously observed.
> Men put off savagery, justice was more puissant than arms,
> citizen thought shame to fight with citizen,
> and he who but now had shown himself truculent would at the sight of an altar
> be transformed and offer wine and salted spelt on the warm hearths.
> [*Fasti*, trans Frazer J.G., *Loeb Classical Library*, 1931]

So, with a little bit of imagination, given that this is a poem, it is not too hard to glimpse the characteristics of an Ur-Nammu, or a Hammurabi, or Shulgi or Moses; all the characteristics of the great reformers who have founded something. So far, so good. But things take a different turn straight afterwards because Jupiter flies into a rage and seems far from satisfied by the worthy work of the great new legislator, instead sending a terrible rain accompanied by an impressive thunderstorm because Numa has touched the altar with his hands, however pure they may be (and "Whatsoever toucheth the altar shall be holy"). I'll sum up briefly: on the advice of a nymph, Egeria, who is his wife, Numa goes to a sacred wood and succeeds in binding two demigods, Picus and Faunus, who have become drunk on the libations offered by the faithful; apologizing for his cruel trick, he says he'll set them free if they will make use of powerful magical formulae unknown to mortals to invoke Jupiter and compel him to show himself; with great reluctance they agree and call on Jupiter. In the presence of Jupiter in all his potency, Numa, the new Moses, cannot fail to be terrified at being in the presence of such divinity:

> The king's heart throbbed, the blood shrank from his whole body,
> and his bristling hair stood stiff,

and he hesitantly asks:

> King and father of the high gods
> vouchsafe expiations sure for thunderbolts,
> if with pure hands we have touched thine offerings,
> and if for that which now we ask a pious tongue doth pray.

And now we come to the point, to the difference between the Sacrifice of Avraham-Hammurabi and that of Numa, the difference between East and West.

Some have noted that the Latins and Greeks always had an antipathy circumcision.

It was known in China, but only as a medical practice and by no means carried out frequently.

The same in India, before the Islamic conquests.

This is precisely the fact that makes the enormous difference between East and West and emerges through the meaning which Numa finds in "sacrificing", "cutting", cutting the throat of the victim, cutting the Covenant, cutting the foreskin; and this difference is proved by the dialogue that follows:

> The god granted his prayer, but hid
> the truth in sayings dark and tortuous,

and alarmed the man by an ambiguous utterance.
"Cut off the head," said he. The king answered him, "We will obey.
We'll cut an onion, dug up in my garden."
The god added, "A man's." "Thou shalt get," said the other, "his hair."
The god demanded a life, and Numa answered him, "A fish's life."
The god laughed and said, "See to it that by these things
thou dost expiate my bolts, O man whom none may keep from converse with the gods."

How different Jupiter is from Yahweh, and how different Numa is from Moses!

I think that such a profoundly different relationship with the deity conceals a number of features, among which the fact that Numa – i.e. the Italians and Hellenes – thinks of a smiling god who good-naturedly lets himself be teased a little, and amused. He is a god who would like human sacrifices like his eastern colleagues, but then smiles on humanity, on Numa's quick-wittedness and, contentedly appreciates his respect for human life. Numa doesn't want to sacrifice anything of himself, or of other people, neither a head nor a living soul. Numa goes back to his normal life, trusting in a god who is quite a decent chap, and not in the least angry. A god who makes do with mortals as they are and does not demand perfection from them. And so Numa goes back to the old sacrifices, the ancient ones that are sufficient: a heifer will be sacrificed, a heifer which has never known the yolk. How different this sounds from the return of Moses from Sinai.

How different the divine characters are: the God of Abraham rages and punishes, and so does Moses.

And how different Moses' wife Zipporah is from Egeria: she immediately circumcises her son (although some, Graves and Patai among them, say that the circumcision of Moses was also performed by his wife as a Midianite marriage ritual.[1]

---

1   This passage in the Bible which tells of Moses' return after fleeing to Midian is quite bizarrely constructed: immediately after receiving instructions from God about what to say to Pharaoh, Moses will have to say, "Let my son go, that he may serve me: and if thou refuse to let him go, behold I will slay thy son, even thy firstborn." Immediately after this comes the description of Moses' own risk of death. In his Commentary on Exodus 4, 24-26, Rashi writes "24 – During the journey he came to a place where he sought shelter, the Lord struck him and tried to kill him. During the journey – the angel attempted to kill Moses because he had not circumcised his son Eliezer; since he had been negligent in this, he wanted to punish him with death. We learn from one Baraità that R. José said: 'God preserve you. Moses was not negligent in carrying out the circumcision; he was thinking, If I circumcise him straightaway and set off on my journey, the child will be

Blood. Always Blood and Sacrifice.
But with Numa we have blood versus onions. Blood versus a haircut. Blood versus fish.

No, if the coin is based on sacrifice and, moreover, a sacrifice of something very personal and precious, no, the Romans and the Hellenes cannot have invented the coin; at least, not if we give Ovid any credit.

Both of them, in various rituals, will often resort to bloodshed, terrible self-mutilations, unrestrained orgies of individual and collective violence; they will make rivers of blood flow, their own and that of many, many others; but for the coin, no, I'd definitely say no.

> exposed to dangers for three days. However, if I circumcise him and wait three days, I am not obeying the Blessed Holy One who commanded me 'Go and return to Egypt.' Why then was he considered deserving of death?' 'Because he gave priority to the deeds concerning the first stage of the journey.' The angel transformed himself into a serpent which began to swallow Moses from head to hip and, after expelling him then began to swallow him from his feet to his hip (that is, to the level of his member). Zipporah then understood that this had happened because the circumcision had been deferred. 25 – Then Zipporah took a sickle and cut her son's foreskin, and threw it at the feet of Moses, saying, 'A bloody husband thou art.' She threw it at his feet – she threw it at the feet of Moses, saying, referring to his son, 'A bloody husband thou art' – that is, thou art the reason why my husband would have been killed, thou wouldst have been the killer of my husband. 26 – The Lord left him and then she said: 'Thou art a bloody husband to me because of the circumcision." He let him go – the angel let him go and she understood that he wished to kill him because of the circumcision. She said, 'Thou art a bloody husband to me – My husband was about to be killed because of the circumcision.'"

It is a slightly confused passage in which features of great tension and drama are mixed together. Maybe a partial answer can be found to the "forgetfulness" of Moses, which Freud mentioned. It remains an interesting and curious fact that the terrifying transformation of the angel into a serpent, starting from the head and then from the foot in its attempt to swallow Moses, stops at the hip, "at the height of the member," as Rashi says. Hip: again the Hip, like Jacob's Hip described earlier; again, Circumcision and Member and Foreskin as central elements of the story (Rashi's parentheses: baraita, a story from oral tradition.)

22.
AVRAHAM, אברהם, FROM UR OF THE CHALDEES. AVRAM, אברם, THE CHALDEAN FROM UR

I previously listed the lineage of Ham, while Japhet's did not interest me.

Now I'll describe those of Shem, since he too had children:

> This is the lineage of Shem: When he was a hundred years old he begat Arpachshad two years after the flood. After he begat Arpachshad he lived five hundred years (…)Arpachshad (….) begat Shèlach (…) Shèlach (…) begat Ever (…) Ever (…) begat Pèleg (…) Pèleg (…) begat Re' ù (…) Re' ù (…) begat Serugh (…) Serugh (…) begat Nachor (…) Nachor (…) begat Terah (…) Terah (…) begat Abram (*Genesis* 11:10-26, *Hebrew Bible*).

At this point, before continuing, I wanted to introduce Abram's genealogy because it seems important for a better understanding of what I am going to talk about next.

Abram is a direct descendant of Shem: i.e. a direct descendant of Noah.

A few pages before, also drawing on *Genesis*, I listed the descendants of Noah who leads directly from Ham to Nimrod, alias Amraphel, alias Hammurabi.

This time I'm not suffering from an excess of enthusiasm if I claim that Abram and Nimrod are related. Both are direct descendants of Noah.

Since Noah was the only survivor of the Flood, I obviously can't call that breaking news but, in this context, I don't think it's hair-splitting to say that the attempt made by the author(s) of the Bible to ennoble his/their own origins was stronger than biology. And we will come across Avram and Nimrod again, as we'll see, in different, very different positions.

The historical truth of the story doesn't matter very much, but it is an important fact that the author(s) of the Bible considered it necessary to attribute a common origin to these two immense figures, an origin I referred to without explanation in the chapter on the Sacrifice; they are in fact Doubles: Avraham and Hammurabi. And their relationship begins in an unusual way, starting from an amusing and dramatic but little-known event

in their development which I find interesting as it appears in commentaries on *Genesis* by the most important scholars and Rabbis.
From the *Bereshit Rabbah*, XXXVIII (see also *Midrash Rabbah*, Trans. and ed. Freedman, H. and Maurice S. London, Soncino Press. 1939):

> 13. And Haran died in the presence of his father Terah (xi, 28).
> R. Hiyya said:
> Terah was a manufacturer of idols. He once went away somewhere and left Abraham to sell them in his place. A man came and wished to buy one. 'How old are you?' Abraham asked him. 'Fifty years/ was the reply. 'Woe to such a man!' he exclaimed, 'you are fifty years old and would worship a day-old object ! ' At this he became ashamed and departed. On another occasion a woman came with a plateful of flour and requested him, 'Take this and offer it to them/ So he took a stick, broke them, and put the stick in the hand of the largest.
> When his father returned he demanded, 'What have you done to them?' 'I cannot conceal it from you, he rejoined. 'A woman came with a plateful of fine meal and requested me to offer it to them. One claimed, "I must eat first," while another claimed, "I must eat first." Thereupon the largest arose, took the stick, and broke them/ 'Why do you make sport of me,' he cried out; 'have they then any knowledge!'
> 'Should not your ears listen to what your mouth is saying?' Abram retorted. Thereupon [Terah] seized him and delivered him to Nimrod.
> 'Let us worship the fire!' he [Nimrod] proposed.
> 'Let us rather worship water, which extinguishes the fire' replied Abram.
> 'Then let us worship water!'
> 'Let us rather worship the clouds which bear the water. '
> 'Then let us worship the clouds!'
> 'Let us rather worship the winds which disperse the clouds'
> 'Then let us worship the wind.'
> ' Let us rather worship human beings, who withstand the Wind.'
> 'You are just bandying words,' he exclaimed; 'we will worship nought but the fire. Behold, I will cast you into it, and let your God whom you adore come and save you from it. Now Haran was standing there undecided. If Abram is victorious, [thought he], I will say that I am of Abram's belief, while if Nimrod is victorious I will say that I am on Nimrod' s side.
> When Abram descended into the fiery furnace and was saved, he [Nimrod] asked him, 'Of whose belief are you?' 'Of Abram's,' he replied. Thereupon he seized and cast him into the fire; his inwards were scorched and he died in his father's presence. Hence it is written, And Haran died in the presence of his father Terah in the land of his kindred, in Ur of the Chaldeans (English Standard Edition); (*Genesis*, 11:28).

Besides the "insolence" and "arrogance" of Abram, in this brief and witty account we find elements of *Euphratesian* law, such as recourse to the judgement of the King and the use of the Ordeal which, I emphasize again, is not of Christian and Mediaeval origin. There is the explicit generational conflict between father and son, Haran's opportunist position which is destined to have no place in the emerging monotheism, and a heap of other things that I will look at in the next chapters dedicated specifically to Avraham. And the Furnace is represented as a central element of the story with all its apparatus of magic and wisdom, as Eliade has argued at length. The Koran gives a more concise version of the same situation in the Sura al-Baqarah (Sura of the Ox) verse 258. This is an episode from Avraham's youth and there are other very enjoyable ones (Ginzberg, 1925, vol. 3). But then Avraham grows up, so much so that he becomes a king. Avraham is king; becomes, *de facto, King,* which is what those who aspire to royalty often do. King of a small tribe of *ḥāwiru*, "free" men, i.e. marginal shepherds-bandits-nomads who survived in the interstices of Egyptian, Canaanite, Philistine, Syrian, and Assyro-Babylonian society: but still "King".

An etymon of Avraham would be *Avir hamon*, "King of the multitudes" (in Disegni, *Hebrew Bible*, p. 27, note). In *Avir* there is also an echo of the much-cited *ḥāwiru*.

I was a bit ironic about Professor Jeremias before, and it seems to me that I should repay him for something I took from him in my polemical vehemence. Jeremias reports some possible etymologies for the name Abram as identified by various scholars, connecting him to the cult of the god Sin, the Moon: *Ab* would indicate the Divine Father – in Hebrew *Aba*, father – who is present in many languages. *Ab-ram* would be of Akkadian origin and would stand for *The (divine) Father is sublime/ great. Ab-ram* should also be connected to war, thus becoming Father of Tumult.

Jeremias (*The Old Testament*, vol. 2, p. 8) connects this final meaning with what remains of the *Universal History* by Nicholas of Damascus (30 C.E.), one fragment of which has this to say about Abram:

In Damascus reigned Abram, who came there with an army from the land of the Chaldees, bordering on the upper half of Babylon. And not long after he moved out again from there with his people towards Canaan, which is now called Judea, where he greatly increased.

I don't think Nicholas of Damascus expressed himself in English, but never mind that for now.

Damascus would be the famous Harran (with two Rs this time) I alluded to a little while ago; and it would have been taken *manu militari* and not as a *buen retiro*, and so Jeremias likens Avraham to a new Mahdi militarily undertaking his own personal Hegira. The quotation by Jeremias also appears in Josephus Flavius's *Jewish Antiquities*. Though expressed less effectively, my hypothesis about the fact that Avraham is a king is very close to that of Jeremias. And of Flavius.

In this case at least we think alike; although, in the grip of another polemical fury, I have to observe that in 360 pages of his *Old Testament* Jeremias does not once mention circumcision, which is Abraham's big idea, confining himself to a single reference, obviously to Isaac (vol. 2, p. 4). Instead he reports the news that Terah, Abram's father, beside selling idols, would in fact have been one of Nimrod's generals, a hypothesis also proposed by Graves and Patai. It would actually have been Nimrod, which derives from *mered*, meaning 'rebellion, cause uproar, tumult', and not Abraham, but since they are doubles it comes to the same thing.

Abram marries a princess, Sarai, connected to the Akkadian *šarratu(m)*, queen. Sarai is also called *Yiskah*, which according to Rashi means "see, look", verbs referring to the divinatory skills attributed to her; but others derive it instead from *Nesikha*, princess/noble, and at the moment I don't care whether she is his sister or his half-sister (*Genesis*, 20:12) or how far, and in what way, she collaborated to increase Avraham's wealth through her relations (?) with Pharaoh who, having desired her, contracts a mysterious venereal disease called *Raatan* (*Raatan* is an unknown term today). But Rashi writes:

> He was struck by the plague called Raatan, which made it impossible for him to have sexual relations.

Others speak about *Raatan* as leprosy, and yet others as lupus, or tinea, or scabies, as Graves and Patai (1980, p. 177) indicate; exactly the same thing that happened to the Pharaoh also happens to Abimelech, for not assaulting her but having merely thought about it in a dream. He too becomes mysteriously ill and, in order to recover (Rashi, Commentary on *Genesis*, chap. XX, p. 3), pays 1,000 pieces of silver, *kesef* indeed (*Genesis* 20:16). The striking fact remains that Sarah is sterile and Pharaoh's disease makes it "impossible for him to have sexual relations", therefore "impotent" (?) and therefore sterile (?) as well, whereas the "disease" of Abimelech, who has *not* even assaulted her, or maybe only in a dream, in fact afflicts his whole family (*Genesis* 20: 18); sterility and closure of "every opening of the body" (Rashi, Commentary on Genesis, chap XX: 14-17). Very strange.

Whatever the case, Sarah must have been really beautiful and full of fascination if she is receiving all this ardent attention after the age of seventy! Beauty to die for. Literally *une femme fatale*.[1] And *kesef*, pace Attali, is also the coin which Abraham "strikes" near Mamre in order to buy the tomb for Sarah: (*Genesis* 23:14-17):

> And Ephron answered Abraham, saying unto him,
> My lord, hearken unto me: the land is worth four hundred shekels of silver; what is that betwixt me and thee? bury therefore thy dead.
> And Abraham hearkened unto Ephron; and Abraham weighed to Ephron the silver, which he had named in the audience of the sons of Heth, four hundred shekels of silver, current money with the merchant.
> And the field of Ephron... and the cave which was therein, and all the trees that were in the field... were made sure
> Unto Abraham for a possession in the presence of the children of Heth, before all that went in at the gate of his city.[2]

It seems to be exactly the same ritual as the *mancipatio* with the Roman *Libripens* and exactly the same as the ancient Mesopotamian, *Euphratesian* ritual with the weigher of silver/ scribe who calls publicly for witnesses

---

[1] The song of Hyrcanus about the beauty of Sarah (from Graves and Patai, 1980, p. 244) says, "How beautiful is Sarai/ Her long, fine, glossy hair, Her shining eyes, her charming nose/ The radiance of her face!/ How full her breasts, how white her skin/ Her arms how goodly, how delicate her hands/ How lissom her legs, how plump her thighs! / Of all virgins and brides/ That beneath the canopy walk/ None can compare with Sarai/ The fairest woman underneath the sky/ Excellent in her beauty/ Yet with all this she is sage and prudent/ And gracefully moves her hands."

[2] In connection with this feature, the fact that Abraham weighs/ strikes "money that is current with merchants", we find a brief description of such coins in *Bereshit Rabbah* XXXIX, 11. As far as I know, no such coins have been discovered but the description of them in the *Jewish Encyclopedia*, citing the *Bereshit Rabbah*, is very appealing: "Abraham, Joshua, David, and Mordecai issued their own coinage. The coins of Abraham had the figure of an old man and an old woman on the face of the coin, and those of a youth and a maiden on the obverse, signifying that after Abraham and Sarah had grown old their youth was renewed and they begat a son." Hence, it would not only be Abraham who struck coins, but many others too, given that it is "current with merchants". In other words, he individually strikes a coin that is recognized by others, which means that both he and others can strike coins in reference to a common, shared standard; von Hayek would be happy about this. If we set aside the fact that the representation of human or animal figures would be in direct conflict with the prohibition in *Deuteronomy* 4:15-19 against reproducing images of any kind, just as the meal which Abraham offers G-d would not be kosher, as I noted earlier, it seems to me that such coins once again endorse the element of sacredness as a central aspect of coinage, given that it is precisely around the theme of age/sterility, therefore covenant/ circumcision and hence the miracle of pregnancy to which the figures stamped on that coin make reference.

(compare the beautiful description of the ancient procedure reported by Bulgarelli, 2001, p. 110, so similar to the Roman one I described earlier, and picked up again in his 2009 text, p. 12, n. 4).

Avraham is "very rich in cattle, in silver, and in gold" (*Genesis*, 13: 2). He fights Hammurabi (and only Doubles fight each other); they sound like doubles too [?], with Hammurabi being called Amraphel. Avraham skilfully fights 318 men and defeats Hammurabi, one of the kings of the East (*Genesis* 14:1; 14:15) to rescue Lot; and a bit of silver-*kesef*. Rashi writes that Amraphel is composed of '*amar*,' "he had said", and '*pul*', "thrown down". Which is why this Hammurabi-Amraphel would be that Nimrod, obviously the ugly, filthy, wicked King of *Ur* described in *Genesis* 10:10 who, years before, in the episode I recounted earlier "had said" to Avram, in that genuine ordeal by fire, that he must "throw himself" into the furnace (!); others say he was catapulted into the furnace because of the conflict that had arisen with his own father, Terah, a conflict caused by the fact that Avram himself had destroyed the idols his father sold, and that this would be followed by the death of Haran who tried to be clever, as I described earlier. Overall, then, a destiny of Doubles who are obviously destined to meet again after this episode, to meet again as kings.

Like Enmerkar before him, like Akhenaton after him, like Hammurabi at the same time, Avraham is not only a king but also a *priest*, a *sacer*dos. Or rather he is not a sacerdos by vocation but because he is chosen directly by God. He is a sacred person. God, even God goes to him, chooses him, and promises him a son, on certain conditions. He is not only *sacer*, he is the quintessence of Enmerkar and Pharoah at the same time.

He is *sacer* because his investiture comes directly from heaven, and still more *sacer* because heaven itself descends to him. He is fully *Homo religiosus* in Eliade's sense, and he glimpses and creates a genuine, new *hierophany*.

Avraham does not climb the mountain or suffer hardships in the desert or return home defeated.

The first time "in a vision" (*Genesis* 15:1), then after a deep sleep and a dark horror (15:12), then prostrating himself (17:3) and lastly wide awake (18:1-6), he argues with God: he negotiates a contract, and even bargains (!) with God (18:24) about the fate, Fair and Just here too, of Sodom and Gomorrah (there is a curious book about the cause of their destruction by Bond and Hempsell which agrees with a sentence from Jeremias (*Old Testament*, vol. 2, p. 84):

> Instead of the fire-flood, sometimes a rain of stones appears

which could represent the fall-out from the impact of a meteorite on the Earth's crust.³ Naturally Alfred Jeremias does not speak about it in those terms.

And some time before, Sarah happened to "laugh" at God. In her own tent! When did Adam dare do that? When did Noah ever allow himself anything like it? When did Moses behave in such a way towards God? When did a Pharaoh go so far?

Maybe one of them, Amenophis IV, had tried to a little while previously, with Aton ("cousin" of Shamash and Marduk) and with the Hittites, *qui paulo ante exierant*, XV-XVIII dynasty, but it hadn't gone very well. When did Enmerkar talk to Inanna like that, eating meat and drinking milk with her, though it wasn't *kasher*? At the very most he invoked her and then had to wait to interpret her obscure messages.

Hammurabi (another coincidence?), King of Ur, or rather *šarrum... ana mat šumerim u akkadim* and sovereign of Babylonia, stands in the presence of the god Shamash who is seated, however (Ur-Nammu actually does the same thing, but here I am considering as superimposed onto Hammurabi, although Ur-Nammu was in the presence of the god Sin).

But Avraham prepares and attends God's meal, actually participates in it. Participates in the "Sacrifice", the concrete celebration of the rite itself. What's more, Avraham is seated. And God is standing!

In Rashi's comment (chap. XVIII, 1) we read:
He was sitting (*yoshev*) [the verb is written *yashav*, he sat; in Akk. *wašabum*].

---

3   Alan Bond and Mark Hempsell studied a round clay tablet, an unusual shape for clay tablets. This tablet seems to be an Assyrian copy from around 700 B.C.E. of an identical but older Sumerian tablet; it was a common exercise in the scribal schools to copy older material for practice. According to Bond and Hempsell, despite the damage to many parts of the tablet, it describes the observation by night of a heavenly body which crossed the sky like a ball of flame as it plunged towards the Earth. This observation seems to have been made at great speed by a Sumerian astronomer who records various stars, constellations, and reference points in the celestial vault on the tablet, and the authors have reconstructed how it would have presented itself to an observer positioned in the middle of Mesopotamia. According to their hypothesis, a meteorite entered Earth's atmosphere from the north-west, struck a glancing blow against a village in the Tyrol, and after a series of bounces, ended up in the Indian Ocean, beyond Somalia, having passed the area of the Dead Sea. The disastrous impacts of this ball of fire generated explosions and falls of varied material which contaminated the environment and destroyed life over large areas. One of these areas of impact would have affected the area of Sodom and Gomorrah. In their reconstruction this would have happened on 29 June 3123 B.C.E.

In fact, Abraham wanted to stand up, but the Holy One, blessed be He, said to him, "Remain seated while I stand! You are a sign for your children, that in future I will stand in the assembly of the judges while they remain seated, as it is written: God stands in the assembly of the judges."

Doubles again: both of them, Avraham and Hammurabi, look into the eyes of God.

And the Double of Doubles: Avraham is seated in God's place, assuming his position, and God stands in Avraham's position. Doubles to whom Sarah is added, eavesdropping, laughing at God, being assailed by Pharaohs and Kings, taken or maybe not, moving angels and miracles. Just like another soap opera!

And in this context, with these elements as I was saying before, it may seem an enormous operation of cultural integration, circum-incision, precisely because it is conceived, if only at the outset, as a striving towards complete monotheism; Caetani (1914, p. 71) astutely observes:

True monotheism is an ideal of theorists or religious dreamers, not a reality existing in the consciousness of peoples.

Circumcision which, in the context into which it is inserted, and with the purpose it assumes at that moment, becomes something new.

I've mentioned this already: if dedicating, sacrificing, donating, offering one's time, one's physical prowess, sweat, devotion (*corvées*) to the gods is already an act of great faith, an act so great as to generate an empire, writing, and money, what should we call the gesture, the act for which one deprives oneself of one's own flesh? It is a sacrifice, a great one. A gift, a great one. An *A/ Oblation*, a great one; which, while still presupposing an apparatus for mediation (the clergy do not disappear), establishes the priority of a direct relationship between the worshipper and God. There are no more common *corvées*, group efforts for building dykes and tilling fields, masses of cultivators-harvesters on lands owned by the temple, troops of shepherds driving and managing enormous flocks of livestock destined for "redistribution", gigantic, orgiastic, collective rituals (the Golden Calf which Moses will make his people swallow is a good representation of the concept). And when the *corvées* reappear in the Biblical account, they take on the connotation of slavery.

In the new faith it is the faithful who individually make an offering to God of something that is their own, personal and intimate. Theoretically there is no further need of a King and Priests to represent, mediate, and ritualise enormous, sumptuous collective celebrations. It is a covenant between the single individual, albeit a member of an elite chosen people, and God; von Rad (1972, p. 27) calls it:

a religion which emphasizes the relationship between God and man, between God and the human community, without ties to fixed localities, but all the more agile in following all the variations in the destinies of the faithful.

Bottéro (1986) also observes that with Abraham we come "to establish... a personal spiritual life... an ideal of life that fills the heart and governs thoughts and deeds" (pp. 90-91). And God, confirming this feature of total uniqueness, though reluctantly, against His will, and a long time later, will grant Israel a King to govern it, but he will be a disaster (1 *Samuel* 8:14-17):

> And he will take your fields, and your vineyards, and your olive-yards, even the best of them, and give them to his servants.... And he will take the tenth of your sheep and ye shall be his servants,

clearly underlining the key idea about the *ḫāwiru*: if you are so foolish as to want him, you'll get him; and indeed, *corvées* and taxes will first be sketched out by David and then endorsed by Solomon. But before all this happens there is an epoch-making change, a revolution. And what is the symbol of this revolution? It is a piece of flesh. A foreskin.

And what should we say about the hundred foreskins of enemy Philistines which Saul demands from David as a bride-price (1 *Samuel* 18:25-27)

An absolutely intimate gift offered, vertically, to a single God.

An old symbol, perhaps, but certainly a new covenant.

And the new covenant is round, empty, a folded string, a ribbon.

A ring. Of flesh. A little circular piece of skin, *circum cæsum*.

I believe that *this* is the soul, the *Mana*, of the Coin.

Certainly this is the Founding Sacrifice, not the near throat-slitting of Isaac (von Rad, 1977) which comes much later and seems to be the regurgitation or compulsion/ re-emergence of ancient, pagan customs – i.e. the ritual sacrifice of the first-born I've mentioned before, also practised by the Canaanites (see for example *Solomon* 106:28 and 37, with Ba'al Peor; but also *Leviticus* 18:21, "and thou shalt not let any of thy seed pass through the fire to Molech," "after the doings of the land of Canaan" (18:3); or 2 Kings 3:27 where the King of Moab "took his eldest son, that should have reigned in his stead, and offered him for a burnt offering upon the wall") and see also Graves and Patai (1980, chap. 34); a sacrifice that God requires to be changed so that only the first-born are dedicated to Him, but "redeemed", replaced, by a sacrificial animal that takes his place as a burn offering.

And the much celebrated Sacrifice of Isaac can be seen as the "simple" repetition, slightly shifted in time and organ, of that first Founding Sacrifice – the Circumcision which happened thirty-seven years before, in which the knife has already "cut off" something from the "child"; a repetition as a kind of regression which reprises archaic gestures, given that the "sublimation"/ "symbolic shift" of circumcision, for some reason, is no longer holding sway. (In any case, it makes me smile a bit to think of Isaac, a man of thirty-seven, looking around, puzzled and astonished, and asking, "But Father, where is the animal to be sacrificed?" It seems that Isaac, not yet a Patriarch, wasn't the sharpest knife in the drawer.)

From eating the children (as Chronos did? He too was destined to be overthrown, as he had overthrown his father), to letting them live but emasculated ("You're not dangerous any more," i.e. the Horde in Freud), to Circumcision ("I could kill you or castrate you, but I'll just give you a wound that won't destroy you; remember, though, I could have done it if I'd wanted to!").

"Cutting off with the knife" and "Child" are two elements that Freud would have little difficulty interpreting (*On Transformations of Instinct Exemplified in Anal Erotism*, SE vol. XVII – p. 133; but also *Analysis of a Phobia in a Five-Year-Old Boy*, SE vol. X, pp. 7-8 and 36n; *Totem and Taboo*, SE vol. XIII, p. 153n).

How strange then, that Freud had so little interest in Abraham... Who knows, maybe Karl (Abraham) was enough for him. (Though I don't know Carlo Bonomi's work well, from the little I've been able to read, it seems that he develops some interesting observations about this.) And on the subject of redemption (and substitution) in the sacrifice, my old friend Jeremias, in the second volume of the same text (*The Old Testament*, p. 49), quotes a Babylonian religious text taken from Zimmern:

The lamb, the substitute for man

> The lamb, he gives for their life
> He gives the head of the lamb for the head of man
> The neck of the lamb for the neck of man
> The breast of the lamb he gives for the breast of man.
> (*Agnus Dei qui tollis peccata mundi...*)

To say nothing of the *Pidyon Haben*, the Ransoming of the First-Born (male) which is effected in modern Judaism thirty days from the child's birth by payment of a coin, 5 shekels, to a descendent of Aaron, i.e. a Cohen. If I wanted to propose a bizarre hypothesis, 5 shekels could be considered a multiple (tenfold) of the half-shekel (for further details, see the *Jewish Encyclopedia* entry on 'Primogeniture').

Back to the subject. Given that earlier on I expatiated at length about elements connected to the business of weighing the Coin in the time of Hammurabi-Avraham and which were necessary for bringing it into being, and given that the half-Shekel was the most widespread coin, and given that, as we shall see in a moment, it was the custom to donate a half-Shekel to the Temple in homage to a tradition the origin of which is lost (in any case, the Torah calls every half-shekel "the ransom of the soul" – see Pacifici, *Commento alla Parashat Ki Tissà*); well then, I ask a "naïve" and humorous question: how much does a foreskin weigh?

Answer: I don't know. (There! I said it, I said it!)

I've asked Jewish friends, who smile and point out that the *Mohel's* responsibilities do not include the task of weighing the child's foreskin. I asked a couple of friends who are surgeons, and who also smiled, telling me that the intervention is a paediatric responsibility, except where it is a matter of pathology (and more rarely, a question of religious conversion) in adults, all of which are cases where there is little room for, or medical interest in, establishing a weight. And so: how much does the baby Isaac's foreskin weigh? or Abraham's? or Hammurabi's? I should say, Abraham-Hammurabi's.

In the literature, medical or otherwise, I've found nothing.

I've let relatives and friends loose on the subject but not received reliable answers. Nothing, for the moment, beyond the slightly ironic and surprised smile with which they look good-naturedly at me. However, it is still the custom today at the Jewish festival of Purim to give a half-shekel to charity, *Tzedaka*, in memory of that ancient custom which prevailed until the second destruction of the Temple, a custom in which every Jew, whether rich or poor, used to give a half Sheqel to the Temple ("the rich shall not give more, and the poor shall not give less, than half a shekel," *Exodus* 30:13- 16; sometimes three half shekels are given, this step being cited three times in the Scriptures).[4]

---

4 Hendin (2010, p. 22) makes attractive and interesting observations on the half Sheqel to be paid to the Temple; some scholars of the Talmud observe that if the offering was made with a small, whole coin worth half a sheqel, this was good and no questions were asked, but if an offering was made, even by two people, with a one sheqel coin, then it had to be changed (this was the function of the moneychangers in the Temple who made Jesus so angry) into two half-sheqel coins; and every one of these coins changed for the offering incurred the Kolbon: a corresponding rate of interest for moneychangers in the Temple as payment for their work; the rate of the Kolbon/ Kolbonot hovered at around 4.2 %. This percentage also applied when offering smaller coins and, of course, coins with values other than those of Israel: in other words, it had to be a single half Sheqel.

[*Exodus* 30-16] And thou shalt take the atonement money of the children of Israel, and shalt appoint it for the service of the tabernacle of the congregation: that it may be a memorial unto the children of Israel before the Lord, to make an atonement for your souls.

(Second destruction of the Temple or destruction of the second Temple? Whichever it was, the first was the work of Nebuchadnezzar in 586 B.C.E., while the second happened under Titus in 70 C.E.) This "offering" was made by the first day, *Rosh Codesh*, of the first month of the new year, Nisan, which was the Jewish New Year's Day which recalled the exodus from Egypt ("this month shall be unto you the beginning of months: it shall be the first month of the year to you," *Exodus* 12:2) and not the one marking the birth/ creation of Adam.

Alongside this, I would like to point out the fact that for the Catholic Church, the first day of the first month of the new year has always been celebrated – officially, I believe from the ninth century – as an anniversary which was only "abrogated", i.e. made optional (!), by Vatican Council II in 1969, and I think it is still preserved unaltered in the Eastern Orthodox Church; this feast is the Circumcision of Jesus. And, another distinctive detail, in the Vespers for this festival, the Vespers of the 'new month's day' of the new year, the reading, which we have encountered before, is from St Paul, Letter to the Galatians, 4:4:

> But when the fulness of the time was come, God sent forth his Son, made of a woman, made under the law.

"Made under the law" refers explicitly to the Law and the Alliance, i.e. *Genesis* 17:1-14, Abraham and the covenant with God, circumcision, set beside the reading of the Gospel about the circumcision of Jesus (*Luke* 2:21), the *Brit Milah*, and the presentation at the Temple (*Luke* 2:42), the *Bar Mitzvah*. Once again: Isaac is circumcised eight days after his birth, and the anniversary of the circumcision of Jesus falls on the first day of the year, eight days after his birth.

On the subject of the *Brit Milah*, which is carried out eight days after birth, an interesting point for reflection which I have not considered in this book since I wanted to keep away from clinical practice, could be the fact that while primary repression, thank God, intervenes to "delete" the trauma of the early wound inflicted by circumcision, it is equally obvious

---

However, not even Hendin can tell us exactly why one half-Sheqel coin, not fractions, not multiples, not in other currency, had to be given to the Temple; he simply reiterates once again the importance of the half sheqel.

that the effects of the wound, though shared and accepted by the group of the circumcised, and representing an important element of the group's very cohesion, are concretely, daily, physiologically present to the eyes of the growing child; and though this fact is a long way from the general concept of part-object, it must introduce some part-element.

There will be other details (the devil!) but in Liturgy, as in all ritual and sacred matters, details are important. The *Brit Milah* and the *Bar Mitzvah* are still present, remembered, and sanctified in the Christian liturgy. Two Alliances and two Baptisms.

Apart from the many deviations which have led to the adoration of the Holy Foreskin and its presence as an "authentic" relic in as many as eight European cathedrals (though others have counted 15), I don't think we should underestimate the symbolic potency of the temporal and symbolic "coincidences" I have mentioned: the first day of the first month of the new year. Another form of that "new covenant", that "new" beginning, that "new" *Brit*, that is so ancient today; and with Abraham and his Covenant once again as witness. From a 1957 Festive Missal for Vespers:

> First of January. The Lord's Circumcision. Shedding his own Blood in the Jewish rite of circumcision, an image of the Baptism, Jesus warns us that this gift that He has given us with his birth in Bethlehem, and which makes us children of God, "comes at a heavy price."

In the same jocular vein, having mentioned Purim, I could paraphrase Géza Róheim and say that the problem is deeply repressed and that this increases its evident importance.

With regard to what I have said so far, I do not believe that the real weight of an adult man's foreskin, that of Avraham-Hammurabi-Ur-Nammu, is an element of great concrete importance, but I don't deny that if an adult man's foreskin turned out to weight about 4.25 grams, i.e. a *half-Siqlum*, I wouldn't resist a *half*-smile, from Manishtushu onwards.

In any case, I think no one can miss the symbolic significance of the "totemic meal", for example (as a metaphor or as a concrete fact); and from a clinical viewpoint it is of no great importance whose thesis is more plausible: Girard's, reconstructing the birth of the royal function as the expression of a sacrificial murder of a scapegoat which generates Royalty (a thesis also arrived at, though indirectly and along a different route, by Mauss when he derives the word *Rex* etymologically from *Res* and *Reus*),[5]

---

5 Mommsen indicates that *Rex* would be descended from *Rector* but, unlike Girard, he doesn't go into the features which lead to the genesis and investiture of the

or Freud's reference to the Horde and the Oedipus conflict; or whether we prefer the observations by Money-Kyrle, or Girard, or Lopez about sacrifice.

It's always a matter of homicides, bloody "ritual" events.

It's always a matter of foundational elements. And they are good to reflect on.

The fact is that any given action leaves a symbolic "precipitate" of that action which is then slightly transformed and deposited, taking new directions, always mobile and changing in the way it is expressed, though remaining connected to its beginnings and continuing to preserve its initial characteristics (remember the word *Ur*) which go on living and working deep down.

And acting.

---

figure of the *Rex*. In the same way, Semerano holds that *Rex* is connected to the Akkadian *rē'ûm*, i.e. Shepherd, title of all the sovereigns of Babylonia and the origin of the Sanskrit *rāj-*, hence *raja*, King/Monarch; here too, as with Mommsen, we have the etymological hypothesis about the royal function but not about the formation and birth of the royal institution; Dumézil also shares Mommsen's opinion. I will add that in Akkadian, we also find *rēštum* (beginning/first), *rēštûm* (pre-eminent, primordial) and *rēšum* (head, chief, commander/leader). The term Shepherd is still used to indicate one of Christ's functions, and it makes me smile to think of Professor Jeremias, who is a Shepherd himself, as the ministers of his cult often call him. The term is present, with this meaning, both in the Hebrew Bible and in the Gospels, but even more so across the whole of Mesopotamian tradition, beginning with those very *Euphratesians* disliked by Ministrer-Shepherd Alfred Jeremias.

23.
ARE WE REALLY SURE ABOUT THE CUNEUS?

Before continuing I want to point out a "little", "secondary" question.

אלהים ברא אתו זכר ונקבה
"*elohim bara oto zakar u neqevah*" (*Genesis*.1:27)

Obviously, with all the variables and semantic, linguistic, poetic, or political possibilities offered by all the world's languages – and fortunately in this case, at least, we can perhaps avoid Gender theory – the translation (with apologies for the transliteration) of the first sentence could be:

"the Lord/G-d created him male and (her) female."
(Oh God, maybe questions of Gender aren't absent after all!)

In this case too, I set off from a long distance away in order to arrive at a point of reflection which, albeit "marginally", has heavily conditioned and still conditions a certain part of our way of thinking.

So, G-d creates man, creates him male (and in his own image and likeness). In this passage from the Torah, this Man is called *zakar* (in others he is *ish*, or *adam*). According to my Hebrew teachers, *zakar*, more Male than Man, is connected to *zikrun*, root z.k.r., which goes back to *lizkor*, remember/ leave a sign; memory, *zikaron/zeker*. Nothing strange about this so far, since the male tangibly leaves a "sign", a little memory, nine months afterwards of himself. All this is very nice and poetic.

However, I also need to add that *zikaru(m)* is also the Akkadian term used to indicate the male/virile (Sum.: URUM) and, specifically, also the penis (Sum.: ĜEŠ/ PIL).

In relation to the woman, *neqevah*, the term is definitely a bit brutal, and sounds like it: *the well/hole inside her*. Here we don't have a Gender problem but I would say do have some little problems of respect; at least we do if we address the subject with a modern sensibility but when all's said and done it's a bit ridiculous to accuse the ancients of not being modern. If they were, they wouldn't be ancient!

That said, we do have to make some reflections on *neqevah*. Like *zakar*, *neqev* (root n.q.v.) also has its direct predecessor in Akkadian: *naqābu(m)*, penetrate sexually, deflower. (B and V very often replace each other, as they do in modern Spanish where V and B are pronounced almost identically. The Spanish have a saying that warns of the potential confusion between the sounds: '*la V de* vaca *no* es la B de burro' (The V of vaca – 'cow', pronounced 'baca' – is not the B of burro – 'ass').

I won't go into the developments of *zakar* since I will speak about it in more detail later, and in a different context. Instead I'll concentrate on the term *neqev*.

Love of my country stops me once again confronting the difficulty in translating the term *Moneta* (from Latin) into English, given that Coin, Coins, Money, Banknote, and Currency do not always have exact equivalents in other languages. However, I would like to start with the definition given by the Online Etymology Dictionary for Coin (the Oxford English Dictionary is more vague, confining itself to a brief report of the possible Latin etymology from *cuneus*; nothing more; as is the case with all the etymological dictionaries I have been able to consult):

> coin (n.)
> c. 1300, "a wedge, a wedge-shaped piece used for some purpose," from Old French *coing* (12c.) "a wedge; stamp; piece of money;" usually "corner, angle," from Latin *cuneus* "a wedge," which is of unknown origin (my emphasis). The die for stamping metal was wedge-shaped, and by late 14c. the English word came to mean "thing stamped, piece of metal converted into money by being impressed with official marks or characters" (a sense that already had developed in Old French). Meaning "coined money collectively, specie" is from late14c.
> Compare quoin, which split off from this word 16c., taking the architectural sense.
> Modern French coin is "corner, angle, nook."
> The custom of striking coins as money began in western Asia Minor in 7c. B.C.E.; Greek tradition and Herodotus credit the Lydians with being first to make and use coins of silver and gold. *Coin-operated* (adj.), of machinery, is attested from 1890. *Coin collector* is attested from 1795.

I thought of quoting this since in Spanish, French, and *de facto* in Italian, we have the same description (in German too, *Münze* is derived from Lat. *moneta*): in other words, Coin would derive from the use of a *Cuneus* by an engraver to stamp an image onto a metal disc with blows from a hammer; hammer and *cuneus* stamp images onto the two faces of the coin.

This term would derive from the Old and Middle French of the 12-16th

centuries, B.C.E. naturally (?), given that metals were being smelted as early as the Calcholithic Period, and still more in the Bronze Age, which would in turn derive (?) from the Latin, *Cuneus*.

I was joking about the 12-14th centuries B.C.E.; even though the Grandeur de la France is over the top and sometimes even comical. It's clear that only the French could think they'd founded the Roman Empire, or the Egyptian Middle Kingdom, or Classical Greece, not to mention Minoan Civilisation, originating the Iron Age and defeating the Akkadians, conquering the Lebanon and Palestine (something they really did, but a "month or two" after the 12th century B.C.E.). And only an Italian could, out of an ancient spite, make jokes of this kind.

Anyway, I apologise to my French colleagues, but it was too good an opportunity to miss poking a bit of fun.

I won't even mention the worn-out, and now threadbare, reference to Lydia since it bores me to death (see the chapter on Ur-Nammu).

The appealing thing about etymology is that, in the all-too-frequent absence of "proofs", we follow our noses or the *ipse dixit* of certain scholars (as a psychoanalyst, I had better be very cautious about this): seeing that some people have decided that, since Coining really is carried out with a Wedge and a Hammer, then the ancients certainly did it this way, and the word Coin can only be derived from that practice. In Italian we have the noun *Conio* and the verb *Coniare*, with the same root as Coin. Therefore Coin must derive from the Latin *Cuneus* (but then what became of the yet more ancient Lydia?).

This is patent silliness: every numismatist is aware of the fact that, long before coins were minted, it was common practice to smelt metal (Iron, Copper, Bronze, Silver, and Gold) and to pour it into appropriate moulds.

The first example that comes to my mind is *Aes Rude* at the start of the Iron Age, which has homologues at various times in Asia/China and in northern Europe (this term, *aes*, specifically *militare*, remains in use in Europe until the 16th century to define the pay given to troops, both mercenary and otherwise) and is certainly not the most recent example. In fact, Coining only begins in the sixth century B.C.E., or the seventh (Lydia again!). Naturally the Assyrians' "drops" of molten silver, cited in Parise (2000), of a standard weight, are irrelevant!

And so I believe that the idea of the *Cuneus* as the origin of Coining must be completely reviewed.

The most ancient coins, or rather – given what I will maintain at the end of this book – the second most ancient, were smelted and not "Coined".

Anyone who knows even a tiny amount about the technique of smelting

knows that at least five elements are needed before anything can be smelted:
1. The ability to know about, extract, and transport the specific material to be smelted;
2. Knowledge of the chemico-physical characteristics of the material to be smelted;
3. A furnace with a crucible: an open fire reaches 600-700°C, insufficient even for copper; melting copper needs 1,083°C; bronze (formed from copper and tin, together with small quantities of various elements/oxides depending on the alloy) from 880°C to 1,150°C; and iron 1,538°C along with bellows and suitable structures for forcing the air;
4. A ladle;
5. A mould.

On the first and second elements, I refer you to the splendid works of M. Eliade on alchemy and the melting of materials. In any case, "Chapeau!" to our ancestors.

With regard to the use and construction of furnaces, I refer to Bottéro. Once again, "Chapeau"!

As far as the ladle is concerned, I refer to the distinctively delightful valley I come from.

Which leaves the Mould.

What is a mould? Stone, clay, plaster, metal, with or without "lost wax", with or without holes, it doesn't matter for the moment.

> mold (n.1) (again from the Online Etymology Dictionary)
> also mould, "hollow pattern of a particular form by which something is shaped or made," c. 1200, originally in a figurative sense, "fashion, form; nature, native constitution, character," metathesized from Old French modle «model, plan, copy; way, manner» (12c., Modern French moule), from Latin modulum (nominative modulus) "measure, model," diminutive of modus "manner" (from PIE root *med- «take appropriate measures»). By c. 1300 as «form into which molten metal, etc., is run to obtain a cast.» By 1570s as «a form of metal or earthenware (later plastic) to give shape to jellies or other food. Figurative use of break the mold «render impossible the creation of another» is from 1560s.

As we now know, the techniques of smelting begin in the Calcholithic Period (the Copper Age) around 5,000 B.C.E., in Mesopotamia; from 3,000 al 1,000 (approximately) the Bronze Age, then passing on to the Iron Age. Obviously I'm simplifying, maybe too much, but it doesn't matter for now. So we have a temporal arc of more than 4,000 years before the arrival of Coining. As I noted before, *sheqel*/sickles were spoken of in the time of the goddess Nanshe, and this was around 2,500 B.C.E.; these *sheqel* were

NOT coined but moulded.

I will also pass over some strange dating drawn from the Dictionary recorded above.

Now, if we choose to believe the Biblical chronology, which I wouldn't recommend, it has forgotten the Assyrian captivity and only recalls the Babylonian (not exactly forgotten, but the accounts of the two captivities are profoundly different: on the one hand we have the edifying accounts of Daniel and Esther, and on the other the dramatic anticipation/ restatement in *Exodus* of the Babylonian captivity) in that very broad lapse of time which runs from at least 5,000 B.C.E. to the sixth century B.C.E. (official beginning of "coining"), we have Sumerian which indicates Woman by referring to her female genitals, also using the term PU, and Akkadian which also uses the term *pû*, and then the term *naqābum* appears, with the same meaning. We have seen the Hebrew with *neqev*, hole/well; while Greek uses κολεος, *koleos* first (but does Renault know?) and then κολπος, *kolpos*, crack/split.

Latin is interesting from this point of view because it brings about a change of perspective. Whereas *kolpos*, crevice, and *neqev* have the implicit and explicit meaning hole/ crevice/ ravine/ tunnel/ orifice to be "explored/ penetrated" exactly like the Akk. term *naqābum*, the Latin shows an "evolution" of the concept, like the Hebrew. We don't know exactly when and why it happened but, from indicating the female genitals with *Cunnus*, it moves to *Vagina*. (I won't go into the anatomical questions of Vulva-Vagina-Uterus, since the ancients rarely went in for subtlety.) And in Hebrew we pass from *neqev* to *pot*, from hole to fountain/ spring (almost recovering the Sumerian/ Akkadian meaning; in Brescian, for what it's worth, we have *pota*). In Hebrew we have also the words, so similar: כוס, cos/glass, and כוס, cus/cunt.

In ancient Latin, Vagina was used to indicate, for example, the scabbard which "protected" the sword and in which the sword was replaced after use; or it indicated the film which enveloped the kernel of fruit, the chaff/ glume which envelops the kernel. From a symbolic viewpoint these could mean the same thing if it weren't for the fact that, for the ancient Latins, that was the vagina, whereas the female genital organ was a *Cunnus*, a hole. This was long before Asterix. In the time of Asterix, Obelix, and Getafix, with Catullus, Juvenal, Martial, Ovid, Lucretius, and Pliny, we witness the appearance of the term Vagina to indicate the female genitals, and the Cunnus continues to be the hole used by the Cuniculus, Coney/ Rabbit for hiding and procreating in its cuniculus, burrow/ tunnel/ den.

What is the point of all this discussion?

"Nothing; everything," Ridley Scott makes Ṣalāḥ al-Dīn say in his very beautiful film *The Crusades* on the value for the Muslims of conquering Jerusalem.

All this discussion achieves "nothing" because those who, for the most diverse reasons, will want to go on believing in the *Cuneus* will go on doing so. This discussion achieves "everything" because a new view may come out of it, one unlike the established view that is regarded as "solid".

If we include "swearwords", which often indicate deep meanings much better than gracious and official language, given that smelting technique entails the use of a Hole/ Shell/ Mould/ Container that has to receive the poured liquid metal, giving it the desired shape, and if we consider swearwords like the English *Cun*-t/ *Conch* and the French *Con* (here we see the subject coming into focus) and add that in French, *Moule* is the shell which serves as a mould, but also acts as a vulgar term for vagina; *Moule* which is at the origin of Mold (v. sopra); *Moule* which is also the mussel, bi-valve/bi-vulva, which well represents both the shell and the vulva; we come to the Spanish where, together with concha, Conchita, valve, Her Majesty, is enthroned the **coño**[1], again to indicate the vagina.

At this point I think that Cunnus/Coño has every right to be considered at least as a possible, quite different, explanation from the much-trumpeted Cuneus.

---

1 Coño: Dal Lat. cunnus.
1. m. (malsonante). Vulva y vagina del aparato genital femenino.
[(vulgar). Vulva and vagina of the female genital apparatus.]
Diccionario de la Real Academia Española (actualizado en 2019).

With regard to the Tilde (~), it should be noted that it is an element, a glyph, of Mediaeval origin, initially used by amanuenses to save space when indicating a "double" letter: i.e. instead of writing, NN, they wrote Ñ. Over time, the Tilde became autonomous, no longer representing a double letter but a sound in its own right, NY. Hence, *Coño* would originally have been written *Conno*, extremely close to the French *Con* and almost identical to the Latin *Cunnus*.

24.
THIS UNKNOWN FORESKIN

Prepuzio: s.m., anat. Piega cutanea che riveste il glande, su cui scorre liberamente in tutta la circonferenza, eccetto che in corrispondenza del meato urinario esterno, dove è fissato dal frenulo o filetto; viene reciso con la circoncisione (Grande Dizionario Utet).

And in the Encyclopaedia Britannica:
The foreskin (or prepuce), extends forward to cover the glans. At birth or during early childhood, the foreskin may be removed by an operation called circumcision (q.v.).

I find the English terms interesting: Prepuce, Foreskin, Fore-Skin. It's the same in German: *Vorhaut*, *Vor-Haut*. Again "Pre-Skin".

The Hebrew term, in the book's title, is very curious: *Orlah*, which it's amusing to imagine translated into Italian as *orlo*; (in French *orle*; in Spanish *orla*; 'hem', 'fringe', or 'border' in English.)

Orlo: Parte estrema, laterale o inferiore che termina una stoffa, una veste, un capo d' abbigliamento (Grande Dizionario Utet).

[Hem: furthest point to the side or the lowest point which terminates a piece of cloth, a garment, or an item of clothing.]

Before continuing, I would like to make a brief digression: given that, if only for the sake of survival, agricultural activity must precede the practice of circumcision, I think it is interesting to report an entry from the *Jewish Encyclopedia*:

> 'ORLAH ("Foreskin" [of the trees]): Name of a treatise in the Mishnah, Tosefta, and Yerushalmi, devoted to a consideration of the law, found in Lev. xix. 23-25, which ordains that the fruit of a newly planted tree shall be regarded as "'orlah" (A. V. "uncircumcised") for the first three years, and that therefore it may not be eaten. This treatise is the tenth in the mishnaic order Zera'im, and is divided into three chapters, containing thirty-five paragraphs in all.
> Chapter 1.: The conditions which exempt trees from or subject them to the law of 'orlah (§§ 1-5); mixing of young shoots of 'orlah or "kil'ayim" with other young shoots (§ 6); parts of the tree which are not considered fruit, such as leaves, blossoms, and sap, and which are therefore not forbidden, either as

'orlah in the case of a young tree, or to the Nazarite in the case of the vine; it is noted in passing, however, that in the case of a tree dedicated to idolatry (the Asherah) the use of these parts in any way is likewise forbidden (§ 7); the parts which are considered fruit in reference to 'orlah, but not in reference to "reba'i" (the fourth year); so that, although these parts may not be eaten during the first three years, it is not obligatory to take them to Jerusalem in the fourth year (§ 8; comp. Lev. xix. 24); concerning the planting of 'orlah shoots (§ 9).

Foreskin. This word always seems to be composed of two terms: Pre/ Fore and Puce/Skin. Let's see if Giovanni Semerano can help:

> The Latin is *Pæputium, i...*
> *Prae-*, in antiquity *prai* (cfr. *praifectos*, "praefectus"), before, because; Oscan *prai* (with a temporal meaning), Umbrian *pre-* etc., v. παραι and Latin *pris ... Parai*, παραι, comes to be connected with *peri*, περι, and *pris* with *prior* and *primus*.

Indeed, circumcision in Greek is *Peri-tomé*, περιτομή.
So this would give us a *Prae-*, a περι, with shades of 'around', 'before', and 'first'.
So far so good. But before what? *Præ-* (Pre-) doesn't seem hard to interpret. But *-puce*? (Here the *Puce* is not the *flea*).
In Latin *Puto,-as,-avi,-atum,-āre* (Semerano).

Puto,-as,-avi,-atum,-āre: besides meaning Consider and Believe, it has the meanings Clean, Tidy, Cleanse, Purge, Prune/Trim, Remove Leaves; therefore, it could be connected to hygiene practices concerned with Cleansing the smegma, which would obviously be easy to connect with *Putor,-ōris*: Decay, Stink, Fetor. Puteoli, the modern Pozzuoli, is directly connected to *Putidus*, 'rotten', 'stinking'. But we also:

> need to go back to *putus*, which originally meant 'open', adj. from the Akk. *putu* (open), Sem. *ptḥ*; in Akkadian there are already all the semantic values that will develop in Latin: *to open a path, way, or well*.... The same original base gives the Lat. *puteus*, -i: 'cistern', 'well', 'hole', and obviously the Lat. *pateō: patet* 'to be open'; and the Gr. Πετάννυμι, 'to be open' (Semerano).

Hole-Well-Cavity-Source. Hence the element of hygiene, the need to clean the cleft of the glans by pulling back the foreskin: in other words, the requirement to keep the penis clean would have generated, perhaps in an excess of zeal, a radical form of advance cleaning, *Putare*, to 'amputate' – i.e. *Am-Putare*, "make clean/ prune/ trim all around"; and *Amphi*, hence Am, has a specific sense of Round, Circular, Around, as in *Amphi*-theatre.

Agreed, smegma *olet*, but not everything that *olet* is Qaqi. Some flowers *olent*, but they aren't Qaqi. Garlic *olet*, but isn't Qaqi. Almost every single essential oil for perfumes *olet*, but these aren't Qaqi. Most spices *olent*, but aren't Qaqi. In other words, *Olēre* isn't just a Stink. Semerano goes on to write:

> If *olōs* (from Müller) can be trusted, it is a cross with the root corresponding to the Akkadian *elēṣu* (to be joyful); *ulluṣu* (to bring joy).

And *Olōs*, connected to Olfactory, is the root of Olezzo, a pleasant odour, fragrance (Pianigiani), closely connected to joy, as the perfumiers know (and as Patrick Süskind does too, though for purely dramatic purposes; and Puccini in *Madame Butterfly*, "*piccina, mogliettina, olezzo di verbena*"!).

On the subject of the Foreskin, Semerano writes:

> (Præ-) putium (…) must have been heard in the same way as *-puttus*, whose obscene meaning is reconnected to the root corresponding to Hebr. pot (female pudenda).

That is: *Potta*, Vagina/Cunt.

Although it is not much prized by many enlightened linguists, the Pianigiani *Etymological Dictionary*, stresses the Sanskrit root *pu* in relation to Putto, Putta, and Potta, i.e. the sense of 'to generate', from which derive other terms such as *puer* and *puella*:

- Putto
   1. Sp. and Port. *puto*: from the Lat. PUTUS from a root PU- inflected as PUT, which has the sense 'generate', 'procreate', whence Sanskr. *pu-tràs* (Celt. *Poatr*) son, *putrî* daughter, *pô-tas, pô-takas*, young (animal); as well as the Lat. *puer* boy, *pùsus* and *pùsio*, little boy, *pu-p-us/a* boy/girl, *pûllus*, any young animal, po'- mum*pomo* and the Gr. PAIS [παῖς, παιδός] boy.
   cf. Paggio/Page, Piccolo/Petit/Petty, Pedagogo/Pedagogue, Pollo/Pollone/ Pullet, Pollulare/Pullulate, Pomo/Pomme, Popa/Poppet, Pre-puzio/Pre-Puce, Puerile, Puledro/Foal/Filly, Pupilla/Pupil, Pube/Pubis, Pusillo/Pusillanimous, Rampollo/Offspring. (Others derive from the Sanskrit root. PU- passive form of PA- feed (v. Padre/Father): whence *putràs* son, would be 'the one who is fed'] The same as Bambino/Baby.
   Deriv. Puttana/Prostitute; Puttello/Little boy.
   2. in the sense of Abject, Luxurious, which may derive from PUTIDUS, which means 'stinking'.

My observation: setting aside the hygienic aspect, which will be present nonetheless, the element of generation present in the Sanskrit root *pu* is definitely related and correspondent to the Sumerian PU, 'source', 'good', 'well', 'cavity', 'spring', 'vessel', which in Akkadian becomes explicitly *Pû(m)*, vagina (literally mouth/orifice), later becoming *Pot* in Hebrew (and I won't dwell on the symbolism of the English *Teapot* or the French *pot*, but there must be something in their shared etymology); and Eliade (1937 p. 43) makes the same connection, though via other routes and in reference to Foce/River mouth, Pozzo/Well, Fonte/Spring.

From Pianigiani again:

- Potta from the barb. Lat. PUTA, which used also to mean foreskin[1] and is perhaps linked to the Root PU, generate, procreate: see also PUTTO. Dien

---

1   We are told that Potta "used also to mean foreskin"; it is also the fold of mucous tissue which covers the prepuce of the clitoris [Pianigiani, revised by Professor Vincenzo Gazich, derives clitoris from Kleis, Κλείς, Κλειδός, 'key' and Kleio, Κλείω, 'to close', associated with Dorà, Δορά, 'skin', in turn connected to Dero, Δέρω, 'to peel' (interesting that Doris, Δορίς, is the sacrificial Knife!); by contrast, the DELI identifies the etymon of Kleitoris, Kleitoridos, Κλειτορίς, ρίδος, from Kleitor [?], 'hill', Klino, Κλίνω the and Klinein, Κλίνειν, 'to incline', and attributes to Falloppio (1523-1562) the introduction of the term which replaced the hitherto current one, tentigo; Havelock Ellis clarifies the matter: "The Greeks called it myrton [μύρτον], the myrtle-berry; Galen and Soranus called it nymphê, νύμφη because it is covered as a bride is veiled, while the old Latin name was 'tentigo', from its power of entering into erection, and columella, the little pillar, from its shape" (Studies in the Psychology of Sex, vol. 5, p. 130)]. It is also possible that this physiological and semantic similarity to the male foreskin is not unconnected to the various types of excision, partial removal, and amputation to which many young girls are unfortunately and brutally subjected in various parts of the world in the name of "tradition". Until I am better able to understand the matter, I will maintain the perspective I have been illustrating up to now, with "the Foreskin of Abraham-Hammurabi" as my specific point of reference: male circumcision which, in any case, seems to open up interesting scenarios for investigation. So, if a further investigation of the clitoridal Prae-putio and the clitoris (in Hebrew, dagdegan, instrument for tickling; a bit like in German where Kitzel, clitoris, is so close to kitzeln, tickle) leads to further developments, it will be interesting to see what direction they take. It will be one coincidence among many, but when at the start of this work I wrote that a little boy had bought a shell to listen to the sound of the sea, I was in fact referring to a Cypraea tigrata. Very beautiful. The Cypraeidae are the conches, also called Kauri, Cauri, Cowries. A great many species of conch have been used across the world as Coins: with three sheep I buy/barter a certain quantity of conches, accepted by my social group, and so I make use of a third means which enables me to buy other goods. This happens/ happened in many parts of Africa, Oceania, America and China. Many writers have studied economies based on conches and one of the greatest is Karl Polanyi. I won't dwell on his ideas here, since this isn't a book about economics. What is important for me is the fil(s) rouge(s) that unite(s) apparently distant things. And

connects it to the Spanish and Portuguese POTE; Provençal and French POT, of Germanic origin; Dutch POT, Swedish POTTA, Danish POTTE; also in the Cimbrian Celtic language POT, Scots Gaelic POIT, Irish POTA, PUITE;

even among the Cypraeidae we find very interesting threads a long, long way from Shit. Threads about sexuality, Latin, fields, Linnaeus, porcelain. Two definitions from Pianigiani: "Porcelain: 1 - .... Portulàca or Porculàca.....[flower] porcelain: 2 – Fr. porcelline, Sp. porcelana, Ger. Porzellan. This term was used in the middle ages to designate a tigrata conch or shell of Venus, Concha Veneris, so called because of a certain similarity of form, from PORCUS or PORCA (via the diminutive PORCELLA), to the vulva of the prostitute." [Here Pianigiani seems to confuse the conch, Cypraea tigrata, with the scallop shell, the coquille Saint Jacques; however, it is still clear that the gleams of light associated with Porcelain are coming from the Cypraea; Pecten Jacobaeus is a bi-valve, thereby having an equal symbolic importance of its own.] Again in Pianigiani, we find two different hypotheses about Porca: "Pòrca = Lat. PORCA which the ancients said was a contraction of PORRIGIA and derived from PORRIGERE or PORGERE, 'to stretch out' (see Porgere); but Pictet, Fick and Duden more plausibly consider it cognate to Old Germ. FURH, FURUH = mod. furche, A.O. furchî, Middle Eng. forwe, mod. Eng. furrow (cf. Forra). Any stretch of raised earth between two furrows in a field, so called because it stretches up from the level of the field." (In other texts, we find more or less decorous references very similar to this. Semerano cites Festus, 244,6: "Porcae appellantur rari sulci, qui ducuntur aquae derivandae gratia, dicti quod porcent, i.e., prohibent aquam frumentis nocere" (the term porcæ [nominative singular porca] means furrows a long way apart, which divert water, so called because they porcent [porcĕo, -es, ēre, to distance, restrain, impede], that is, they prevent water from damaging the crop) going on to underline the link between Porca and Porcus; Porcus, domestic pig, as opposed to Aper, the wild pig.) There seem to be no direct links to the Lira. So I go and look up Forra (furrow): "Fòrra from the Old Fr. FEURRE (= mod. FOURRÉ) furrow, channel: which some see as connected to FOURRE, sheath, from the Germanic. FODR sheath (v. Fodero); and Tobler derives it more probably from the Ger. FURRE, another form of furche, ant. furuh, furh [whence Eng. furrow] furrow = ant Scand. for canal (cf. Porca). Opening, Cavity or similar, long, narrow and steep, between mountains and hills, covered for the most part by thick scrub, and also a deep, narrow Bòtro, ditch, with steep sides, ordinarily carved out by waters in muddy terrain." (A sentence that could have been taken from the Song of Songs.] So I take a quick look at Botro: "bòtro, cf. dial. berg. e bresc., buder, sinkhole; Rum. butura depth; from Gr. BOTH- ROS (= BOTH- YNOS) ditch, from the European root BADH or BAD, to dig, hence also the Lith. BED-IT, dig, BED- RE, cave, and the Lat. FOD- ERE (for PHODERE, BODERE) dig (cf. Fosso [drain], Bottino [cesspit]?). Others see a match with Gr. BUTHOS (= BYSSOS, BATHOS) depth (v. Abisso [abyss]).– Sinon. of Borro [trench]; Fossa [ditch] and Fosso [drain] which is a rustic term". Since my distinctively attractive home valley is indeed in the vicinity of Brescia, I can add that, in contemporary language, what the Author indicates as "buder" – though being Siennese rather than Brescian it's not a given that he'll report it faithfully – has in any case been turned into "buren".

the last of which, like the Latin CONCHA, i.e. CONCA, reunites the two meanings of vessel and vulva. Female Pudenda. [I'll add the English POT, and the Brescian POTA.]

Looking at the term Prepuce/Foreskin through the lens of these etymons, I think we can glimpse a small change of register which makes the *praeputium* shift from being an element that simply *precedes* the glans, and so is centred on the penis itself with a purely anatomical, static, and self-referential connotation, to a *praeputium* which becomes *prae-liminare* to the encounter with the Pot. For the change of vowel from U to O, I think we can view this in the same way as we do Ur and Or.

Pre-Puce: understood not only and not so much as Pre-Glans, to be cleaned by washing or removing it, but Pre-Puce understood also as Pre-Putio, i.e. *Pre-Pot*, i.e. Pre-Vagina. And maybe also vice versa. A shift from a simple body-part to the indicator of a relationship with an Other. Being concerned with circumcision of the penis, this is clearly a male-oriented permutation of the discussion. The pre-liminary, *prae-limen*, to an encounter.

Limen, not *Limes*, which in a suggestive and evocative image Semerano invites us to read as "entry into the interior". And the entry, the encounter, the offering, the gift of a *prae* and a pot, represents better than anything else, the mutual giving, the offering oneself to the Other, losing oneself in the passing of the *limen*, entering into Other territories, the territories of Others.

There is no longer a Gift and no longer a Trade. Only Desire. Mutual.

And the sacrifice to the deity of something that is part of the most delicate, complex, elaborate, and mature thing in existence, being an important generative, reproductive, amorous, and erotic feature, one that is, or could/ should be, the peak of Genitality, cannot only be seen as a pure and simple hygienic exercise, as Philippe Haddad rightly observes. It is not only a little piece of flesh or skin.

It is indeed a little piece of Genitality with all the profound implications that this term, Genitality, has assumed in the history of psychoanalysis.

And turning to Folklore, which I think would Freud like, I'll add that there are countless words available in the Italian language, especially in its dialectal roots, beginning with POT, varying the sound of the "O" in the most bizarre ways, but usually remaining faithful to the initial concept. One final observation, going back to something I mentioned earlier: *Kesef*, coin, has the same root as *Kasaf*, third person masculine past tense, 'he (ardently) desired,' from *Lichsof*, 'to yearn/long for'; in Hebrew it is the third person, past tense, that gives you the root of a verb.

Lichsof, sometimes understood in the sense of heartrending nostalgia; it is not clear to me how far and in what way it indicates Envy, as Jacques Attali suggests.

And, as Freud taught, few things are as strong as desire. Therefore, while it is certainly true on the one hand that the Coin is often to be considered an object of desire in itself, on the other hand it seems equally clear, given the story I've told so far, that it can be considered the expression of a burning desire, of a yearning that comes to take material form, is rendered concrete and tangible. And thus it becomes a Symbol which is once again the offering of a desire that represents the culmination of individual sexual maturity and genitality. A mutual Gift. And a concrete and symbolic offering of such a gift to God Himself.

A little ring, a decidedly important little Circle.

It seems that, as we near the end of this work, some features relating to Gaia may re-emerge, revitalised, features which, when I wrote about Enmerkar and writing, I had taken to be on the wane in favour of a world where the Father made his overbearing appearance as Third, as Writing, condemning what preceded him to oblivion.

Maybe that's not how it is. Perhaps there is a parallel path along which the two roads will meet again in the end. Maybe Gaia anticipated many of the "new" paternal intuitions, developing them in her own way, with different, but no less important, sensibilities.

Perhaps the Cypraea[2] was the start of this whole new development;

---

2   Since Linnaeus (c. 1758) Cypraea has been given the specific name Cypraea moneta, and within its classification into sub-species and forms, I have found at least six which refer to Cypraea as C. Monetaria Moneta (Cypraea M.M. barthelemyi, M.M. erosaformis, harrisi, icteriana, rhomboides, tubercolosa). This appealing gastropod enjoys at least 17 synonyms which recall its "nature" as a Coin, which should not astonish us, since many scholars have advanced their own classifications and, especially, since it was used as coinage until the early nineteenth century over an extremely wide area, even if we ignore the exchange of Chinese and Melanesian Cypraeids. I'll add that in Latin, the Conch, in Spanish Concha, Concita, Chica and so on, are synonyms, not always offensive, sometimes frankly endearing, others only slightly vulgar, of Vulva and Vagina, as well as Seashell. Vagina in Latin isn't Vagina, but Cunnus. In Latin, Vagina indicates a scabbard, for example, and the husk that covers a grain of wheat. Vulva is Vulva in Latin, and its meaning extends over many other aspects of nature, but always with the function of "Contain", "Protect", "Be a gate/ rind/ veil"; to say nothing about the evident symbolism between the seashell, the sea, and Venus. If I add that a couple of the little fairies who buzz around "The Sleeping Beauty in the Wood", i.e. the archaic Roman gods, are called Imporcitor, the deity who steadies the

maybe it was just a trial run; maybe it just something set in motion that later made its own way. Perhaps Gaia created, or tried to create, her own fund, her own monetary system which dried up in the end, for internal reasons, and maybe those reasons those reasons still have something to do with symbiosis. Perhaps Gaia is the Source, PU, of desire itself.
Perhaps.
Who knows?

---

movement/ action of ploughing the furrows – the Porca – and Insitor, the deity who places the seed in the furrow (!), I think it's clear what a "sexual" picture emerges from the Seashell. Genital, not Anal; and in relation to the Cypraea too. In China, in the thirteenth century B.C.E. according to some, but only from the ninth according to others, the first minted forms of coin had the shape of a shell, Cypraea again, with no particular inscription – i.e. no Shamash – as if the shell shape in itself had value, almost a magical element, or maybe not; we could think that the shape of that first coin, on the wave of the sentences I quoted from Laum, assumed value in itself as representing something valuable, precious: the female genital organ. Another curious feature is the term used to indicate each of the parts into which the husk, rind, or shell is divided in pulses, for example: i.e. Valva; and in malacology, from the Greek Malkòs: 'soft', i.e. the study of molluscs, each of whose two parts that form the shell of bivalves is indeed a valve; and the Pianigiani Etymological Dictionary does in fact connect Valve to Velare (to veil) and to the Vulva, from the Sanskrit ùlvam = vurvam, connected to the root VAR- i.e. to cover, to wrap, to veil.

25.
THE INGOT

Some time ago I was wandering through the rooms of the British Museum trying to recover the memory of myself as a boy many years before, a bit lost, passing astonished and curious through those same rooms, when I noticed, to my surprise, that an exhibition was being held: "Money, a History", and very good it was too.

Anyone who wants to take a look at what I'm talking about, now that the exhibition is no longer there, can click on the British Museum and type "The El Amarna Hoard". But I think that exhibition has become a permanent gallery.

El Amarna is the ancient capital, called Akhetaten, founded by Amenhotep IV, i.e. Akhenaton, grandfather of Tutankhaton who then passed into history as Tutankhamon, testifying in this change of name to Akhenaton's attempt, which failed soon after his death, to set a "monotheistic" religious cult (one God, abolition of idols, abolition of magical practice and incantations, tombs without material goods, monogamy, etc.), the cult of Aton, replacing the old cult of Amon, in line with what I have been illustrating so far. An operation in which it was Avraham who succeeded instead. I recall that the famous "El Amarna" letters are tablets written in Babylonian cuneiform characters, which nevertheless show all the academic training of scribes based on Sumerian cuneiform, which was the lingua franca of the period, at least for diplomatic purposes, rather like Latin until the last century, 1400 years after the fall of Rome, which was no longer spoken by any nation, apart from the Vatican, but was used as a common element all over Europe.

And the use of cuneiform would continue until the first century C.E., before disappearing, supplanted by the now omnipresent Phoenicio-Latin alphabet brought by Roman conquest.

On the British Museum website there are only a couple of photographs of the finds that composed that exhibition; only a meagre pair of photographs, but very interesting if you look at them carefully; the finds are coins, Egyptian in this case, from the fifteenth century B.C.E., Eighteenth Dynasty, historically not so far from the era of Hammurabi-

Avraham and immediately following the Hittite domination, which are a good illustration of what I'm saying. They are silver ingots (*lingotti* in Italian) shaped like little sausages. Ingot/*lingotto* is a term of uncertain derivation, so I'm wronging no one if I point out that in Bengali the penis is called *Lingo*, which means Precious Thing, and is exactly equivalent to the famous *Lingam-Linga* of Hindi culture and language, necessarily connected to *Yoni*.

Dr. Anna Bruna Maffessoli suggests (and I thank her) that in Devanagari ingot is called इनगट, pronounced *inagat*, that sounds so close to ingot.

Here I'll ignore – though, in saying this, I'm clearly not doing it – *Joshua* 7:21, 24 which tells of *leshon zahav*, 'tongue of gold', 'lingua d'oro', i.e. 'ling-otto d'oro', *Leshon*, a descendant of the Akkadian *lišanu*, which in *Isaiah* 13:12 (King James Bible) is linked to the land of Ophir that is placed by many scholars, following Schwennhagen, in Peru, Mexico, or Brazil (www. ancient-hebrew.org); while, for Caetani, Ophir is Rhodesia, the Zambesi and Malaysia (funny the book written by Brook Wilensky-Lanfort).

And the union of *Lingam* and *Yoni* could also be seen as the concrete representation of the concept of genitality expounded earlier through my "revisiting" of the term Foreskin (and maybe also the proof that Brescia and Bergamo have Chaldean, if not Sumerian, roots!).

The *Lingam*[1] is considered an attribute of Shiva and some believe that

---

[1] De Gubernatis (1899, pp. 90-93): "The great Purusha, creator of the world in Indian myth, had been an enormous linga, a colossal phallus (...) The legend of the origin of Priapus is not only Eastern, but attributes the monstrous Priapus to a journey made by Bacchus to India, where the cult of the linga is still so widespread, and in Udayapura a special form of the god Shiva is worshipped under the name Eka-linga (a word that means 'he who alone has the linga, who is all linga').... and who knows whether in that Greco-Latin name Priâpos we do not have a contraction of the Indian names Pragiâpa or Pragiâpati (...) which is the name given to the creator God, the male progenitor, the parusha, the cosmogonic phallus." Here De Gubernatis, with the information available to him and in the cultural climate of his time, is making a sweeping summary connecting the Indian Aryans and the Latin Aryans as descendants of the same stock, and not the Latins as humble shepherds of the Palaeolithic era. Therefore, even if we don't know Enki, who in any case emerges from the Eka-linga (given that it is also true of Enki that "His penis became erect, he ejaculated/ And filled the river with ever-changing water/ As if it were a cow at pasture/ Lowing for her calf left in the stable.../ The Tigris instantly submitted to him," Bottéro, p. 174, lines 250-260) and even though his discussion is aimed at supporting the concept of "Indo-European", he very intelligently adds, "The Aryans already knew, from Vedic times, talented craftsmen of every race; and the very notion which Vedic myth

it is indeed, symbolically, Shiva himself, with no concession to Priapus (although it was the custom to place little statues of Priapus in the gardens of ancient Rome for good luck, and the our thoughts can hardly fail to turn to the gnomes who so often appear in gardens; and the gnomes...); and it is interesting to note, as Graves does, that Priapus, another gardener, is often represented accompanied by his trusty knife, or sickle, or billhook, for *pruning* (!).

It seems that India – but not Sanskrit India, which will come many centuries later – may be the original homeland of the SAG.GI.GA, the Black Heads (Akk. *ṣalmāt qaqqadim*), an epithet which the Sumerians gave themselves, the people of *Shine'ar*, who lived in *ki-en-gir*, the land of the civilised Lords, the ancestors of King Enmerkar, Manishtushu, Ur-Nammu, Sargon, Gilgamesh, Hammurabi, and Avraham – that is, the people who founded Ur. Other scholars maintain that their land of origin was the Caucasus mountains; others point to the Zagros mountains, others Eritrea, and yet others Dilmun, which they locate in modern Bahrain. Whatever the case, they are all lands to which the Sumerians extended their gifts, as Semerano observes in the introduction to his *Etymological Dictionary of the Greek Language*, vol. 1, p. XIII:

> It is via Mesopotamia that, in the eighth century B.C.E., India receives a form of Semitic script, the best-known variant of which will be *Brāhmī*,

thus neatly dismissing the "fable of the Indo-European", despite having devoted his life to it. As I was saying, in that exhibition there were *Lingam-Ingots* with an established weight in multiples of *Shat*, from which flakes of silver were "peeled off" according to equally precise and established criteria of value and weight, and shavings-bands-rings that were "curled", forming the *Gur-Gaggaru-Kikkar-Shat* that was "sliced" directly with a knife from the silver sausage of the *Lingam-Ingot*. Indeed, we could consider them to be Circumcised. One small observation: a hobby-horse of those who maintain the concept of the Pre-Coin is that the minted coin is the only one to have characteristics that can be immediately identified

---

gives us of Vr'itra, the great enemy of the God Indra, a son of Tvashtar, lets us suppose that, under the myth of a heavenly combat, there lies the shadow of the combat which the first Aryans of the Vedic era had with non-Aryan tribes, who could have been Heteans." [De Gubernatis does not know the Euphratesians but intuits that something preceded the Aryans, which for him means the Hittites.] Other interesting references to the Linga can be found in the 1887 text by De Gubernatis.

and recognized without the need to weigh it, and this would constitute the point of transition from the pre-coin to the coin proper, stamped or moulded. From the high-resolution photographs on the British Museum website we can observe that the Shat/Shaty, the little rings, show signs of intentional, precise incisions made with exceptional skill. In other words, they are intended incisions with a specific purpose, and certainly cannot be interpreted as the product of a chance abrasion. And so, if, long before minting, those responsible for "simple weighing" took the trouble to carve "decorations", different ones on different rings, I think it's hard to see this as a matter of chance. Even today, 20 cent Euro coins are distinguished from 10s or 50s both by weight and size, but I very often distinguish them by the "design" on recto or verso. I'm not in a position to say if the different designs corresponded to different values among the various rings, but it is still striking that so much attention, care, and skill were devoted to the chiselling and differentiation of each ring.

Taking account of all that has been accumulating so far, I wonder: what on earth would Bernhard Laum say?

26.
FREE ASSOCIATIONS: THE BUTTER CURLER AND OTHER MATTERS

I've almost come to the end of what I wanted to present. From all the material that's around me and all the material I have in mind, I've "isolated" an ingot and some flakes of silver[1] (*en passant*, since my drafting of this book inevitably caused her quite a bit of torment, my wife gave me a little pair of pharmacist's scales for my birthday: wrapped with it were some little flakes of metal which represent, in this case sealed, the values of the fractions of grams needed for weighing, and which can be picked up and manipulated only by means of tiny pincers that accompany the scales) and I have focused my attention on some "curls", some little metal rings that are "peeled" off the ingot. Now I find a memory coming into view: a Christmas lunch when I was a child, before which my mother took out from a drawer a strange instrument I had never seen before, or perhaps had never noticed in other circumstances: I asked her, "What's that thingy?" and my mother, smiling, told me it was a "butter curler". I have always found this word *arricciaburro* funny. And even today I have no idea if this is what it's really called…. Yes, it seems this really is its name; I've searched for "Arricciaburro" on Google and up came a load of pictures of this comical instrument. At my tender age, I now know that it really is called *arricciaburro*, butter curler, and this is funny too. Anyway, that strange object, whether because of the Christmas atmosphere, because I was a child, because I found the very idea of curling butter amusing in itself, or maybe because of the sound of the word with all those Rs, because of the utensil's bizarre shape, or because of the idea of holiday, I don't know, but for me "that thingy" brings with it a feeling of happiness,

---

1 *Unstruck coin flans* is the name to various types of coin by Hendin (2010, p. 47). I've not been able to reproduce his photographs of these "coins", but the caption reads, "Unstruck coin flans of various size, these are referred to in the Talmud as 'uncoined metal which bears no imprint', and underline the fiduciary nature of bronze in ancient Judaea." It is astounding how closely these little flakes of metal resemble those present in the British Museum photograph. And it is astounding how much they are like the little flakes of my pharmacist's scales.

holidays, and joy. I don't know if Hammurabi's mum and Avraham's also used a "butter curler" for the Sacred Feast of Christmas, but similar bizarre devices were in some way the first shape, i.e. the first little ring, in the story of the Coin. Not, however, with butter; or maybe yes, given that no trace of them remains.

I'm joking again, perhaps to lighten the mood of what I'm writing, and maybe also to "lighten" the farewell that I can glimpse not far ahead, but it seemed an appealing association and I wanted to share it.

But now, more seriously, I'd like to finish with three quotations, two from the Bible and one the product of Mediaeval mysticism not so very far from delirium, with all due respect to the mystic in question; but I've just remembered that I'm a psychoanalyst and that I often want to "translate", apparently "betraying" what I hear, given that it's often only in this way that we can understand hidden things.

I could call it an unsaturated conclusion. Intentionally unsaturated. Although it might perhaps introduce an explanatory hypothesis about what Menger (1909) calls "unintentional action", the kind that would be behind the birth of the coin.

The first quotations is about Eliezer (no thigh, though) when he encounters Rebecca at the well, which (again !!!) is in Harran, the same Harran from which Avraham set out (*Genesis* 24:22-23):

> When the camels had finished drinking, the man [Eliezer] took a gold nose-ring weighing a half shekel, and two bracelets for her arms weighing ten gold shekels, and said, "Tell me whose daughter you are..."
> (*New Revised Standard Version, Catholic Edition*. After 'a half shekel...', the Bible of the Conferenza Episcopale Italiana has "and hung it from her nostril, as was the custom then.")

The gold ring hung from the nose is in fact called *Nezem*, nose ring. And it weighs, perhaps by coincidence with what I have presented, exactly half a shekel.

The second quotation is taken from Ezekiel (16:6-14) through whose voice speaks God Himself who, like a lover/beloved, addresses Jerusalem, which He will later curse:

> I passed by you, and saw you flailing about in your blood.
> As you lay in your blood, I said to you, "Live! and grow up like a plant of the field."
> You grew up and became tall and arrived at full womanhood;
> your breasts were formed, and your hair had grown; yet you were naked and bare.

I passed by you again and looked on you; you were at the age for love. I spread the edge of my cloak over you, and covered your nakedness: I pledged myself to you and entered into a covenant with you, says the Lord GOD, and you became mine.

Then I bathed you with water and washed off the blood from you, and anointed you with oil.

I clothed you with embroidered cloth and with sandals of fine leather; I bound you in fine linen and covered you with rich fabric.

I adorned you with ornaments: I put *bracelets* on your arms, a *chain* on your neck, a *ring on your nose, earrings* in your ears, and a beautiful *crown* upon your head.

You were adorned with gold and silver, while your clothing was of fine linen, rich fabric, and embroidered cloth. You had choice flour and honey and oil for food. You grew exceedingly beautiful, fit to be a queen.

Your fame spread among the nations on account of your beauty, for it was perfect because of my splendor that I had bestowed on you, says the Lord GOD.

(*New Revised Standard Version, Catholic Edition*; my italics)

The third quotation is taken from the letters of St Caterina da Siena (1347-1380), which are written in an obviously old form of Italian, specifically letter 201, *Saint Catherine of Siena as seen in her letters*, Trans. and ed. Vida D. Scudder. London, Dent. 1906] where Caterina alludes to a wedding ring invisible to others, with which Jesus would marry her, and the nature and composition of which is made clear by the text:

To Sister Bartolomea della Seta, nun in the Convent of Santo Stefano, Pisa:

Now thus I tell thee, dearest my daughter, that I want thee to do. And be for me a mirror of virtue, following the footsteps of Christ crucified. Bathe thee in the Blood of Christ crucified, and so live, as is my will, that thou nor seek nor will aught but the Crucified, like a true bride, bought with the Blood of Christ crucified. Well seest thou that thou art a bride, and that He has wedded thee and every creature, not with a ring of silver, but with the ring of His Flesh. O depth and height of Love unspeakable, how didst Thou love this Bride, the human race! O Life through which all things do live, Thou hast plucked it from the hands of the devil, who possessed it as his own; from his hands Thou hast plucked it, catching the devil with the hook of Thy humanity, and hast wedded it with Thy flesh. Thou hast given Thy Blood for a pledge, and at the last, sacrificing Thy body, Thou hast made the payment. Now drink deep, my daughter, and fall not into negligence, but arise with true zeal, and by this Blood may the hardness of thy heart be broken in such wise that it never may close again, for any ignorance or negligence, nor for the speech of any creature. I say no more. Remain in the holy and sweet grace of God.

Sweet Jesus, Jesus Love.

Myst-icism sometimes seems really "Myst-erious" (!). Mysterious like the subtitle of this book. Mysticism, the discipline itself, and not St Catherine, is often very close to De-lirium – i.e. "out of the furrow" as the term's etymology tells us, close to an outright Psychosis: but sometimes it is the result of a fully worked through preconscious process; some say it's even a manifestation of the Divine, others of the Unconscious; still others think of frankly and dramatically pathological alterations caused by degenerative states. It's hard to say. I don't know. Certainly Mysticism (see also Freni, 2000), from Enoch to Meister Eckhart, is sometimes very curious indeed.

And it's curious that rings, nose rings, earrings, arm rings, bracelets, anklets, necklaces, torques, diadems, crowns that are gifts, expressions of affection, devotion, dedication, are all round, empty: little round strips of metal. Like a Gur. Like a Kikkar. Like a Nezem.

Like a half שקל, a half Sheqel.

Reserves of value, precious materials, expressions of feelings, elements of vanity, representations of desire, instruments of seduction and beauty, jewels that are so different from one another. So many things at once. Hard to say.

But always Blood; and Sacrifice; and Covenant; and Desire; and Love. Like a little Coin.

CONCLUSIONS

I think that when we discover a new expression in the eyes of our beloved, they are often the eyes of someone who is looking with astonishment and joy at what is before her, and it is the eyes themselves that have changed: so changed that they can see something that may not have been there before but, even if it had been, those eyes would not have been able to see it. And this happens for many other reasons. The new thing we can observe is in the eye of the observer; only the change in the way I look permits me to something in a different way. Even an author like Proust, maybe because he was rather idle, says something along these lines when he writes that "The true voyage of discovery does not consist in finding new lands but in having new eyes." If I find a lot of little pieces of silver wrapped up in a bag near a furnace, I might easily think that they are a residue from the production of jewels, possibly even intended to undergo a further smelting; but they could be the Unstruck coin flans I wrote about before; like small pennies, *argent de poche*, Coins for everyday life. Or if I come across a pile of little rings I might conclude that they are old earrings of the kind that votive statues are full of, or merely symbols of an all too human vanity; but they could be Anaq /Torquis, precious like a Gold Full Sovereign Victoria 1841, Money for substantial transactions. And I will mentally think of them, classify them, regard them according to convictions which have gone into making me who I am. If I am convinced that no monetary system could have existed in antiquity because I am convinced that the Coin was invented in Lydia during the 8th-7th centuries BCE and absolutely convinced that the Coin must always and only be that round, compact thing, that little disc with something stamped on it, then obviously I will not be able to recognise as a Coin anything that departs from what my "eyes" are inclined to recognise. Eyes which have forgotten the Attic standard and the first *denarius*. Eyes which had already forgotten the Latin Monetary Union which established the basis – a singular coincidence – of 4.5g of silver for a Franc and a ratio of 1:15 to gold; almost exactly as so, so long ago, when NANNA-Sin was in a relationship with UTU-Shamash. Eyes which will be unable to see in the Sumerian *"Assayer of silver"/scribe* the "grandfather" of the Roman *Libripens*; eyes which will be unable to see the authority of the British Royal Mint in the certification of UTU-Shamash; eyes which will be unable to see the frenetic activity of weighing metal as the forerunner of Justice, which is in turn the daughter of a successful attempt to escape from a dynamic of Theft-Counter-theft, the concrete expression of a Predator-Prey "dialectic" in inexorable alternation, in turn generating an infinite Talion with the trails of blood which characterise it.

In the absence of something new in the eyes of the beholder, how is it possible to see the new which we can sometimes encounter outside?

My dissatisfaction with dusty old geographical maps has stirred a new curiosity which has made me glimpse something I'll have come across goodness knows how many times without even thinking it might be something to recognise.

In these pages I have tried to look in a slightly different way, trying to keep as far as possible from clinical practice, Trade, Theft, Money, Coin, and a few other things. I think the old story about barter turning – hey presto! – into coins no longer has a leg to stand on, at least for me. And I also think I can start to reflect more calmly and clearly on the meanings of Money and Coin. And I think that Faeces have as much of a relationship with Coin as fish and chips do with the tectonic plates.

I also think, affectionately I'd say, about a coinage in circulation in everyday life in antiquity, that of the ancient Near East; a fairly generalised affection, of course, felt for those people of flesh and blood who rush about trying to manage everyday problems, and work, pray, love, laugh, hate, and do deals; who assess, wait, sign contracts, and weigh everything; humble, unknown actors and authors of a circulating coinage which involves artisans, courtesans, gardeners, princes, housewives, the *ḫāwiru*; and the interpreters of dreams.

An everyday life as old as it is human and present, and of which I have no claim to ownership. An everyday life and a Humanity so ancient but so like the humanity I know; a Humanity which, day by day, silently builds its own destiny. I have played, as well as I can and trying not to get pebbles in my shoes with Words, Languages, History, Symbols, and Ideas I have encountered, "in dialogue" with some giants and dwarves who, for good or ill, have been actors and authors of a representation which cuts across and coincides with History itself; a History made of misunderstandings, thefts, frauds, lies, bold innovations, bad faith, flashes of genius, wretchedness, courage, meanness, brutality, and beauty. I have come across little pieces of silver and gold which represent little pieces of simple physicality, small poor body parts not so very different from those which represented the first units of measurement, such as a thumb, or a span/palm, a hand, foot, arm, forearm/*Cubitus*, or a King's Foot, or the legend of the Yarda which features a nose and a finger, or simply a tear: but which at the same time are elements of great symbolic significance, still fragments of "simple" humanity, but now seen with slightly different eyes.

I believe we must acknowledge our debt to the Sumerians, Akkadians, and Babylonians, and to Ur, for many things including Money and the Coin; these represent the impulse towards an idea of freedom, respect, and reciprocity that is much more modern than I could ever have expected. Money and the Coin are part of the enormous and poorly known debt of acknowledgement and gratitude which modernity owes them.

I think that much of what I have been writing, considered from a strictly historical perspective, may be properly be considered little more (or little less) than a fable: Abraham cannot have got into a fight with Hammurabi for the simple reason that the name of his home city, "Ur of the Chaldees", refers directly to a civilisation, the Chaldean, from at least a thousand years later than Hammurabi's time. But if the link between these figures has just been a narrative expedient which I have made use of in order to stimulate a little curiosity – an expedient of which I am certainly not the originator but that is no reason for what I have been recounting to lose even a single *gUr* or a ŠE of its true meaning, since *Ur (u) Kasdim* neatly summarises everything which has preceded it and generates a great deal of what came after it.

Playing seriously, I have invented a story whose aim was to enter into the mechanisms which led to the birth of Money and the Coin; a story which, dispensing with the individual characters to whom I have assigned a part for my own purposes, intended to try and recreate new links and stimuli around an old problem. A story that is certainly not History, at least not entirely; but another story, one which aims to tell and refer to yet other stories.

But without a funny butter-curler connected to the memory of a far-off and joyful Christmas lunch; without a lovely shell, that *Cypraea* which I bought on distant beaches so long ago to "hear" the murmur of the sea; without a curious return to London, a London which I so loved and so hated; without an odd, finely honed Finnish knife, a Martiini Bronze Bird (and I realise, now this entire text which revolves around the Bronze Age, has also passed from a Knife, Bronze Bird, with all the significance that these terms – Age-Bronze-Bird-Knife – can evoke) with which I risked a small, not religious or ritual, circumcision of my left index finger (not the little finger as it was for the Wolf Man), I do not think I would have been able to look at the things I've looked at in the way I have looked at them.

And without the work done by so many others on whose shoulders I have allowed myself to jump, sometimes kicking a bit, without even asking permission or saying "please", but to whom I have to say "thank you", the ingots, the chips, the scales, and the rings would not have told me much more than they had said before.

In any case, this long voyage of mine in the "South Seas", with all its lacunae, errors, imprecisions, badly typed quotations, and all the things I have not been able to say; this crossing of dangerous, fragile, wet, wobbly Tibetan rope bridges, relying on the legs of Dr. Lemuel Gulliver, after having scampered around among dwarves and giants, comes to an end. Here.

BIBLIOGRAPHY

Abraham K., *Opere*, Boringhieri, 1975

Adams C., *For Good and Evil: The Impact Of Taxes On The Course Of Civilization*, Madison Book- Lanham, Maryland-USA, 2001

Aelianus, *De animalium natura libri* (200 AD), BUR, 1998

Aglietta M., Orléan A., *La violence de la monnaie*, PUF, 1984

Akerlof G. A., R. J. Shiller, *Spiriti animali*, Rizzoli, 2009

Alary P., *Du troc à la monnaie; le troc a-t-il donné la naissance à la monnaie?*, Un. Orleans, 2011

Albanese B., *Le situazioni possessorie nel diritto privato romano*, Palumbo, 1985

Albertani Brixiensis, *Liber Consolationis et Consilii* (1246), Thor Sundby, 1873 [Reprint 2012]

Alberti D., *La nascita della moneta*, lamoneta.it, 2015

Albright W. F., *Magan, Meluḫa and the synchronism between Menes and Narâm-Šin*, Journ. Egypt. Arch., pag 85-86, 1921

Alighieri D., *La Divina Commedia*, Salani, 1936

Ambrosoli S., Ricci S., *Monete greche*, Hoepli, 1917

Apuleius Lucius, *The Golden Ass*, Abbey Library, 1935

Aranjo-Ruiz V., *Istituzioni di Diritto Romano* (1994), Ed. Jovene, 2002

Aristotle, *Etica nicomachea*, Laterza, 1999

Aristotle, *Politica*, Laterza, 1983

Aristotle, *L'amministrazione della casa*, Laterza, 1995

Aristotle, *La costituzione degli ateniesi*, Rizzoli, 1999

Arnaud G., *Money as Signifier. A Lacanian Insight into Monetary Order*, in Free Association, Vol 10 Part 1 (No. 53): 25-43 ; 2003

Arrow K. J., *Razionalità individuale e collettiva nei sistemi economici*, (in

Egidi M. e Turvani M.)

Ascalone E., Peyronel L., *Balance weights from Tell Mardik–Ebla and weighing sistems in levant during the middle bronze age*, in Weights in context, Ist. It. di Num., Studi e Materiali, 13 – Roma, 2006

Ascalone E., *Mesopotamia; assiri, sumeri e babilonesi*, Mondadori Electa, 2005

Astori D., *The seven Noachian Precepts between Monotheistic Religions and Human Liberties*, in Atti del Convegno "Monotheistic Religions and Human Liberties", Costanza, 2010 (in prep.)

Attali J., *Les Juifs, le Monde et l'Argent*, Fayard, 2002

Attali J., *Dizionario innamorato dell'ebraismo*, Fazi Editore, 2013

Augustinus, *Commentary on John*, Homily 40, 9: ed. Fitzgerald AD

Augustinus, *The Works of Saint Augustine: a Translation for the 21st century*, New City Press, New York, 2009

Augustinus, *De Civitate Dei*, Mondadori, 2011

Bausani A., *L' alfabeto come calendario arcaico*, Oriens Antiquus, vol. XVII, 1978

Bentham J., *Defence of Usury* (1787), HardPress, 2019

Benveniste E., *Il vocabolario delle istituzioni indo-europee*, Einaudi, 1981

Bereshit Rabbah, UTET, 2000

Bernal M., *Atena nera* (1987), Il Saggiatore, 2011

Bettelheim B., *Ferite simboliche*, Sansoni, 1973

Bianchi E. et al., *Il dio denaro*, Rizzoli, 2010

Bibbia Ebraica, *Pentateuco e Haftarot*, Giuntina, 1995

Bibbia, Ed. Paoline, 1965

Bibbia di Gerusalemme, Ed. Dehoniane, 2005

Bibbia (La Sacra), Società Biblica di Ginevra, 1994

Bible (The Holy), English Standard Edition, Crossway, 2001

Bindman Y. Rabbi, *The seven colours of the rainbow*, Resources Pubblication, 1995

Bion W., *Attention and Interpretation*, Rowman and Littlefield publisher Inc., 1995

Bobokhyan A., *Identifying balance weights in bronze age Troia*, Istituto Italiano di Numismatica, Studi e materiali – 13, 2006

Bond A., Hempsell M., *A Sumerian observation of the Kofels' Impact Event*, Alcuin Academics, 2008

Bonfante P., *Le leggi di Hammurabi*, S.E.L., 1903

Borneman E. (edited by), *The Psychoanalysis of Money* (1973), Urizen Book, 1976

Bottéro J., *Le problème des ḫabiru*, Imprimerie National, 1954

Bottéro J., *The Birth of God*, University Park, University of Pennsylvania Press. 2000

Bottéro J., *Mesopotamia. La scrittura, la mentalità e gli dei* (1987), Einaudi, 1991

Bottéro J., Kramer S. N., *Uomini e Dei della Mesopotamia* (1989), Einaudi, 1992

Bottéro J., *L'Épopée de Gilgameš*, Gallimar, 1992

Bottéro J. (a cura di), *L'Oriente antico* (1992), Edizioni Dedalo, 1994

Bressett K., *Money of the Bible*, Whitman Publishing, LLC, 2007

Buccellati G., *Quando in alto i cieli*, Jaca Book, 2012

Buccellati G., *Alle origini della politica*, Jaca Book, 2013

Bulgarelli O., *Il denaro alle origini delle origini*, Spirali, 2001

Bulgarelli O., *Moneta ed economia nell'Antica Mesopotamia*, Rivista trimestrale di Diritto dell' Economia, n. 3, 2009

Caesar G. J., *De bello gallico*, Einaudi, 1991

Caetani L., *Studi di storia orientale* (1911), Hoepli (Reprint)

Caetani L., *Altri studi storia orientale* (1914), L' Erma di Bretschneider, 1997

Cagni L., *L'epopea di Erra*, Istituto di Studi del Vicino Oriente, 1969

Calcagno C., *Circoncisione, dalla selce al bisturi*, arabA Fenice, 2009

Camia F., Privitera S., *Obeloi*, Pandemos, 2009

Campanella T., *The city of the Sun*, Outlook Verlag, 2018

Catherine (Saint) of Siena as seen in her letters, Trans. and ed. Vida D. Scudder. London, Dent. 1906

Cassirer E., *Simbolo, mito e cultura*, Laterza, 1985

Cassirer E., *Linguaggio e mito*, Garzanti, 1975

Catarzi M., *Il simbolo denaro fra norma e rappresentazione* [in N. Parise (a cura di), Bernhard Laum, Istituto Italiano di Numismatica, Studi e materiali, 5 - Roma, 1997]

Chaucher G., *Canterbury Tales* (1386), J.Thor Sundby, 1823

Chiodi S., *Eracle fra Oriente e Occidente*, pp. 93-116, Melammu Synposia 4, 2004

Cicero M. T., *De Divinazione*, UTET, 2013

Cicero M. T., *De Natura Deorum*, UTET, 2013

Croce B., *Materialismo storico ed economia marxista* (1899), Bibliopolis, 2001

Croce B., *Ciò che è vivo e ciò che è morto della filosofia di Hegel* (1907), Bibliopolis, 2006

Coe M. D, Stone M. van, *Readings the Maya Gliphs*, Thames & Hudson, 2001

Corano (a cura di A. Bausani), Rizzoli, 1999

Dallen M. E., *The rainbow covenant*, Lightcatcher Books, 2003

Damasio A. R., *L'errore di Cartesio* (1995), Adelphi, 2012

Dash M., *La febbre dei tulipani*, Rizzoli, 1999

Décarie T. G., *Piaget e Freud*, Armando, 1976

De Gubernatis A., *Zoological Mythology* (1872), Ayer Co Pub, 1978

De Gubernatis A., *Mitologia Comparata* (1887), Hoepli, [Reprint 2012]

De Gubernatis A., *Roma e l'Oriente* (1899), Soc. Ed. Dante Alighieri, Kissinger Reprints, 2012

De Martino E., *Il mondo magico* (1948), Bollati Boringhieri, 2007

De Santis S., *I sogni*, F. lli Bocca, 1899

Dimen M., *Money, Love, and Hate: Contradiction and Paradox in Psychoanalysis* (1994), Psychoanal. Dial. 4:69-100

Diodorus Siculus, *Bibliotheca Historica*, Sellerio, 1986

Descartes, *The Passions of the Soul and Other Late Philosophical Writings*. Oxford, OUP; 2015

Desmonde W. H., *On the Anal Origin of Money* (1953), Am. Imago, 10:375-378

Desmonde W. H., *Magic, Myth, and Money*, The Free Press of Glencoe, Inc., 1962

Dumezil G., *La religione romana arcaica* (1964), Rizzoli, 1997

Eagleton C., William J., *Money, a history*, The British Museum Press, 2007

Egidi M., Turvani M., *Le ragioni delle organizzazioni economiche*, Rosenberg & Sellier, 1994

Bibliography

Einzig P., *Primitive Money* (1949), Eyre and Spottiswoode, 1951

Eliade M., *Il mito dell' alchimia* (1935), Bollati Boringhieri, 2001

Eliade M., *Cosmologia e alchimia babilonesi* (1937), Sansoni, 1992

Eliade M., *Il sacro e il profano* (1956), Bollati Boringhieri, 2006

Eliade M., *Alchimia asiatica* (1978), Bollati Boringhieri, 2001

Ellis, H., *Studies in the Psychology of Sex* (1906), Forgotten Books, 2012

Engel F., *L' origine della famiglia, della proprietà privata e dello Stato* (1884), Newton Compton, 1977

Erasmus of Rotterdam, *The prise of Folly*, Penguin, 2004

Eusebius of Caesarea, *Chronicon Bipartitum*, Bapt. Aucher Ancyrani (Kissinger Legacy Reprints)

Fenichel O., *The Drive to Amass Wealth* (1938), Psychoanal. Q., 7:69-95

Fenichel O., *The Psychoanalytic Theory of Neurosis*. Hove, Routledge, 1990

Ferenczi S., *Opere*, Cortina, 1990

Ferrarese L., *Dottrina Frenologica*, Stamperia dell' AQUILA, 1836

Freni S., *La dimensione mistica nell' esperienza psicoanalitica*, Relazione al Centro Milanese "Cesare Musatti", 2000

Freud S., *The Standard Edition*, Vintage, The Hogarth Press (2001)

Freud S., Salomé L. A., *Eros e conoscenza, Lettere 1912-1936*, Bollati Boringhieri, 2010

Freud S., *Lettere a Wilhelm Fliess*, Bollati Boringhieri, 2011

Freud S., *Epistolari, Lettere fra Freud ed il Pastore Pfister, 1906-1913*, Bollati Boringhieri, 1990

Friedman D. M., *Storia del pene*, Castelvecchi, 2001

Fromm E., *To Have or To Be* (1976), Bloomsbury, 2013

Furnham A., Argyle M., *The Psychology of Money*, Routledge, 1998

Galbraith J. K., *Storia dell' economia*, Rizzoli 1987

Galiani F., *Della Moneta* (1750), COFIDE, 1991

Galiani F., *Dialogue sur le commerce des Blés* (1768), Fayard, 1984

Gallo F., *Studi sul trasferimento della proprietà in diritto romano*, Giappichelli, 1955

Gallo F., *Studi sulla distinzione fra "res mancipi" e "res nec mancipi"*, Riv. Dir. Rom. IV - 2004

Garbini G., *Il Poema di Baal di Ilumilku*, Paideia, 2014

Garbini G., *History and Ideology in Ancient Israel*, SCM Press,2011

Garbini G., *Mith and History in the Bible*, Continuum-3pl, 2009

Garbini G., *I Filistei*, Paideia, Brescia, 2012

Gaur A., *La scrittura* (1984), Edizioni Dedalo, 1997

Gelb I. J, *A study of writing*, The University of Chicago Press, 1952

Gelb I. J., *Ebla and Kish civilization*, Istituto Universitario Orientale, 1981

Genovesi A., *Lezioni di Commercio o sia d'Economia Civile* (1769), Remondini

George A. R., 2013, *The Poem of Erra and Ishum: A Babylonian Poet's View of War*, in H. Kennedy (ed.), *Warfare and Poetry in the Middle East* (London: I.B. Taauris, 39-71

Ginzberg L., (1909, 1937), *The Legends of the Jews*. Baltimore, Johns Hopkins UP. 1998, I

Ginzberg L., *Le leggende degli ebrei, II, Da Abramo a Giacobbe* (1925), Adelphi, 2010

Girard R., *Il Sacrificio* (2002), Cortina, 2004

Girard R., *Delle cose nascoste sin dalla fondazione del mondo*, Adelphi, 1983

Gironde S., *La Neuroeconomia*, Il Mulino, 2010

Glover E., *I fondamenti teorici e clinici della Psicoanalisi* (1939), Astrolabio, 1971

Goethe J. W., *Faust e Urfaust* (1808), Feltrinelli, 2000

Gramsci A., *Letteratura e vita nazionale*, Editori Riuniti, 1973

Graves R., *I miti greci*, Longanesi, 1989

Graves R., Patai R., *I Miti Ebraici*, Longanesi, 1980

Haarmann H., *Modelli di civiltà a confronto nel mondo antico*, in *Origini della scrittura* (a cura di Bocchi G., Ceruti M.), Bruno Mondadori, 2002

Haddad P., *L'ebraismo spiegato ai miei amici*, Giuntina, 2003

Haight D. F., *Is Money a Four-Letter Word ?* (1977), Psychoanal. Rev., 64:621-629

Hayek F. A. von, *Legge, Legislazione e Libertà* (1973-1982), Il Saggiatore, 1982

Hayek F. A. von, *La via della schiavitù* (1949), Liberilibri, 2011

Hegel G. W. F., *Opere*, UTET, 2002

Heidegger M., *History of the Concept of Time: Prolegomena*, Indiana University Press, 1992

Heidegger M., *Il concetto di tempo* (1924), Adelphi, 1998

Heidegger M., *Essere e tempo* (1927), UTET, 1996

Hendin D., *Guide to Biblical Coins*, AMPHORA, 2010

Herodotus, *Storie*, UTET, 2006

Hesiod, *Teogonia*, UTET, 1977

Hill P. and Keynes R.(Edit. By), *Lidia and Maynard*, Charles Scribner's Sons, New York, 1989

Hinshelwood R. D., *Dizionario di psicoanalisi kleiniana*, Cortina, 1990

Hobbes T., *Leviatano* (1651), Editori Riuniti, 1976

Hughes R., *The Culture of Complain* (1994), Harvill, 1999

Hume D., *Discorsi Politici* (1752), Boringhieri, 1959

Hume D., *Of Money* (1752), Laterza, 1982

Isidore of Seville, *Etymologiae* (625 d.C.) UTET, 2004

Jacobs T., *Money: Some considerations on its impact on psychoanalytic education and psychoanalytic practice* (in *Money Talks*, ed. by Berger B. and Newman S., Routledge) 2012

Jeremias A., *Babylonian conception of heaven and hell* (1902), La Vergne USA, 2011

Jeremias A., *The Old Testament in the light of the ancient East (voll.I e II)* (1904), reprint from the collection of the University of Michigan Library, 2011

Jeremias A., *Babilonisches im Neuen Testament* (1905), reprint from the collection of the University of Michigan Library, 2011

Jevons W. S., *Money and mechanism of exchange*, D. Appleton and Company, 1875

Jevons W. S., *Economia politica* (1871), Hoepli, 1924

Josephus Flavius, *Antichità Giudaiche*, UTET, 2006

Jung C. G., *Opere*, Boringhieri, 1990

Jung C. G., *Collected Works*, Routledge, 1991

Kahneman D., Tversky A., *Choises, Values and Frames*, Cambridge University Press, 2000

Kahneman D., *Pensieri lenti e veloci*, Mondadori, 2012

Kant I., *Opere*, UTET, 1993

Kant I., *Groundwork for the Metaphysics of Morals*, ed. Allen W. Wood,

Cambridge, Mass. Yale University. 2002.
Kant I., *La falsa sottigliezza delle quattro figure sillogistiche* (1762), Ist. Edit. e Polig. Internaz., 2001
Kardiner A., *My analysis with Freud*, W.W. Norton & Company Inc., 1977
Keynes J. M., *The End of Laissez-Faire* (1926), Prometheus Book, 2004
Keynes J. M., *Teoria generale dell'occupazione, dell' interesse e della moneta* (1936),UTET, 1971
Khan M. M. R., *Alienation in Perversions*, London,Hogarth 1979
Klebanow S., *Power, Gender, and Money* (1989), J. Amer. Acad. Psychoanal.,17:321-328
Klebanow S., Lowenkopf E., *Money and Mind*, Plenum Press, 1991
Klein M., *Envy and Gratitude*, Vintage Classic, 1997
Klitsche de la Grange T., *Funzionarismo*, liberilibri, 2013
Kramer S.N., *The Sumerians: their History, Culture and Character*, Chicago and London, University of Chicago Press, 1958
Krueger D. W., *Money Meanings and Madness: a Psychoanalytic Perspective* (1991), Psychoanal. Rev., 78:209-224
Krueger D. W. (Edited by), *The last taboo*, Brunner/Mazel, INC., 1986
Kuhn T. S., *La struttura delle rivoluzioni scientifiche* (1962), Einaudi, 1980
Kutra V., *I Celti e il Mediterraneo*, Jaca Book, 2004
Laitman M. (a cura di), *Zohar*, URRA, 2011
Lasky, E., *Psychoanalysts' and Psychotherapists' Conflicts About Setting Fees* (1984), Psychoanal. Psychol., 1:289-300
Laum B., *Origine della Moneta e teoria del sacrificio*, in N. Parise (a cura di) Istituto Italiano di Numismatica, Studi e Materiali, 5 – Roma, 1997
Laum B., *Heiliges Geld* (1924), Semele, 2006
Le Goff J., *La borsa e la vita* (1986), Mondadori, 2004
Le Goff J., *Lo sterco del Diavolo*, Laterza, 2010
Leoni B., *La libertà e la legge* (1961), liberilibri, 1995
Levi-Strauss C., *Le strutture elementari della parentela* (1968), Feltrinelli, 1978
Levi-Strauss C., *Il pensiero selvaggio* (1962), Il Saggiatore, 1976
Lichtenstein A., *Le sette leggi di Noè*, Lamed, 1986
Liverani M., *Oltre la Bibbia*, Laterza, 2007

Liverani M., *Uruk, la prima città*, Laterza, 1998

Liverani M., *Antico oriente. Storia società economia*, Laterza, 1988

Liverani M., *Le lettere di el–Amarna; 1 e 2*, Paideia Editrice, 1999

Locke J., *The Two Treatises of Civil Government* (Hollis ed.),1689

Locke J., *The Works of John Lock in nine volumes*, C. and J. Rivington, London, 1824

Locke J., *Essay Concerning Human Understanding* (1690), Hackett Publishing Co,1996

Locke J., *Ragionamenti sopra la moneta* (1691), ICEB, 1986

Lopez D., *Il desiderio, il sacrificio, il capro espiatorio*, Colla, 2008

Lopez D., *La strada dei Maestri*, Colla, 2011

Lopez D., *La psicoanalisi della persona*, Boringhieri, 1983

Lowenkopf E., *The Almighty Dollar*, iUniverse Inc., 2003

Luther M., *Table Talk*, The Lutheran Publication Company, Philadelphia

Luther M., *Discorsi a Tavola* (1531-1546), Einaudi, 1969

Luther M., *Degli ebrei e delle loro menzogne* (1543), Einaudi, 2008

Madden F. W., *Coins of the Jews* (1881), Georg Olms, 1976

Maffei S., *Dell'impiego del Danaro* (1744), Stamperia Lazzarini

Malthus T. R., *Principi di economia politica* (1820), ISEDI, 1975

Mandel G., *L'alfabeto ebraico*, Mondadori, 2000

Mander P., Notizia P., *L'uso dell'argento nell'economia del regno della III dinastia di Ur*, Rivista di Storia Economica, a. XXV,n. 1, 2009

Martin P., *Argent et psycanalyse*, Navarin, 1984

Martin F., *Money: The Unauthorised Biography*, Bodley Head, London, 2013

Marx K., *Il Capitale*, Editori Riuniti, 1974

Marx K., *Manoscritti economico-filosofici del 1844*, Einaudi, 1976

Marx K., *Marxism and Art: Essays Classic and Contemporary*, ed. Solomon, M. Detroit, Wayne State University Press, 1979.

Marx K., *History of Economic Theories*, ed. Kautsky K. Papamoa Press, 2018

Matthiae P., *Ebla*, Einaudi,1977

Mauss M., *Les origines de la notion de la monnaie*, Antropologie, 25, pagg. 14-19, 1914

Mauss M., *Saggio sul dono* (1923), Einaudi, 2002

Menger C., *Principii fondamentali di Economia Politica* (1909), Laterza, 1925

Midrash Rabbah (Trans. and ed. Freedman, H. and Maurice S. London) Soncino Press, 1939.

Mill J. S., *Principles of Political Economy*, Cosimo, 1848

Mises L. von, *Teoria della moneta* (1912), Edizioni Scientifiche Italiane, 2012

Mises L. von, *Burocrazia* (1944), Rusconi, 1991

Mises L. von, *Teoria e Storia* (1957), Rubettino, 2009

Mises L. von, *Libertà e Proprietà* (1958), Rubettino, 2007

Mises L. von, *Politica economica* (1979), Liberilibri, 2007

Miyazaki I., *China's Examination Hell: Civil Service Examinations of Imperial China*, Yale University Press; New Ed edition (1 July 1981)

Mommsen T., *Storia di Roma* (1856), Dall' Oglio, 1971

Money- Kyrle R., *Il significato del sacrificio* (1929), Bollati Boringhieri, 1994

Money- Kyrle R., *Man's Picture of His World*, Karnak Boook, 2005

Motterlini M., Piattelli Palmarini M. (a cura) *Critica della ragione economica*, Saggiatore, 2005

Motterlini M., *Economia Emotiva*, Rizzoli, 2006

Motterlini M., *Trappole mentali*, Rizzoli, 2008

Murdin L., *How money talks*, Karnac, 2012

Myers K., *Show me the money: (the "Problem" of) the Therapist's Desire, Subjectivity, and Relationship to the Fee* (2008), Contemp.Psychoanal., 44:118-140

Neusner J., *Mekhilta according to Rabbi Ishmael*, vol. 2, Brown University, 1988

Nietzsche F., *Complete Works of Friedrich Nietzsche*. Hastings, Delphi Classics, 2015

Oppenheim A.L., *Ancient Mesopotamia: Portrait of a Dead Civilization*, The University of Chicago Press, 1964

Orwell G., *Politics and the English Language* (1946), Oxford City Press, 2009

Ovid P. N., *Le Metamorfosi,* Garzanti, 1992

Ovid P. N., *Fasti*, (trans Frazer J.G., *Loeb Classical Library*, 1931)

Parise N., *La nascita della moneta*, Aspis, 2000

Bibliography

Parise N., *Lineamenti di preistoria monetaria greca*, in "La moneta greca e romana", a cura di F. Panvini Rosati, "L'Erma" di Bretschneider, 2000

Parry J., Bloch M., *Money and the morality of Exchange*, Cambridge University Press, 1996

Pettinato G., *Babilonia centro dell'universo*, Rusconi, 1994

Pettinato G., *La scrittura celeste*, Mondadori, 1998

Pettinato G., *Mitologia Sumerica*, UTET, 2001

Pettinato G., *I miti degli inferi assiro- babilonesi*, Paideia, 2003

Pettinato G., *I re di Sumer I*, Paideia, 2003

Pettinato G., *La saga di Gilgameš*, Mondadori, 2004

Pettinato G., *Mitologia assiro-babilonese*, UTET, 2005

Pettinato G., *I Sumeri*, Bompiani, 2005

Peyronel L., *Storia ed archeologia del commercio nell'oriente antico*, Carocci, 2008

Piaget J., Inhelder B., *La psicologia del bambino* (1966) Einaudi, 1975

Piaget J., *Lo sviluppo mentale del bambino* (1967), Einaudi, 1973

Plato, *Opere*, UTET, 1992

Pliny the Elder, *Naturalis Historia* (77 AD), Einaudi, 1986

Popper K., *La società aperta e i suoi nemici* (1945), Armando 1973

Popper K., *Logica della scoperta scientifica* (1934), Einaudi,1981

Proudhon P. J., *Cos'è la proprietà*, (1840), Laterza,1974

Pseudo Xenophon, *Contro la democrazia, La costituzione degli ateniesi*, Le Mani, 2012

Quesnay F., "Analyse de la formule arithmétique du tableau économiqueu de la distribution des dépenses annuelles d'une Nation agricole", *Journal de l'Agriculture, du Commerce et des Finances*, 2:3 (1766), 11–41.

Rad G. von, *Teologia dell'Antico Testamento, vol I* (1962), Paideia, 1972

Rad G. von, *Il sacrificio di Abramo*, Morcelliana, 1977

Ramorino F., *Mitologia Classica*, Hoepli, 1979

Rank O., *La nascita del mito dell'eroe* (1909), SugarCo, 1991

Rashi de Troyes, *Commento alla Genesi* (1090 c.a), Marietti, 1985

Rashi de Troyes, *Commento all'Esodo* (1090 c.a), Marietti, 1985

Reich W., *Analisi del carattere* (1933), Sugarco, 1973

Reik T., *Dogma and Compulsion*, Greenwood Press, 1951

Ricardo D., *Sui principi dell'economia politica e della tassazione* (1817), ISEDI, 1976

Rousseau J. J., *Il contratto sociale* (1762), Einaudi 1973

Rousseau J. J., *Emile* (1762), Armando, 1995

Sahr R., *Inflation Convertors Factors for Dollars 1665 to Estimated 2014*, Oregon State University Site

Saint-Exupery A. de, *Il piccolo principe* (1943), Bompiani, 2010

Salomé L. A., *Anal und Sexual*, Guaraldi, 1977

Saporetti C., *Le leggi della Mesopotamia*, Le Lettere, 1984

Saporetti C., *La storia del siciliano Peppe e del poveruomo babilonese*, Sellerio, 1985

Saporetti C., *Come sognavano gli antichi*, Rusconi, 1996

Saulcy F. de, *Recherches sur la numismatique Judaique* (1854), Kessinger Publishing, 2010

Saussure F. de, *Corso di linguistica generale* (1916), Laterza, 1974

Schama S., *The Embarassment of Riches* (1987), Harper Press, 2004

Schama S., *The Story of the Jews*, Vintage, 2014

Schimmel I., *Vous êtes bien payée pour savoir ce que ça m'a coûté pour en arriver là*", Revue française de psychanalyse, 54(2) 533-552

Schimmel-Reiss I., *La Psychanalyse et l'Argent*, Odil Jacob, 1993

Schmandt-Besserat D., *Before Writing*, University of Texas, 1992

Searle J.R., *The Rediscovery of the Mind*, Cambridge, Mass; MIT Press,1992

Searle J.R., *Mind: a Brief Introduction*, New York, Oxford University Press, 2004

Sedillot R., *Histoire, morale & immorale, de la monnaie*, Bordas, 1989

Semerano G., *Il popolo che sconfisse la morte*, Bruno Mondadori, 2003

Semerano G., *L'infinito: un equivoco millenario*, Bruno Mondadori, 2004

Semerano G., *La favola dell'indoeuropeo*, Bruno Mondadori, 2005

Semerano G., *Le origini della cultura europea* (Voll. I-II), Olschki, 2007

Semi A. A. (a cura di), *Trattato di Psicoanalisi*, Cortina, 1988

Seminara S., *Guerra e pace ai tempi di Hammu-rapi, voll. I e II*, Paideia, 2004

Seminara S., *Immortalità dei simboli*, Bompiani, 2006

Serafini S., *La Scrittura Celeste: nell'alfabeto un'antica testimonianza archeoastronomica?*, Rivista Italiana di Archeoastronomia, II, 2004

Bibliography

Sermonti G., *L'alfabeto scende dalle stelle*, Mimesis, 2009

Servet J. M., *Le troc primitif, un mythe fondateur d'une approche économiste de la monnaie*, Revue Numismatique, 6a série, Tome 157, 2001

Simmel G., *The Philosophy of Money* (1900), London, Routledge. 2004

Simmel G., *Il denaro nella cultura moderna*, Armando, 2005

Simmel G., *Denaro e vita*, Mimesis, 2010

Simon H. A., *La razionalità nella psicologia e nell'economia*, (in Egidi M. e Turvani M.)

Simon H. A., *Behavioural Economics*, Carnegie Mellon University, 1994

Sismondi S. de, J.-C-L., *Nuovi principi di economia politica* (1815), ISEDI, 1975

Smith A., *La ricchezza delle nazioni* (1776), Cugini Pomba, 1851

Smith A., *Teoria dei sentimenti morali* (1759), Rizzoli, 1995

Smith A., *Considerazioni sulla formazione originaria delle lingue* (1763), in R. Carotenuto, Quaderni Facoltà Scienze Politiche, 26, Giannini Editore, 1987

Sofocle, *Tragedie e frammenti*, UTET, 2006

Spinoza B., *Trattato teologico-politico* (1670), Bompiani, 2001

Spinoza B., *Etica* (1677), Bollati Boringhieri, 2006

Sraffa P., *Production of Commodities by Means of Commodities*, Cambridge, Cambridge University Press, 1960

Steiner G., *Il libro dei libri* (1996), Vita e Pensiero, 2012

Stuart Mill J., *Principles of Political Economy*, 1848, Cosimo: New York,

Swift J., *Gullivers's Travels* (1667), Andesite Press, 2015

Temin P., *A Market Economy in the Early Roman Empire*, Oxford University Press, 2001

Thomas Aquinas (saint), *I vizi capitali*, Rizzoli, 1996

Toaff G. (a cura di), *Sefer Yezirah*, Carucci, 1988

Toaff A., *The Bloody Satanic Sacrifice Rituals of the Jewish Race: Blood Passover* (2007), Independently published, 2020

Tögel C., *Sigmund Freud's practice: visits and consultation, psychoanalyses, remuneration*. The Psychoanalytic Quaterly, Vol. LXXVIII, N. 4, 2009

Tuckett D., *Minding the markets*, palgrave macmillan, 2011

Turri M. G., *La distinzione fra moneta e denaro*, Carrocci, 2010

Tversky A., Kahneman D., *Scelta razionale e rappresentazione delle decisioni*, (in Egidi M. e Turvani M.)

Vaughn K. I., *John Locke and the Labor Theory of Value*, Journal of Libertarian Studies, Vol. 2, No. 4, pag. 311-326, Pergamon Press, 1978

Venturi Ferriolo M., *Aristotele e la Crematistica. La storia di un problema e le sue fonti*, La Nuova Italia, 1983

Vergili P. M., *Eneide*, Istituto Enciclopedia Italiana, 2000

Verri P., *Opere filosofiche e d'economia politica*, Silvestri, 1818

Viderman S., *De l'argent, en psychanalyse et au-delà*, PUF, 1992

Voltaire, *Candido* (1759), Bietti, 1973

Voltaire, *Trattato sulla tolleranza* (1763), PGreco, 2010

Voltaire, *Lettere filosofiche* (1778), Barbera Editore, 2007

Voltolin A. (a cura di), *L'ideologia del Denaro*, Bruno Mondadori, 2011

Warner S. L., *Sigmund Freud and Money* (1989), J. Amer. Acad. Psychoanal., 17:609-622

Waddell L., *Egyptian Civilization, Its Sumerian Origin and Real Chronology*, Luzac and Co, 1930

Weber M., *Economia e società; comunità religiose* (1922), Donzelli, 2006

Weber M., *The Protestant Ethic and the Spirit of Capitalism* (1907), Oxford University Press, 2010

Wechsler E., Schoffer D., *La metàfora milenaria*, APM Biblioteca Nueva, Madrid, 1998

Wilensky-Lanford B., *Paradise Lust*, Grove Press, 2011

Williamson G. C., *The money of the bible* (1894), Reprint Bibliolife, 2011

Winckler H., *Altorientalische Forschungen*, E. Pfeiffer, 1895

Winckler H., *The history of Babylonia and Assyria* (1907), Michigan Library, 2011

Winckler H., *La cultura spirituale di Babilonia* (1907), Editori Riuniti, 2004 (Postfazione di P. Mander)

Wittfogel K. A., *Il dispotismo orientale* (1962), Sugarco, 1980

Xenophon, *Economico*, Rizzoli, 1991

Yerushalmi Y. H., *Il Mosè di Freud* (1991), Einaudi, 1996

Zimmern H., *The Babylonian and the Hebrew Genesis* (1901), David Nutt Reprint Kessinger Publishing, 2010

Bibliography

Sitography

ePSD, The electronic Pennsylvania Sumerian Dictionary

ETCLS, The Electronic Text Corpus of Sumerian Literature, Faculty of Oriental Studies of Oxford

cdli, Cuneiform digital library initiative, Los Angeles, Oxford, Berlin

CAD, The Assyrian Dictionary of the Oriental Institute of the University of Chicago.

the vatican website www.vatican.va

the S.Agostino.webarchive.

Jewish Encyclopedia

Biblos.com

www.moneymuseum.com

Languages

for French:
- *Le nouveau Petit Robert*, Le Robert,
- *Ghiotti*, Petrini; and thanks to M.me Pascale Bernasconi;

for English:
- *The Oxford Dictionary and Thesaurus*, Oxford University Press,
- *Grande dizionario Hazon*, Garzanti,
- *Irish - English Dictionary*, Oxford University Press; and thanks to Andrew Harwood, Miss Niki Balani, Miss Susan Cumming, Miss Jenny Hayes and especially to Dr. Gillian Clayton;

for German:
- *Dizionario di Tedesco*, Giacoma L. e Kolb S., Zanichelli; and thanks to Frau Evelyn Rachbauer;

for Latin and Greek:
- *IL Castiglioni- Mariotti*, Loescher,
- *Georges-Calonghi*, Rosenberg & Sellier,
- *Rocc*i, Società Editrice Dante Alighieri,

- *Vocabolario Greco-Italiano e Italiano-Greco*, F. Fontanella, Ed. Molinari, 1833;
and thanks to Prof. Vincenzo Gazich;

for Italian:
- *Grande Dizionario della lingua italiana*, UTET,
- *Enciclopedia Italiana*, Treccani,
- *Devoto- Oli*, Le Monnier,
- *Zingarelli*, Zanichelli;

for Spanish:
- *Diccionario de la lengua Española*, Real Academia Española,
- *Dizionario moderno Spagnolo- Italiano*, Frisoni G., Hoepli; and thanks to Prof. Isabel Manzanares;

for Sanskrit :
- *Dizionario Sanscrito*, Vallardi;

for Sumerian and Akkadian words and grammar: Giovanni Pettinato, Mario Liverani, Giovanni Semerano,
and also:
- *Introduction to Akkadian*, Caplice R., Editrice Pontificio Istituto Biblico, 2002,
- *Sumerian Lexicon*, Halloran J. A., Logogram Publishing, 2006;
- *A grammar of Akkadian*, Huehnergard J., Harvard College, 1996;
- *Sumerian Grammar*, Edzard D. O., Soc. of Biblical Literature, 2003
- *A Concise Dictionary of Akkadian*, Black J., George A., Postgate N., Harrassowitz Verlag, 2000;
- *Manuel d'épigraphie akkadienne*, Labat R. e F., Geuthner, 1994;
- *L'origine del cuneiforme*, Mander P., Aracne, 2005
- *Appunti di grammatica sumera*, D'Agostino F., Mander P., Aracne, 2007; and thanks to Dr. Alvise Matessi of Università di Pavia and Prof. Davide Astori of Università di Parma;

Bibliography

for many etymology:
- *Vocabolario etimologico della lingua italiana Pianigiani*, on line,
- *DELI*, Cortellazzo M., Zolli P., Zanichelli,
- *Dizionario etimologico della lingua greca*, and
- *Dizionario etimologico della lingua latina*, di Giovanni Semerano, Olschki Editore;

for Hebrew language:
- *Dizionario Italiano-Ebraico-Italiano*, Ed. Prolog, Giuntina,
- *Etymological Dictionary of the Hebrew language*, di E. Klein, Carta Jerusalem, 1987; and thanks to Giovanna Ruggeri Kaufmann, Dr. Baruch Avezov, Dr. Reny Icin and Prof. Davide Astori (Universty of Parma)

for Ancient Roman Laws:
thanks to Avv. Federico Biemmi.

NAME INDEX

Abraham K., 12, 72, 270
Avraham (from Ur), 261-274
Acrisius, 28
Adams C., 207
Aelianus, 27, 28
Ænea, 212, 213
Agesilaus, 134
Aglietta M., 58, 62n, 76, 84, 231n
Akerlof G.A., 51
Alary P., 40
Albertanus Brixiensis, 49
Alberti M.E.., 231
Albright W. F., 232
Alighieri D., 49, 186n
Apuleius, 72n
Andronicus Livius, 201
Aranjo-Ruiz V., 210
Aristotle, 12, 37, 38, 44-45, 61n, 127, 194
Argyle M., 52
Arnaud G., 73
Arrow K. J., 52
Ascalone E., 30, 32, 103, 228n, 229
Asterix, 279
Astori D., 158
Attali J., 12, 60, 107, 129n, 153, 206, 217, 237, 238, 247n
Augustinus (saint), 42, 58, 59, 67, 154, 155

Basello G.P., 88
Bastiat F., 194
Bausani A., 143

Bentham J., 46, 207
Benveniste E., 119, 203
Bernal M., 60
Berossus, 27, 100
Bettelheim B., 246
Bindman Y. Rabbi, 158
Bion W.R., 243
Bismarck O. von, 88, 136
Blesio G., 17
Bloch M., 167
Bobokhyan A, 189
Bond A., 266, 267n
Borneman E., 79
Bottéro J., 21, 24, 29, 30, 32, 101, 103, 104, 108, 109, 112-113, 115, 153, 157n, 197n, 254, 269, 290n
Bresciani-Turroni C., 84
Brennus, 21, 200
Bressett K., 224n, 226
Broome J., 144
Buccellati G., 103, 171
Bulgarelli O., 169, 175, 189, 191, 212, 266

Caetani L., 79, 151, 242n, 244, 268
Cagni L., 104
Calcagno C., 127, 245n
Calvin J., 36, 67, 69, 169
Camia F., 231
Campanella T., 42, 66n
Carducci G., 115, 233
Cassirer E.,119
Catarzi M., 129n, 145, 236

Cathars, 69
Catherine (saint) of Siena, 295
Catullus G.V., 279
Celts, 69n
Cesar J., 185
Chaucer G., 48
Chiodi S., 104
Chomsky N., 119
Christie A., 20
Chrysostom J. (saint), 58
Cicero M.T., 45
Coe M. D.,140
Cremerius J., 71
Croce B., 69, 136, 193
Crusoe R., 13
Cyprian (saint), 59
Cyrus (the Great), 29, 115.

Dallen M. E., 158
Dal Pozzolo A., 253
Damasio A. R., 53
Dash M., 54
De Gubernatis A., 61, 63, 71n, 72n, 155, 174, 290n, 291n
Delitzsch F., 93, 97
De Martino E., 150
De Santis S., 152
Descartes, 40, 41, 51, 53, 65, 65n
Desmond W.H., 80, 94, 233
Dimen M., 80
Diodorus Siculus, 110
Dionysius the Elder, 36
Disegni R., 263
Disney W., 12
Donne J., 67
Dracon, 37
Dumezil G., 154, 170

Eckart Maister, 296
Egidi M., 52
Einzig P. 41, 68n, 167n, 191-192
Eissler K.R., 70
Eliade M., 30, 79, 97, 101, 150, 236, 244, 266, 284

Ellis, H., 284n
Engel F., 38
Enmerkar, 23, 24, 25, 26, 28, 31, 97, 129
Erasmus, 20
Erikson E.H., 66n
Etchegoyen H., 70
Eusebius of Caesarea, 100

Fachinelli E., 21
Fenichel O., 79
Ferenczi S., 63, 86
Ferrarese L., 66
Flavius Josephus, 254, 264
Fliess W., 74, 77, 81, 86, 87, 89, 90
Freni S., 296
Freud S. 11, 28, 30, 33, 60, 70, 72, 74, 75, 76, 77, 81-115, 245, 270, 274
Friedman D. M., 251
Fromm E., 29, 57-62, 136, 206
Furnham A., 52

Galbraith J. K., 61
Galiani F., 38, 39, 43, 49, 121
Galilei G., 12
Gall F., 66
Gallo F., 210
Garbini G., 21, 70n, 99
Gaur A., 139, 142
Gedo J., 71
Geertz C., 150
Gelb I.J., 160
Genovesi A., 39, 124
George A. R., 104
Gernet L., 233
Gessel B., 54
Getafix, 279
Ginzberg L., 23n, 216n
Girard R., 62, 62n, 157, 236, 273n
Gironde S., 53
Giulia, 17
Glover E., 70
Goethe J. W., 12, 219
Graeber D., 21

Name Index

Gramsci A., 61
Graves R., 112, 152, 250, 251, 258, 264, 269, 291
Grotefend G. F., 29

Haarmann H., 141
Haddad P., 286
Haight D. F., 67
Halloran A., 225
Harari Y. N., 21
Hayek F. A. von, 40, 43, 49, 137, 167
Hegel G. W. F., 19, 134,136
Heidegger M., 149, 174
Heller A., 50
Hempsell M., 267n
Hendin D., 219, 225, 229, 271n, 272n, 293n
Herodotus, 12, 30, 40, 41, 121, 245
Hesiod, 112, 174
Hill P., 19
Hincks E., 29
Hobbes T., 39, 64, 128
Hughes R., 46
Hume D., 35, 51
Hycarnus, 265n

Iliad,114
Ingham G., 19
Innes A. M., 21
Isidore of Seville, 23, 174n

Jacobs T., 80, 138
Jeremias A., 12, 29, 30, 81-115, 169, 263, 266, 274n
Jevons W. S., 226
Jung C. G., 70n, 81, 225
Juno, 200, 201, 203

Kahneman D., 21, 52
Kant I., 37, 73, 76, 169
Kardiner A., 83n, 84
Keynes J. M., 12, 19, 20, 51, 64, 137, 167n, 201
Keynes R., 19

Khan M. M. R., 126
Khan R.F., 19
Klebanow S., 80, 141
Klein E.,102, 107, 173
Klein M., 200, 201
Klitsche de la Grange T., 194
Kramer S.N., 21, 29, 159, 165, 216n, 245n
Krauss F.R., 89
Krugman P., 39
Krueger D. W., 52, 75
Kuhn T. S., 12
Kutra V., 69n

Lacan J., 21
Lasky, E., 80
Laum B., 54, 94, 129n, 145, 169, 181-192, 199, 233, 235-237, 255
Law J., 53n
Layard A. H., 245n
Lea S., 52
Le Goff J., 48
Leoni B., 168
Levi-Strauss C., 119
Lichtenstein A., 158
Linneus, 185, 285n, 287n
Liverani M., 21, 60, 152, 193, 195, 196, 215, 220, 240
Locke J., 38, 39, 43, 51, 123n, 168n
Lopez D., 28, 135, 137, 274
Lowenkopf E., 34, 141
Lucretius T. C., 205
Luther M., 66, 66n, 97, 206-207
Lycurgus, 37

Madden F. W., 225, 227n
Maffei S., 39, 42-43, 45, 46, 124, 203
Maffessoli A.B., 290
Malighetti R., 150
Malinowski B., 21
Malthus T. R., 39
Mamon, 81-115
Mandel G., 252n
Mander P., 88, 153n, 226

Marcuse H., 21
Marduk, 100
Martin P., 124
Martin F., 40, 180
Marx K., 19, 38, 39, 40, 45, 54, 59, 118, 119, 135
Matthiae P., 160
Mauss M., 134, 182, 194, 203-205, 273
Menger C., 49, 294
Meskiaggasher, 26
Mill J. S., 35, 51, 73
Milton J., 119
Mises L. von, 49, 84, 188
Miyazaki I., 47
Mommsen T., 188, 214, 255, 273n
Money- Kyrle R., 235, 274
Morgan J.J.M. de, 88
Motterlini M., 53
Murdin L., 47, 219
Musatti G. C., 17
Müller M., 97
Myers K., 80
Mørkholm O., 219

Nebuchadnezzar, 113, 184
Neuser J., 108
Newton I., 53, 65n
Nicolas of Damascus, 91, 263
Nietzsche F., 43
Noah, 32, 157, 158
Nungalpiriggal, 26

Oannes, 27
Obelix, 279
Oppert J, 29
Orléan A., 58, 62n, 76, 84, 231n
Orwell G., 46
Ovid P. N., 110, 112, 213, 256

Pareto V., 21
Parise N., 148, 181, 182-183, 192, 231-233, 277
Parpola S., 88
Parry J., 167

Pasiphaë, 186, 186n
Pettinato G., 21, 23, 24, 26, 27, 32, 88, 98, 103, 104, 111, 139, 183, 189, 190, 196, 232, 240
Peyronel L., 32, 160, 161-162, 169, 229
Piaget J., 76
Pianigiani O., 117,120n, 155n, 201, 214n, 285n, 288n
Piketty T., 21
Plato, 19, 36, 38, 42
Pliny the Elder, 47, 50-51, 110
Plutarch, 157
Polangy K., 54, 284n
Ponzi C., 54n
Popper K., 18, 19, 41, 168
Pound (£), 214
Privitera S., 231
Proudhon P. J., 61n
Proust M., 297
Pseudo Xenophon, 194
Puccini G., 283

Quesnay F., 40, 124

Rad G. von, 244, 268
Rank O., 30
Rashi de Troyes, 60, 162, 259n, 264, 266, 267
Rawlinson E., 29
Reich W., 70
Reik T., 253
Ricardo D., 39
Róheim G., 273
Rossini G., 86
Rousseau J. J., 132, 165

Sahr R., 83, 84
Salomé L. A., 70, 82, 83, 84, 137
Santoni Rugiu P., 245n
Saporetti C., 109, 245
Saraval A, 21, 70
Sargon (the Great), 29, 231
Sargon II, 231
Saulcy F. de, 225, 230

Name Index

Saussure F. de, 119
Say J.B., 118
Schama S., 53, 214, 220
Schimmel I., 12, 28, 61n, 73, 74, 75, 76, 81, 129n, 247n, 217
Schmandt-Besserat D., 144-145
Schoffer D., 99
Schumpeter J., 21
Schwennhagen L., 140, 290
Scott R., 280
Searle J. R., 53, 147
Sedillot R., 167, 225
Segal H., 74
Seipel I., 82
Semerano G., 14, 15, 21, 29, 57, 61, 120, 130n, 172, 233, 282, 283, 286, 291
Seminara S., 147, 216n
Seneca L. A., 46
Serafini S., 143
Sermonti G., 143, 143n
Servet J. M., 40, 203
Seyffart G., 144
Simmel G., 35, 216n
Simon H. A., 52
Sismondi S. de, J.-C-L., 39, 47, 219
Smith A, 36, 39, 48, 54
Smith G., 29, 157n
Solms M., 53
Solon, 37
Sophocles, 29
Spallanzani L., 205
Spielrein S., 81
Spinoza B., 19, 37
Sraffa P., 124
Stecchini L., 234n
Steiner G., 173
Stone M. van, 140
Stucken E., 92
Stuart Mill J., 35
Swift J., 34, 55

Tahler R., 21
Temin P., 32
Thomas (saint) Aquinas, 41, 45, 49, 64
Thureau-Dangin F., 115
Toaff A., 107, 108
Tolkien J.R.R., 227n
Tögel C., 82, 84
Troisi M, 209n
Tuckett D., 52
Tversky A., 52

Vaughn K. I., 38, 193
Venturi Ferriolo M., 36
Verri P., 38, 39
Vespasian T. F., 63
Viderman S., 36, 108
Vergilius P. M., 47, 138, 213
Voltaire, 25
Vuk T., 52

Waddell L., 232
Wagner A., 194
Warner S. L., 82
Weber M., 55, 59, 67, 69, 172, 175
Webley S., 52
Wechsler E., 99
Whately R., 181n

Wilensky-Lanfort B., 290
Williamson G. C., 224n, 225
Winckler H., 87, 89, 91-92, 94-95, 97, 143n, 196
Winnicott D., 137
Wittfogel K. A., 167
Woolley L., 194, 217

Xenophon, 12, 44, 134, 213

Yerushalmi Y. H., 93, 196, 244

Zimmern H., 87, 89, 99, 100, 115, 270
Zucconi S., 77

MIMESIS GROUP
www.mimesis-group.com

MIMESIS INTERNATIONAL
www.mimesisinternational.com
info@mimesisinternational.com

MIMESIS EDIZIONI
www.mimesisedizioni.it
mimesis@mimesisedizioni.it

ÉDITIONS MIMÉSIS
www.editionsmimesis.fr
info@editionsmimesis.fr

MIMESIS COMMUNICATION
www.mim-c.net

MIMESIS EU
www.mim-eu.com

Printed by
Geca Industrie Grafiche – San Giuliano Milanese (MI)
October 2020